An Introduction to
Probability
With MATHEMATICA®

An Introduction to

Probability

With MATHEMATICA®

Edward P. C. Kao

Department of Mathematics
University of Houston, USA

World Scientific

NEW JERSEY · LONDON · SINGAPORE · BEIJING · SHANGHAI · HONG KONG · TAIPEI · CHENNAI · TOKYO

Published by

World Scientific Publishing Co. Pte. Ltd.

5 Toh Tuck Link, Singapore 596224

USA office: 27 Warren Street, Suite 401-402, Hackensack, NJ 07601

UK office: 57 Shelton Street, Covent Garden, London WC2H 9HE

Library of Congress Control Number: 2022011701

British Library Cataloguing-in-Publication Data
A catalogue record for this book is available from the British Library.

MATHEMATICA® is a computational software available at www.wolfram.com with a trademark registered to Wolfram Group, LLC.

AN INTRODUCTION TO PROBABILITY
With MATHEMATICA®

ISBN 978-981-124-543-5 (hardcover)
ISBN 978-981-124-678-4 (paperback)
ISBN 978-981-124-544-2 (ebook for institutions)
ISBN 978-981-124-545-9 (ebook for individuals)

For any available supplementary material, please visit
https://www.worldscientific.com/worldscibooks/10.1142/12505#t=suppl

Desk Editors: Vishnu Mohan/Rok Ting Tan

Typeset by Stallion Press
Email: enquiries@stallionpress.com

Printed in Singapore

To Connie

Preface

In the past several decades, advancements made in computer hardware and software have been phenomenal. This opens up challenges for incorporating them effectively in classroom teaching. This text is one attempt to answer the challenge.

The goals of this book are as follows: to (i) develop model building and problem-solving skills dealing with uncertainties, (ii) build a solid foundation for subsequent studies in subjects, shown below, requiring probability as a pivotal prerequisite, (iii) steer towards a setting more conducive to explore heavy tailed distributions, and (iv) present subject matters in an accessible and coherent manner for *students*.

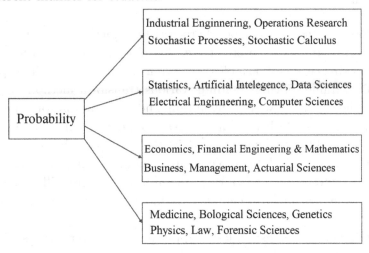

The unique features of this book are summarized below.

- It introduces the use of Mathematica® as a part of a toolbox for working with probability. Consequently, it extends the range of problems easily solvable by making use of the power of modern software.

As an example, a common theme in probability is the application of the "law of unconscious statistician"

$$E(g(X)) = \int g(x) f_X(x) dx,$$

where X is a random variable with probability density $f_X(x)$ and $g(x)$ is a function of x. For an application in finance, we let X be the price of a stock and assume X follows a lognormal distribution with parameters $(\mu, \sigma) = (3, 1)$ and $g(x) = (x - 10)^+$ is the payoff function of a call option. Then, the two alternative ways to do the computation are

```
∫₁₀^∞ (x - 10) PDF[LogNormalDistribution[3, 1], x] dx // N
```

```
24.0593
```

```
NExpectation[Max[x - 10, 0], x ≈ LogNormalDistribution[3, 1]]
```

```
24.0593
```

We see that the syntax makes the idea underlying the computation transparent.

- It introduces the use of the characteristic function without assuming complex variable as its prerequisite. The few relevant ideas used in the text involving complex variables are extremely rudimentary. This provides an unified framework for handling various transform methods — particularly when dealing with convolution and their asymptotic behaviors. With the aid of Mathematica, its usage renders itself to many serious applications. This is demonstrated in the example relating to variance gamma — a modern-date staple in financial mathematics. Once students are being exposed to the utility of the approach, more will be motivated towards the learning of complex variables and analysis in an early stage.
- It contains numerous examples in applications in diverse areas of general interest where nontrivial computations are needed. Moreover, Mathematica provides an experimental platform where students can explore

alternative solution procedures with comparatively minimal programming efforts.

Our endeavor in learning probability should be a collaboration between us and the software. It is not easy for machines to catch up with a human being's ingenuity. But we should never underestimate the immense power the computer possesses. Without actually doing the computation, many times brilliant ideas can contain hidden errors. To expound the point, I will use Problem 28 in Chapter 5 as an example. The problem and its solution are originally given in Jim Pitman's wonderful book *Probability* (p. 504, Springer-Verlag, 1993). Both the problem and its solution are intriguing and ingenious. However when carrying out a Mathematica implementation, it immediately points out a problem involving 0^0 that calls for intervention — namely, a modification of the solution procedure.

Quoting Martin Gardner (p. 39, *The Colossal Book of Short Puzzles and Probability*, edited by Dana Richards, Norton, 2006), "Probability is the one area of mathematics in which experts have often made mistakes. The difficulty is not with the formulas but is figuring out what exactly is being asked!" While I fully agree with the first sentence, the second one only tells a part of the story. It is also imperative that students should attempt to acquire a *thorough* understanding of the logics and rationales underlying the formulas — moreover their interrelations. Merely *searching a formula* to solve a problem, while a popular choice by many students as I have observed in the past, is not a productive way to proceed. The best hint to get a problem solved is shown on the surface of an IBM coffee mug with a logo stating "THINK", and then, if needed, proceed with experimenting on a computer.

There are many excellent software packages available to meet the need for computation. One may ask why I have chosen Mathematica for this text. The simple reason is that for its "WYSIWYG" (what you see is what you get) feature. For *teaching*, this feature is most attractive. I will make my point in the following two illustrative examples.

Example 1. This is the well-known coupon collection problem. Consider the case in which there are 52 distinct types of coupon. One coupon is randomly placed in each cereal box. We want to compute the expected number of boxes to be purchased in order to obtain a complete set (Example 3, Chapter 4 using the inclusion–exclusion formula, and Example 21, Chapter 8 using Kolmogorov's Poissonization approach). The following displays all the codings involved to solve the problem:

Approach 1: Use the inclusion-exclusion formula

```
: BN = 52;
```

```
: fn[i_, n_] = Binomial[BN - i, i] * ((BN - i) / BN)^n * (-1)^(i+1);
```

```
: pn[n_] := Total[Table[fn[i, n - 1], {i, 1, (BN - 1)}]] -
      Total[Table[fn[i, n], {i, 1, (BN - 1)}]];
```

```
: EN = Total[Table[n * pn[n], {n, 52, 1000}]]  // N
```

```
: 235.978
```

Approach 2: Use Kolmogorov's Poissonization

$$\text{cpo} := 1 - \text{PDF}\left[\text{PoissonDistribution}\left[\frac{t}{52}\right], 0\right]$$

$$\text{EN} = \int_0^\infty \left(1 - \text{cpo}^{52}\right) dt \ // \ N$$

```
: 235.978
```

Example 2. The well-known airline overbooking problem is one that treats the interactions involving Poisson arrivals, booking limit (BL), and plane capacity (PC). The following shows the coding and computation for coming up with the answers:

```
pmf[x_] = PDF[PoissonDistribution[175], x] // N;
```

```
BL = 185;  PC = 162;
```

$$\text{ES} = \sum_{x=0}^{300} \sum_{k=0}^{\text{Min}[x,\text{BL}]} k \ \text{Binomial}[\text{Min}[x, \text{BL}], k] \ (0.91)^k \ (0.09)^{(\text{Min}[x,\text{BL}]-k)} \ \text{pmf}[x]$$

```
157.655
```

$$\text{EB} = \sum_{x=0}^{300} \sum_{k=\text{PC}+1}^{\text{Min}[x,\text{BL}]} (k - \text{PC}) \ \text{Binomial}[\text{Min}[x, \text{BL}], k] \ (0.91)^k \ (0.09)^{(\text{Min}[x,\text{BL}]-k)} \ \text{pmf}[x]$$

```
2.21417
```

The coding clearly signals the types of conditional expectation involved in the computation.

During the pandemic of COVID-19, homework submissions by students have been done by emails. The Wolfram Language (that powers Mathematica) has an unique feature that enables students to do text with intricate math symbols as well as serious computation under a *single* platform (by toggling between ALT-7 and ALT-9). This indispensable feature makes the choice of the software desirable albeit its demand on users for close adherence to its syntax.

There will be instructors who prefer the use of other software. The simple resolution is to replace the last section of each chapter — entitled "Exploring with Mathematica" by their own computer implementations without losing continuity.

I took my first *dose* of probability (using a term in this age of vaccine) in a course called STAT 116 at Stanford in the fall quarter of 1965. The course was taught by late Professor Rupert G. Miller, Jr. The course left a strong impression on me and has kept me interested in the field every since. One of the four teaching assistants for the course was Professor Sheldon M. Ross whose book *A First Course in Probability*, Pearson, 2019 (now at its tenth edition) I have used fondly as the required text for many years. Readers may notice omnipresent influences of Professor Ross's book in the pedagogical development of the current text. My enthusiasm in adding computation as an indispensable ingredient to the study of probability was kindled by the work of Professor Ward Whitt of Columbia — particularly those related to numerical inversions of transform. My sojourn as the Area Editor in computational probability and analysis during 1996–2007 for *INFORMS Journal on Computing* has also contributed to my predilection with regard to computation and applications.

I would like to thank my colleagues Professors Giles Auchmuty, Robert Azencott, Jeffrey Morgan, Claudia Neuhauser, and Mathew Nicol for their comments about the Preface of the book. Their suggestions have contributed to my revisions of the manuscript. The staff at Wolfram Research had given the needed help in the use of Mathematica. Paige Bremner, the Managing Editor of Publishing Projects along with Dave Withoff, the Senior Technical Content Editor at Wolfram Research, provided me with feedbacks and encouragements. My daughter Candice Kao of Google reviewed "A Terse Introduction to Mathematica" from a vintage point of an user and made many insightful suggestions. Finally, I would also like to express my thanks to the staff at World Scientific Publishing, particularly Shreya Gopi, Ann Lai, Vishnu Mohan, for their time and effort in this joint venture. As the author, I bear full responsibility for any errors and would appreciate hearing about them.

Edward P. C. Kao
Department of Mathematics
University of Houston
Houston, TX 77204-3008

Contents

Notations and Abbreviations

Ω	sample space
ω	an element of the sample space
CDF	cumulative distribution function
PDF	probability density function
PMF	probability mass function
$P(E)$	probability of event E
$O(E)$	odds of event E
$X \perp Y$	random variables X and Y are independent
$E(X)$	mean of random variable X
$Var(X)$	variance of random variable X
$X_{(i)}$	the ith order statistic of $\{X_i\}$ (the ith smallest of $\{X_i\}$)
$X\|Y$	random variable X conditional on random variable Y
$\otimes$	element-by-element product of two vectors
J	Jacobian matrix
i.i.d.	independently identically distributed random variables
$x \equiv y$	y is defined as x
$x \leftarrow y$	x is to be replaced by y
$x \rightarrow y$	x converges to y
$X \sim dist$	random variable X dollows distribution $dist$
$\Phi(\cdot)$	CDF of $N(0,1)$

$\phi(\cdot)$	*PDF* of $N(0,1)$
$erf(\cdot)$	error function
$\beta(a,b)$	beta function
$\Gamma(\alpha)$	gamma function
$p_X^g(\cdot)$	probability generation function of random variable X
$\lambda_X(\cdot)$	hazard rate function
$M_X(\cdot)$	moment generation function of random variable X
$\varphi_X(\cdot)$	characteristic function of random variable X
$\mathcal{M}_X(\cdot)$	Mellin transform of random variable X
Λ	covariance matrix of a random vector
BVN	bivariate normal distribution
MVN(μ, Λ)	multivariate normal with mean vec μ and COV matrix Λ
$\xrightarrow{p}$	convergence in probability
$\xrightarrow{d}$	convergence in distribution
$\xrightarrow{r}$	convergence in $r-$ mean
$\xrightarrow{m.s.}$	convergence in mean-squares (i.e., $\xrightarrow{x}$ convergence)
CLT	central limit theorem
WLLN	weak law of large numbers
SLLN	strong law of large numbers

Discrete Random Variables

$bern(p)$	Bernoulli distribution
$bino(n,p)$	binomial distribution
$geom(p)$	geometric distribution
$hyper(n,m,N)$	hypergeometric distribution
$negb(n,p)$	negative binomial distribution
$pois(\lambda)$	Poisson distribution

Continuous Random Variables

$beta(a,b)$	beta distribution
$C(m,a)$	Cauchy distribution
χ_n^n	chi-square distribution
$expo(\lambda)$	exponential distribution

$Erlang(n, \lambda)$	Erlang distribution
$F(m, n)$	Fisher's distribution
$gamma(\alpha, \lambda)$	gamma distribution
$lognormal(\mu, \sigma^2)$	lognormal distribution
$N(\mu, \sigma^2)$	normal distribution
$Pareto(\lambda, a)$	Pareto distribution
$t(n)$	Student's t distribution
$unif(a, b)$	uniform distribution
$VG(\alpha, \lambda, \gamma)$	variance-gamma distribution
$weibull(\nu, \alpha, \beta)$	Weibull distribution

Chapter 1

Permutation and Combination

1.1. Introduction

We start this journey through probability with an exploration of various approaches to do counting. Why? To make a long story short, say we would like to find the likelihood of the occurrence of a given event, call it event E, of interest. Let m be the number of scenarios that are constituents of E and n be the total number of possible scenarios in the experiment of interest. Assume that all scenarios are equally likely to occur. Then it is reasonable to assert that the probability E occurs is given by $P(E) = m/n$. For example, we consider a simple experiment involving drawing a card from a standard card deck. Let E denote the event of getting a spade. Then we see that $P(E) = 13/52$ as there are 13 spades in a card deck of 52 cards. Probabilistic problems at times can be involved and tricky. If we can master our expertise in counting, then many times finding probability amounts to counting with respect to figuring out the numerator m and the denominator n of $P(E)$.

1.2. Permutations

Assume that there are n distinguishable items. We would like to line them up. How many ways can this be done? So we do this methodically. For the first spot, any one of the n items can occupy the spot. Once that is done, we move to the second spot. There are $n - 1$ remaining items. Any one of the $n - 1$ remaining items can occupy the second spot. We reiterate the process till we reach the last spot. There is only one remaining item. Thus there is one possibility. To summarize, the total number of ways to line these n items up is given by

$$n \times (n-1) \times (n-2) \times \cdots \times 2 \times 1 \equiv n!.$$

Example 1. Suppose there are five boys to be lined up on a stage. How many ways such a lineup can be formed?

Solution. There are $5! = 120$ possible lineups.   □

Variant. Assume that there are n distinguishable items. We would like to select k of them and line them up. There are

$$(n)_k \equiv n \times (n-1) \times \cdots \times (n-k+1) = \frac{n!}{(n-k)!}$$

ways to do this. We see that the order of selection matters and each item can be selected only once. The notation $(n)_k$ is sometimes called $k-permutation$.

Example 2. Suppose we want to use 26 alphabets to make license plates that only display 6 alphabets Assume alphabets are not allowed to be repeated. How many license plates are possible?

Solution. There are $(26)_6 = (26)(25)(24)(23)(22)(21) = 165\,765\,600$ possible license plates. Another way to state the answer is $\frac{26!}{(26-6)!} = \frac{26!}{20!} = 165\,765\,600$.   □

Generalization. Assume that there are k experiments. We do these experiments sequentially. In the first experiment, we have n_1 distinct items and would like to line them up. In the second experiment, we have n_2 distinct items and would like to line them up. We repeat the process till we reach the kth experiment. Then the total number of possible permutations is given by

$$n_1! \times n_2! \times \cdots \times n_k!.$$

Example 3. Suppose there are 3 distinct books in algebra, 4 distinct books in calculus, and 6 distinct books in geometry. Books are to be lined up in the order of algebra, calculus, and geometry. How many arrangements are possible?

Solution. There are $3!4!6! = 103,680$ arrangements possible.   □

When Some Items are Indistinguishable. Assume that there are n items. We would like to line them up. However, there are k distinguishable *groups*. Items in a given group are indistinguishable. Assume $n_1 + \cdots + n_k = n$. How many distinct arrangements are possible?

Assume for the moment all n items were distinguishable. If so, there would be $n!$ ways to permute them. Since items in the same group are not distinguishable, there are $n_1! \times n_2! \times \cdots \times n_k!$ duplicates, i.e., being counted $n_1! \times n_2! \times \cdots \times n_k!$ times as duplicates. Thus, in actuality, the distinguishable arrangements are

$$\frac{n!}{n_1! \times n_2! \times \cdots \times n_k!}.$$

The following example clearly demonstrates the use of above result.

Example 4. Suppose we use three alphabets and two numerals to form a plaque with five places. How many distinct arrangements are there: (a) If alphabets are distinguishable and numerals are distinguishable, and (b) if alphabets are indistinguishable and so are the numerals?

(a) If alphabets and numerals are all distinguishable. Then there are $5! = 120$ ways to permute them.
(b) If the alphabets are indistinguishable and so are the numerals, we remove the indistinguishable duplicates for both cases. Hence

$$\frac{5!}{3!2!} = 10. \qquad \square$$

Example 5. How many different 10-digit passcodes can be formed using three *'s, four \$'s, and three ^'s?

Solution. We can make a total of

$$\frac{10!}{3!4!3!} = 4200$$

different passcodes. $\qquad \square$

Example 6. How many different letter arrangements can be formed from the letters *PEPPER*?

Solution. Assuming for the moment the three P's and two E's were distinguishable. Then we would have $6! = 720$ ways to permute them. Since the three P's and two E's are actually indistinguishable, we remove these two sets of duplicates from the earlier permutations. Therefore, the number of distinct letter arrangements is

$$\frac{6!}{2!3!} = 60.$$

In Illustration 1, we use *Mathematica* to replicate the computation. $\qquad \square$

1.3.   Combinations

Suppose we have n distinguishable items from which we will choose r of them. How many ways can it be done? As an example, we have a class of 32 students and want to select 5 of them to form a basketball teams. How many selections are there? We can come up with an answer by first approaching it from a permutation perspective. We have r spots to be filled from a potential of n candidates. Thus there are

$$(n)_r = n \times (n-1) \times \cdots \times (n-r+1) = \frac{n!}{(n-r)!}$$

ways to fill the r spots. Since in actuality, the *order* with which a spot gets filled with a candidate is irrelevant. Thus the above count contains $r!$ of duplicates. Hence, the number of distinguishable choices is given by

$$\frac{(n)_r}{r!} = \frac{n!}{(n-r)!r!} \equiv \binom{n}{r}.$$

The term $\binom{n}{r}$ is commonly used as the notation for combinations.

Example 7. A mathematics department has 32 faculty members. A promotion and tenure committee of size 7 is to be formed. How many possible ways such a committee can be formed?

Solution. There are $\binom{32}{7} = \frac{32!}{7!25!} = 3,365,856$ ways to form such a committee.   □

Example 8. (Binomial Theorem). We want to expand $(x+y)^n$.

The expansion involves the filling of the n spots with x's and y's. Thus, a typical term involves $x^k y^{n-k}$, i.e., x shows up x times and y shows up $n-x$ times. For the term $x^k y^{n-k}$, out of the n spots, k of which will be occupied by x's. Thus, for this particular terms, there are $\binom{n}{k}$ ways to make such a selection. Since x can be selected from 0 to n times, we conclude that

$$(x+y)^n = \sum_{k=0}^{n} \binom{n}{k} x^k y^{n-k}.$$

The above also suggests the term $\binom{n}{k}$ is sometimes called the binomial coefficient. □

1.4. Basic Principles of Counting

There are some basic principles of counting in our tool box to assist us in counting: by *multiplications*, by *additions*, by *exclusion*, and by *nesting*.

Counting by Multiplication. When there are k *independent* experiments. For $i = 1, \ldots, k$, we assume experiment i has n_i distinct outcomes. Then the total number of outcomes from the aggregate of k experiments is given by

$$n_1 \times n_2 \times \cdots \times n_k.$$

Here, we assume that the ordering of occurrences of outcomes is not our concern. We call the above *counting is by multiplications*.

Example 9. Consider a group of 4 women and 6 men. If 2 women and 3 men are to be selected to form a jury, how many possible choices are there?

Solution. There are $\binom{4}{2}\binom{6}{3} = 120$ possible choices. □

There is another form of counting by multiplication. This occurs when there is a *dependency* between successive experiments. Assume that there are k experiments to be done sequentially. For $i = 1, \ldots, k$, we assume experiment i has n_i distinct outcomes *given that* experiments $1, \ldots, k-1$ has already been done. Then the total number of outcomes from the aggregate of k experiments is given by

$$n_1 \times n_2 \times \cdots \times n_k.$$

The notion of permutation follows precisely the above idea.

Example 10. Suppose we draw to cards in succession from a standard card deck. How many ways two diamonds can be drawn?

Solution. There are $(13) \times (12) = 156$ ways two diamonds can be drawn. □

Counting by Addition. We now address the second basic principle of counting: by *additions*. If we can partition the outcomes of an experiment into mutually exclusive and collectively exhaustive components. Them the total number of possible outcomes will be the *sum* of the numbers of outcomes from all its components.

Example 11. Suppose that there are 4 boys and 4 girls to be seated in a row of chairs. Assume that boys and girls cannot sit next to each other. What are the number of seating arrangements possible?

Solution. One way to start a seating plan is to have a boy occupying the leftmost position. Following this, there are 4! ways to seat the boys and then 4! to seat the girls. Similarly, another way is to have a girl occupying the leftmost position. Again, there are 4! ways to seat the girls and then 4! to seat the boys. These two components form a partition of all the possible outcomes. Thus the total number of seating arrangements is

$$(4!)^2 + (4!)^2 = 1152. \qquad \square$$

Example 12. There are 8 people playing in a neighborhood basketball court. Five of them will be chosen to form a team to compete against another team in a different neighborhood. How many choices are there if two of the players are feuding and will not join the team together?

Solution. The possible outcomes can be partitioned into two components: $A = \{$none of the feuding players joins the team$\}$ and $B = \{$exactly one of the feuding player joins the team$\}$.

For component A, there are $\binom{6}{5}$ ways to form the team. For component B, there are $\binom{2}{1}\binom{6}{5}$ ways to form the teams. Thus the number of possible ways a team can be formed is

$$\binom{6}{5} + \binom{2}{1}\binom{6}{5} = 18. \qquad \square$$

Counting by Subtraction. Another principle of counting is by exclusion, or by subtraction. Many times, we may be asked to count events involving a statement that reads "at least some of the outcomes possess the indicated attributes." If the total number of possible outcomes *and* the number pertaining none of the outcomes possessing the indicated attributes are easy to compute, then the difference of the two is what we are looking for. Thus a simple subtraction will yield the needed result.

Example 13. How many 3 digit area codes can be formed while allowing at most two of the digits assuming the same value?

Solution. We see that there are 10^3 possible outcomes without any exclusion. Now there are $10 \times 9 \times 8$ cases with all three digits being different. Thus the difference yields the needed answer, i.e., $10^3 - 10 \times 9 \times 8 = 280$. □

Example 14. Assume a person is either college educated or high school educated. Also assume that a person is either a football fan, or a basketball fan, or a soccer fan (not both). There are 15 people chosen for a study. (a) Of the 15 chosen people, what is the number of possible outcomes? (b) What is the number of possible outcomes if at least one of the people chosen is high school educated?

Solution. (a) We have 15 spots, each spot can assume $2 \times 3 = 6$ attributes. Thus the total number of outcomes is 15^6.
(b) The number of outcomes associated with "none is high school educated" is 15^3. Thus the number of possible outcomes if at least one of the chosen is high school educated is $15^6 - 15^3$. □

Counting by Nesting. The last principle to assist us in counting is by first considering several *collections* of elements with each collection sharing a common feature. We consider the counting at the *collection* level. We then return to the counting within each collection. We multiply results obtained from all these collections to arrive at the needed result. We call this approach as counting by *nesting*, or grouping. This approach is illustrated in the following example.

Example 15. Suppose there are 4 distinct chemistry books, 5 distinct biology books, and 8 distinct physics books. These books are to be line up on the bookshelf with books dealing with the same subject staying contiguous. How many different arrangements are possible?

Solution. We first consider books dealing with the same subject a collection. There are three collections — hence 3! ways to arrange them. Within the three collections, there are 4!, 5!, and 8! ways to permute them, respectively. Therefore the total number ways to arrange the books is

$$3! \times (4! \times 5! \times 8!) = 696\,,729,\,600.$$ □

1.5.   Binomial and Multinomial Coefficients

We first explore more about the properties of the binomial coefficient. Many identities involving the binomial coefficient can be established by a combinatorial argument. To illustrate, we have

$$\binom{n+1}{r} = \binom{1}{1}\binom{n}{r-1} + \binom{1}{0}\binom{n}{r} = \binom{n}{r-1} + \binom{n}{r}.$$

The left side is about the number of ways to select r items from $n+1$ items. To establish the equivalent right side, we argue that a specific item is either being included in the selection, resulting in $\binom{n}{r-1}$, or not being included by the selection, resulting in $\binom{n}{r}$. These eventualities form a partition hence we use counting by addition.

To establish another identity using the combinatorial argument, we consider

$$\binom{n+m}{r} = \binom{n}{0}\binom{m}{r} + \binom{n}{1}\binom{m}{r-1} + \cdots + \binom{n}{r}\binom{m}{0}.$$

Assume that an urn contains n white and m black balls. When we remove r balls from the urn, a part of which, say i, will be white balls and the remaining $r-i$ balls will be black balls where $i = 0, \ldots, r$. The left side of the last equality gives the number of possible such selections when considered in aggregate. The right side of it gives the result when we sum over all possible scenarios leading to the same objective. Other identities can be obtained in similar manner without resorting to a brute-force work involving the factorial notation.

We now introduce the multinomial coefficient. Let $n_1 + \cdots + n_k = n$. We define

$$\binom{n}{n_1, \ldots, n_k} = \frac{n!}{n_1! n_2! \cdots n_k!}.$$

The above is called the multinomial coefficient. It gives the number of possible outcomes obtainable from choosing n items, for which n_1 of them are indistinguishable, n_2 are indistinguishable, $\ldots$, and n_k are indistinguishable, i.e., there are k distinct groups.

By direct expansion, we obtain the following identity:

$$\binom{n}{n_1, \ldots, n_k} = \binom{n}{n_1}\binom{n - n_1}{n_2} \cdots \binom{n - (n_1 + \cdots + n_{k-1})}{n_k}.$$

The above amounts to making the needed selections sequentially and applying the multiplication principle of counting.

Example 16. A high school class has 15 students. (a) If three teams are to be formed with 5 members per team, how many ways the three teams can be formed. (b) Assume for the three teams, each has a name, say Team Alpha, or Beta, or Gamma. How many formations are possible?

Solution. (a) There are $\binom{15}{5,5,5} = \frac{15!}{5!5!5!} = 756,756$ possible selections.
(b) We can first form a group of three teams using the result obtained in part (a). We then assign a team name to each one of the three teams so chosen. There are 3! ways to do so. In conclusion, there are $756,756 \times 6 = 4,540,536$ possible arrangements. We also use of the "Multinomial" function of *Mathematica* in Illustration 2 to do part (a). □

Example 17 (Multinomial Expansion). We want to expand $(x_1 + x_2 + \cdots + x_k)^n$.

The expansion involves the filling of the n spots with x_1's, x_2's, $\ldots$, and x_k's. Thus a typical term involves $x_1^{n_1} x_2^{n_2} \cdots x_k^{n_k}$, i.e., x_1 shows up n_1 times, x_2 shows up n_2 times, $\ldots$, and x_k shows up n_k times. For the term $x_1^{n_1} x_2^{n_2} \cdots x_k^{n_k}$, out of the n spots, n_1 of which will be occupied by x_1's, n_2 of which will be occupied by x_2's, and so on. We use the multinomial coefficients to account for the multiplier associated with each product term. This gives the needed expansion:

$$(x_1 + x_2 + \cdots + x_k)^n = \sum_{\substack{(n_1,\ldots,n_k): \\ n_1 + \cdots + n_k = n \\ n_i = 0,1,2,\ldots,n}}^{n} \binom{n}{n_1, n_2, \ldots, n_k} x_1^{n_1} x_2^{n_2} \cdots x_k^{n_k}. \quad (1.1)$$

□

1.6. Occupancy Problems

One objective of presenting this problem is to fine-tune our combinatorial analysis skill. A version of the occupancy problem can be described as follows: There are n indistinguishable balls to be placed in r urns. How many ways could that be done provided that each urn must contain at least one balls.

Following Polya: look at an easily solvable problem, solve it, and develop some insights for solving the original problem.

We start with a small example. Consider $n = 5$ and $r = 3$. In this case, we can enumerate the solution:

1,1,3	0#0#000
1,2,2	0#00#00
1,3,1	0#000#0
2,1,2	00#0#00
2,2,1	00#00#0
3,1,1	000#0#0

In the above table, we place our enumerated results in the first column. Thus we know the answer is 6, namely, there are 6 ways the 5 balls can be put into the three urns. Now we look for hints that would lead us to the solution. Assume that the 5 balls, each marked as "0", are lined up. There are four "gaps" between to five "0". We choose any two such gaps and fill each one of them with a marker "#". By doing so, the five "0" are being split into three groups. The numbers of 0's in the three demarcated groups corresponds to the numbers of balls in the respective urns. This is clearly illustrated in the second column of the above tally. Thus there is one-to-one correspondence between the placements of markers and the resulting solution. The number of ways this can be done is given by

$$\binom{n-1}{r-1} = \binom{5-1}{3-1} = \binom{4}{2} = 6$$

We have just solve the following general problem: Let $x_i =$ the number of balls in urn i. Assume $x_1 + \cdots + x_r = n$ and we require $x_i \geq 1$ for each i. Then the number of vectors $(x_1, \ldots, x_r)$ that satisfies the stated conditions is

$$\binom{n-1}{r-1}.$$

Variant. Consider a variant that reads

$$y_1 + \cdots + y_r = n,$$
$$y_i \geq 0.$$

We are interested in finding the number of vectors $(y_1, \ldots, y_r)$ that satisfies the above conditions.

We let $x_i = y_i + 1$. Hence $y_i = x_i - 1$. We make the substitution in the above equation:

$$(x_1 - 1) + \cdots + (x_r - 1) = n,$$
$$x_i - 1 \geq 0$$

or equivalently

$$x_1 + \cdots + x_r = n + r,$$
$$x_i \geq 1 \quad \text{for each } i.$$

But the answer for this version is $\binom{n+r-1}{r-1}$. Hence the problem is solved. □

Example 18. A person plan to leave 10 million dollars to her four children. Her favorite child should receive at lease 3 million dollars. How many ways the inheritance can be divided among the four children. Assume that each child should receive at least 1 million dollars.

Using one million as an unit. Let $y_i =$ the amount of inheritance to be received by child i. Then, we have

$$y_1 + \cdots + y_4 = 10,$$
$$y_1 = 3, 4, \ldots,$$
$$y_2, y_3, y_4 = 1, 2, \ldots .$$

We set $x_1 = y_1 - 2$ and $x_i = y_i$, $i = 2, 3$, and 4. Thus, an equivalent problem becomes

$$(x_1 + 2) + x_2 + x_3 + x_4 = 10,$$
$$x_i \geq 1 \quad \text{for each } i$$

or

$$x_1 + \cdots + x_4 = 8$$
$$x_1 \geq 1 \text{ for each } i.$$

Thus, the number of ways the inheritance can be allocated is $\binom{8+4-1}{4-1} = \binom{11}{3} = 165.$ □

1.7. Combinatorial Generating Functions

In this section, we explore the use of polynomials to assist us in doing counting involving permutations and combinations. Stated simply, we would like to make use of the coefficients in a multinomial expansion of a polynomial to generate results for a certain class combinatorial problems. It is in this context, we call a polynomial expansion, accompanied by relevant interpretations of its coefficients, a combinatorial generating function.

Consider a multinomial expansion of $(x_1 + x_2 + x_3)^4$, we try to give interpretations to its resulting coefficients. Using (1.1), we obtain

$$(x_1 + x_2 + x_3)^4 = x_1^4 + 4x_1^3x_2 + 4x_1^3x_3 + 6x_1^2x_2^2 + \underbrace{12x_1^2x_2x_3}_{(*)}$$

$$+ 6x_1^2x_3^2 + 4x_1x_2^3 + 12x_1x_2^2x_3 + 12x_1x_2x_3^2$$

$$+ 4x_1x_3^3 + x_2^4 + 4x_2^3x_3 + 6x_2^2x_3^2 + 4x_2x_3^3 + x_3^4. \quad (1.2)$$

Consider the term marked by $(*)$, $x_1^2x_2x_3$. Assume that we have 4 "1's" to be placed in three slots indexed by 1, 2, and 3, where 4 is the exponent shown at the left hand side of (1.2). We envision that two "1's" are placed in slot 1, one "1" in slot 2, and one "1" in slot 3, For the each "1" we assign a *value* of one to the corresponding x_i. For this specific term, we end up in (1.2) having $x_1^2 = 1 \cdot 1 = 1$, $x_2^1 = 1$, and $x_3^1 = 1$. We notice that for the three *exponents* of the term $x_1^2x_2x_3$ can be represented by an *ordered* triplet $(n_1, n_2, n_3) = (2, 1, 1)$. Now its coefficient in associated with this term in (1.2) represents the number of ways the 4 "1's" can be distributed to form the pattern (2, 1, 1), namely, $\binom{4}{2,1,1} = 12$.

Following the above line of argument, we consider the case where we have 4 indistinguishable balls to be randomly distributed to three urns. If we are interested in the number of cases where we observe the *ordered* pattern (2, 2, 0), then the coefficient associated with the term $x_1^2x_2^2$ gives the answer. It is 6. On the other hand, if we are interested in the number of cases of *unordered* pattern (2, 2, 0), the sum of the coefficients associated with the terms $x_1^2x_2^2$, $x_1^2x_3^2$, and $x_2^2x_3^2$ yields 18. Extracting these coefficients from a polynomial can easily be done using *Mathematica*. This is shown in Illustration 3.

In this example, when we set all $\{x_i\}$ to 1, then it give

$$\sum_{\substack{(n_1,n_2,n_3) \\ n_1+n_2+n_3=4 \\ n_i=0,1,\dots\,\forall i}} \binom{4}{n_1, n_2, n_3} = 81.$$

This is equal to $(1+1+1)^4 = 3^4 = 81$ — the number of ways the four "1's" can land in each one of the three slots. From the result of the occupancy problem, the number vectors (n_1, n_2, n_3) such that the conditions under the summation sign is met is given by

$$\binom{n+r-1}{r-1} = \binom{4+3-1}{3-1} = \binom{6}{2} = 15.$$

This is the number of terms on the right side of (1.2). For example, the first three terms on the right side of (1.2) correspond to the ordered triplets $(4, 0, 0), (3, 0, 1)$, and $(3, 2, 0)$. The coefficient in front of each triplet gives the numbers of ways the "1's" can be randomly distributed to the respective slots. Having given the aforementioned interpretations to the multinomial expansion of a polynomial, we are ready to give examples to demonstrate the applications of combinatorial generating function.

Example 19. Consider there are four identical balls are randomly thrown into three numbered urns. How many ways in which two balls land in one urn and another two land in a different urn.

Solution. We set the polynomial as it has just been suggested $(x_1+x_2+x_3)^4$. We sum over the coefficients associated with the terms $x_1^2 x_2^2$, $x_1^2 x_3^2$, and $x_2^2 x_3^2$ in the expansion. In Illustration 3, we find that it is 18 — as expected. □

Example 20. An elevator starts from the basement of a four-story condo with 7 seven passengers on board. Assume that passengers will leave for the four floors randomly and there are no one enters the elevator for this ride towards the top floor. (a) What is the number ways, denote it by $n(E)$, that all passengers get discharged from the elevator and there are four passengers leave for the third floor? (b) What is the number ways, denote it by $n(S)$, that all passengers get discharged from the elevator?

Solution. We set up the polynomial as

$$(x_1 + x_2 + t\, x_3 + x_4)^7.$$

We note that a tag "t" has been placed along with x_3. The specific event of leaving from floor 3 is being duly acknowledged. Then we can determine n by summing all the coefficients in the above expansion whose terms involving t^4. From Illustration 4, we see that $n(E) = 945$ and $n(S) = 16384$. □

Example 21. Consider the roll of a fair die 6 times. Let E be the event of getting a sum of 20. Find the number of ways this can occur. Denote this number by n.

Solution. Define the polynomial by

$$(x_1 + x^2 + x^3 + x^4 + x^5 + x^6).$$

We envision that in each roll of the die, if it lands a number "i," then an "1" is tosses into slot i whose exponent is i. The exponent counts the *number* resulting from the roll. If we sum up all coefficients associated with the above polynomial expansion with their exponents being 20. It gives the value of n. □

1.8. Exploring with Mathematica

Illustration 1. This following results are related to Example 6:

```
x = Length[Permutations[{P₁, E₁, P₂, P₃, E₂, R}]]
```
720

```
y = Length[Permutations[{P, E, P, P, E, R}]]
```
60

```
z = y * 2! * 3!
```
720   □

Illustration 2. Recall Example 16, we have 15 students and want to form three teams with five members per team. The number of ways the three teams cab be formed is given by the multinomial coefficient:

```
Multinomial[5, 5, 5]
```
756 756   □

Illustration 3. The number of terms involving $x_1^2 x_2^2$, $x_1^2 x_3^2$, $x_2^2 x_3^2$ in a polynomial expansion of $(x_1 + x_2 + x_3)^4$ as stated in Example 19.

```
Total[Coefficient[(x₁ + x₂ + x₃)⁴, {x₁² x₂², x₁² x₃², x₂² x₃²}]]
```
18   □

Illustration 4. The counting of the number of cases, $n(E)$, where four passengers leaving at floor 3 in Example 20. The total number of discharge scenarios is given by $n(S)$.

```
tx = CoefficientList[Expand[(x₁ + x₂ + t x₃ + x₄)⁷], t][[5]];
```

We note that we use [[5]] as opposed to [[4]] because of the offset for the term t is 0.

```
nE = Total[tx] /. {x₁ → 1, x₂ → 1, x₃ → 1, x₄ → 1, t → 1}
```

945

```
tt = Collect[Expand[(x₁ + x₂ + x₃ + x₄)⁷], {t}];
```

```
nS = Total[tt] /. {x₁ → 1, x₂ → 1, x₃ → 1, x₄ → 1}
```

16 384 ☐

Illustration 5. Example 21 about the number of cases, nE, in rolling a fair die 6 times and getting a total of 20.

$$p = \text{Expand}\left[\left(x + x^2 + x^3 + x^4 + x^5 + x^6\right)^6\right];$$

$$nE = \text{Coefficient}\left[p, x^{20}\right]$$

4221 ☐

Problems

1. In how many ways can 5 boys and 5 girls be seated around a round table so that no two boys sit next to each other?

2. In how many ways can a lady having 10 dresses, 5 pairs of shoes, and 2 hats be dressed?

3. In how many ways can we place in a bookcases with two works each of three volumes and two works each of four volumes, so that the volume of the same work are not separated?

4. In how many ways can r indistinguishable objects be distributed to n persons if there is no restriction on the number of objects that a person may receives?

5. In how many ways can r distinguishable objects be distributed to n persons if there is no restriction on the number of objects that a person may receive?

6. Eight points are chosen on the circumference of a circle. (a) How many chords can be drawn by joining these point in all possible ways? If the

8 points are considered vertices of a polygons, how many (b) triangles, and how many (c) hexagons can be formed?

7. (a) In how many ways can 20 recruits be distributed into four groups each consisting of 5 recruits? (b) In how many ways can they be distributed into 4 camps, each camp receiving 5 recruits?

8. Using 7 consonants and 5 vowels, how many words consisting of 4 consonants and 3 vowels can we form?

9. A house has 12 rooms. We want to paint 4 yellow, 3 purple, and 5 red. In how many ways can this be done?

10. How many ways can 5 history books, 3 math books, and 4 novels be arranged on a shelf if books of each type must be together?

11. Twelve different toys are to be divided among 3 children so that each one gets 4 toys. How many ways can this be done?

12. How many ways can 4 men and 4 woman can sit in a row if no two men or two women sit next to each others?

13. (a) In how many ways can 4 boys and 4 girls sit in a row? (b) In how many ways can 4 boys and 4 girls sit in a row if boys and girls are each to sit together? (c) In how many ways if only boys must sit together? (d) In how many ways if no two people of the same gender are allowed to sit together?

14. How many 3 digit numbers xyz with x, y, z all ranging from 0 to 9 have at least 2 of their digits equal? (b) How many have exactly 2 equal digits?

15. We have $20,000 that must be invested among 4 possible opportunities. Each investment must be integral in units of $1,000. If an investment is to be made in a given opportunity at all, then the minimal amounts to be invested in the each one of the four investments are $2,000, $2,000, $3,000, and $4,000, respectively. How many different investment strategies are available if

    (a) An investment must be made in each opportunity?
    (b) Investments must be made in at least 3 of the four opportunities?

16. An elevator starts at the basement with 8 people (not including the elevator operator) and discharges them all by the time it reaches the

top floor, the 6th floor. In how many ways could the operator have observed the 8 people leaving the elevator (e.g., all leave from the 1st floor)? What if the 8 people consist of 5 men and 3 women (e.g., 2 men leave from the 2nd floor, 1 man leaves from 3rd floor, 1 man leaves from the 5th floor, and all 3 women leave from the 6th floor)?

17. Delegates from 12 countries, including Russia, Australia, New Zealand, and the United States, are to be seated in a row. How many different seating arrangements are possible if the Australian and New Zealand delegates are to be next to each other and the Russian and U.S. delegates are not to be next to each other?

18. A dance class consists of 22 students, of which 10 are women and 12 are men. If 5 men and 5 women are to be chosen and then paired off, how many results are possible?

19. A person has 8 friends, of whom 5 will be invited to a party. (a) How many choices are there if 2 of the friends are feuding and will not attend together? (b) How many choices are there if 2 of the friends will only attend together?

20. If 8 identical blackboards are to be divided among 4 schools, how many divisions are possible? How many if each school must receive at least 1 blackboard?

21. (a) In how many ways can n indistinguishable balls be put into n numbered boxes so that exactly one box is empty? (b) In how many ways can n distinguishable balls be put into n numbered boxes so that exactly one box is empty?

22. A girl decides to choose either a shirt or a tie for a birthday present. There are 3 shirts and 2 ties to choose from. (a) How many choices does she have if she will get only one of them? (b) If she may get both a shirt and a tie?

23. There are 3 kinds of shirts on sale. (a) If two men buy one shirt each how many possibilities are there? (b) If two shirts are sold, how many possibilities are there?

24. How many different initials can be formed with 2 or 3 letters of the alphabet? How large must the alphabet be in order that one million people can be identified by 3-letter initials?

25. (a) In how many ways can 4 boys and 4 girls be paired off? (b) In how many ways can they stand in a row in alternating gender?

26. (a) In how many ways can a committee of three be chosen from 20 people? (b) In how many ways can a president, a secretary and a treasurer be chosen?

27. You walk into a party without knowing anyone there. There are 6 women and 4 men and you know there are 4 married couples. (a) In how many ways can you guess who the couples are? (b) What if you know there are exactly 3 couples?

28. A child has 5 blocks with the letter A, 3 blocks with the letter B and 1 block with the letter C. Assuming these 9 blocks are to be randomly lined up to form letters. How many distinct words can be formed?

29. There are 10 indistinguishable balls and 5 numbered urns. These balls are randomly thrown into the urns. (a) How many different distributions of balls into the urns. (b) How many different distributions of balls into the urns provided that each urn must contain at least one ball.

30. Letters in the Morse code are formed by sending a succession of dashes and dots with repetitions allowed. How many letters can be formed using ten symbols or less?

31. Consider an experiment consists of determining the type of job — either blue collar or white collar, and the political affiliation — Republican, Democrat, or Independent — of the 15 members of an adult soccer team (a) How many elements are in the sample space? (b) How many elements are there when at least one of the team members is a blue-collar worker? (c) How many elements are there when none of the team members consider himself or herself an independent?

Remarks and References

Example 6 is based on Example 3d in Chapter 1 of Ross [4]. G. Polya [3] gives a clear road map for solving intricate combinatorial problems. The book by Graham, Knuth, and Patashnik [1], has a whole chapter on binomial coefficients. The exposition on combinatorial generating functions stems from Mood and Graybill [2, pp. 27–31]. A related work of interest on generating functions is the book by Wilf [5].

[1] Graham, R. L., Knuth, D. E. and Patashnik, O. *Concrete Mathematics: A Foundation for Computer Science*, Addison-Wesley, 1999.

[2] Mood, A. M. and Graybill, F. A., *Introduction to the Theory of Statistics*, 2nd edn., McGraw-Hill, 1963.

[3] Polya, G., *How to Solve It*, A Doubleday Anchor Book, 1957. (Originally published by Princeton University Press in 1945.)

[4] Ross, S. M., *A First Course in Probability*, 10th edn., Pearson, 2019.

[5] Wilf, H. S., *Generating functionology*, Academic Press, 1990.

Chapter 2

Axioms of Probability

2.1. Introduction

The term probability is used everywhere. On a daily basis, your local weather person will say something like the probability of precipitation tomorrow is such and such, or the probability that someone will win an election is this or that. Exactly, what is meant by probability? In general, there is no precise definition of probability in terms of events occurring in our daily life. Roughly, probability can be thought of the likelihood whether an event will occur. But it can be made more precise if we are dealing with casino games. Legends have it that probability had its origin from card games played in French royal courts.

2.2. Sample Space, Events, and Sets

We start with the definition of a *discrete* sample space. The set of possible sample points, denoted by $\{\omega_i\}$, of an "experiment" is called a sample space S. We can view sample points are the elements of the set Ω. The notion of sample point and sample space can best be illustrated by the following examples.

Example 1. Consider the flipping of a coin. If the coin is flip only once, then there are only two possible outcomes: $\omega_1 = \{\text{the head turns up}\} \equiv \{H\}$ and $\{\text{the tail turns up}\} = \{T\}$. Thus, the sample space is given by $\Omega = \{\omega_1, \omega_2\}$.

Suppose the coin is flipped twice. Then we keep track of the order with which heads or tails turn up. Then, we have $\Omega = \{\omega_1, \ldots, \omega_4\} \equiv \{HH, HT, TH, HH\}$.

Suppose the coin is flipped repeatedly until heads occur for the first time. Then, we have $\Omega = \{\omega_1, \omega_2, \omega_3, \omega_4, \ldots\} \equiv \{H, TH, TTH, TTTH, \ldots\}$ □

Example 2. Consider an experiment where each trial involves the flipping a coin and rolling a die. Then we have $\Omega = \{\omega_1, \ldots, \omega_{12}\} \equiv \{(H, 1), (H, 2), \ldots, (H, 6), (T, 1), (T, 2), \ldots, (T, 6)\}$. Here each sample point is a vector denoting the outcome of each trial.   □

We now define what is meant by an event. Events are defined on the sample space with some specific prescription. In other words, events are subsets of S where its components meet some defining criteria.

Example 3. Consider the flipping of a coin twice. Recall that the sample space is given by $\Omega = \{\omega_1, \ldots, \omega_4\} \equiv \{HH, HT, TH, TT\}$. Suppose now we define an event being the number of times heads showing up. Then there are three events possible: $\{0, 1, 2\} \equiv \{(TT), (HT, TH), (HH)\} = \{\omega_4, (\omega_2, \omega_3), \omega_1\}$.   □

Since sample space is a set and events are defined on the sample space. It is natural to explore properties of sets and exploit their relations to facilitate our work with probability.

Events are subsets of the sample space Ω. We consider two events E and F and define their union $E \cup F$ and the intersection $E \cap F$. We say sample point $\omega \in E \cap F$, if $\omega \in E$ and $\omega \in F$. We say sample point $\omega \in E \cup F$, if $\omega \in E$, or $\omega \in F$, or $\omega \in E \cap F$. These relations are shown in the following graphs, known as the Venn diagrams:

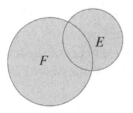
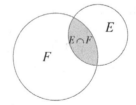

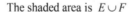

The shaded area is $E \cup F$ The shaded area is $E \cap F$

If $E \cap F = \varnothing$, then E and F are said to be disjoint, where $\varnothing$ is called the *null* set, the set contains no sample points. We use E^c to denote the *complement* of the set E, i.e., $\omega \in E^c$ implies that $\omega \in \Omega \backslash E$, the sample space with the set E removed, where we use "$\backslash$" to denote set exclusion. If $E \in \Omega$ and $F \in \Omega$, the set $E \backslash F$ is the set of sample points ω in E but not in F, i.e., $E \backslash F = E \cap F^c$.

When we write $E \subset F$, we mean E is a subset of F, i.e., if $\omega \in E$, then $\omega \in F$. The converse may or may not true. The following laws are the common rules governing union and intersections:

Commutative laws: $E \cup F = F \cup E$, $EF = FE$.

Associate laws: $(E \cup F) \cup G = E \cup (F \cup G)$, $(EF)G = E(FG)$,

Distributive laws: $(E \cup F)G = EG \cup FG$, $EF \cup G = (E \cup G)(F \cup G)$.

We note that following the convention, we omit the intersection symbol and write AB for $A \cap B$. Moreover, the above three laws follow the duality in the sense when the symbols $\cup$ and $\cap$ are switched, they yield the respective counterparts.

The following laws are known as DeMorgan's laws:

$$\left(\bigcup_{i=1}^{n} E_i \right)^c = \bigcap_{i=1}^{n} E_i, \quad \left(\bigcap_{i=1}^{n} E_i \right)^c = \bigcup_{i=1}^{n} E_i^c.$$

For the case when there are only events E and F, the above laws simplify to

$$(E \cup F)^c = E^c F^c, \quad (EF)^c = E^c \cup F^c.$$

One way to prove DeMorgan's laws is by induction using the above identities as the starting point. The three operations defined so far: complement, union and intersection obey the DeMorgan's law.

We now introduce the fourth operation on sets — namely, the *symmetric difference* between two sets E and F. It is defined by

$$E \triangle F = (E \backslash F) \cup (F \backslash E) = (EF^c) \cup (FE^c).$$

It is shown in the following graph:

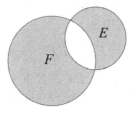

The shaded area is $E \triangle F$

In Illustration 1, we use Mathematica to demonstrate the application of the symmetric difference operator $\triangle$ on sets.

2.3. Axioms of Probability

If the number of sample points in event E is $n(E)$ and the number of sample points in sample space Ω is $n(\Omega)$, assuming that each sample point in the sample space Ω is equally likely to occur, then the probability event E occurs is given by

$$P(E) = \frac{n(E)}{n(\Omega)}.$$

The probability so defined must obey the following three axioms:

Axiom 1: $0 \le P(E) \le 1$;

Axiom 2: $P(\Omega) = 1$;

Axiom 3: Consider events $E_1, E_2, \ldots$. Assuming events $E_i E_j = \emptyset$ whenever $i \ne j$ (known as the events being mutually exclusive), then

$$P\left(\bigcup_{i=1}^{\infty} E_i\right) = \sum_{i=1}^{\infty} P(E_i).$$

The following conditions are consequences of the above three axioms:

1. $P(E) = 1 - P(E^c)$.
2. If $E \subset F$, then $P(E) \le P(F)$.

   Proof. As $E \subset F$, we have $F = F\Omega = F(E \cup E^c) = FE \cup FE^c = E \cup FE^c$ and $E \cap FE^c = \emptyset$. Thus

   $$P(F) = P(E) + P(FE^c).$$

   Since $P(FE^c) \ge 0$, we have $P(E) \le P(F)$.
3. $P(E \cup F) = P(E) + P(F) - P(EF)$.

   One way to see the identity is to examine the graph depicting $E \cup F$. Then we see that $P(E) + P(F)$ implies that $P(EF)$ is being added twice. So we remove one duplicate $P(EF)$.

   A methodical way is to write

   $$E = A\Omega = E(F \cup F^c) = EF \cup EF^c \quad \text{and note} \quad EF \cap EF^c = \emptyset.$$

   Similarly, $F = EF \cup E^c F$ and $EF \cap E^c F = \emptyset$. Thus $P(E) = P(EF) + P(EF^c)$ and $P(F) = P(EF) + P(E^c F)$. Moreover,

   $$P(E) + P(F) = P(EF) + P(EF^c) + P(EF) + P(E^c F) = P(EF) + P(E \cup F).$$

   Thus, the desired result follows.

4. $P(E \cup F \cup G) = P(E) + P(F) + P(G) - P(EF) - P(EG) - P(FG) + P(EFG)$.

Proof. One proof of the above is by induction. We consider $A \equiv F \cup G$ as a set at the moment. We have $P(E \cup A) = P(E) + P(A) - P(EA)$. Now

$$P(A) = P(E) + P(F) - P(EF)$$

and

$$P(E(F \cup G)) = P(EF \cup EG) = P(EF) + P(EG) - P(EFG).$$

Making the substitutions, we find the needed result. $\square$

5. A further generalization of the above to more than three events should be obvious due to its symmetries in the type of terms involved and the alternating of signs involved. But the general result is known as the inclusion–exclusion formula. Let $\{E_i\}_{i=1}^n$ be events. Then

$$
\begin{aligned}
P(E_1 \cup E_2 \cup \cdots \cup E_n) = {}& \sum_{i=1}^n P(E_i) - \sum_{i_1 < i_2} P(E_{i_1} E_{i_2}) \\
& + \cdots + (-1)^{k+1} \sum_{i_1 < i_2 < \cdots i_k} P(E_{i_1} E_{i_2} \cdots E_{i_k}) \\
& + \cdots + (-1)^{n+1} P(E_1 E_2 \cdots E_n).
\end{aligned}
$$

The summation $\sum_{i_1 < i_2 < \cdots i_k} P(E_{i_1} E_{i_2} \cdots E_{i_k})$ is taken over all of the $\binom{n}{k}$ possible sets of size k from a total of n components $\{E_i\}$.

6. Boole's Inequalities. Let $\{E_i\}_{i=1}^n$ be a sequence of events. We have

$$P\left(\bigcup_{i=1}^n E_i\right) \leq \sum_{i=1}^n P(E_i) \quad \text{and} \quad P\left(\bigcap_{i=1}^n E_i\right) \geq 1 - \sum_{i=1}^n P(E_i^c).$$

Proof. We do the proof by induction. For $n = 1$, the inequality on the LHS of the above is clearly true. Assume it is true for $m \leq n$. Then we have

$$P\left(\bigcup_{i=1}^{m+1} E_i\right) = P\left(\left(\bigcup_{i=1}^{m} E_i\right) \bigcup E_{m+1}\right)$$

$$= P\left(\bigcup_{i=1}^{m} E_i\right) + P(E_{m+1}) - P\left(\left(\bigcup_{i=1}^{m} E_i\right) \bigcap E_{m+1}\right)$$

$$\leq P\left(\bigcup_{i=1}^{m} E_i\right) + P(E_{m+1}) \leq \sum_{i=1}^{m+1} P(E_i).$$

For the second inequality, we apply DeMorgan's law and the first inequality. This gives

$$P\left(\bigcap_{i=1}^{n} E_i\right) = P\left(\left(\bigcup_{i=1}^{n} E_i^c\right)^c\right) \geq 1 - P\left(\bigcup_{i=1}^{n} E_i^c\right) \geq 1 - \sum_{i=1}^{n} P(E_i^c). \qquad \square$$

Example 4. Integers are drawn randomly with replacement from the set $\{1, \ldots, 1000\}$. Let E be the event that a number drawn is divisible by 5, 7, or 13. Let D_i denote the event that the number drawn is divisible by i.

Applying the inclusion–exclusion formula with $n = 3$, we have

$$P(E) = P(D_3 \cup D_5 \cup D_7)$$

$$= P(D_5) + P(D_7) + P(D_{13}) - P(D_5 D_7) - P(D_5 D_{13}) - P(D_7 D_{13})$$

$$+ P(D_5 D_7 D_{13}).$$

In Illustration 2, we find that $P(E) = 0.367$. \qquad $\square$

2.4. Sample Space Having Equally Likely Outcomes

As stated in the last section, if we assume that each sample point in the sample space Ω is equally likely to occur. Then the probability event E occurs is given by

$$P(E) = \frac{n(E)}{n(\Omega)}.$$

This is what we mean by *random sampling*, namely, each sample point will have equal probability of being sampled. Many times this is obvious from the type of "experiments" in which we are engaged. But at times, we must be cautious whether this assumption holds.

Example 5. A card is drawn randomly from a well-shuffled standard card deck of 52 cards. What is the probability that the card drawn is a heart or an ace or both?

Solution. Let $E = \{$the card drawn is a heart$\}$ and $F = \{$the card drawn is an ace$\}$. Then we have

$$P(E \cup F) = P(E) + P(F) - P(EF)$$
$$= \frac{13}{52} + \frac{4}{52} - \frac{1}{52} = \frac{16}{52}.$$

An interesting alternative to solve the same problem is using

$$P(E \cup F) = 1 - P(E^c F^c).$$

How many non-hearts are there? There are 39 of them. Of the 39 non-hearts, there are 3 aces. So there are 36 non-hearts and non-aces. Hence

$$P(E \cup F) = 1 - \frac{36}{52}. \qquad \square$$

Example 6. If 5 balls are thrown randomly into 10 boxes, what is the probability that no box will contain more than 1 ball?

Solution. We see each ball has 10 places to go. Thus the number of ways the 5 balls can go is 10^5. Hence we have $n(\Omega) = 10^5$. Now the event E that no box will contain more than 1 ball implies that the 5 balls will be distributed to 5 boxes with each contains exactly one. There are $\binom{10}{5}$ to choose these 5 balls with each occupies one box. But there are $5!$ ways to distribute these 5 balls to their respective destinations. Thus $n(E) = \binom{10}{5}5!$. Hence

$$P(E) = \frac{n(E)}{n(\Omega)} = \frac{\binom{10}{5}5!}{10^5} = 0.3024.$$

Alternatively, we can find the above by placing each ball sequentially over the remaining boxes. This gives

$$P(E) = \frac{10 \cdot 9 \cdot 8 \cdot 7 \cdot 6}{10^5} = \frac{(10)_5}{10^5} = 0.3024. \qquad \square$$

Example 7. You are playing blackjack against the dealer. A blackjack consists of one ace together with one card from the 16 cards consisting of tens, jacks, queens, and kings. What is the probability that neither you nor the dealer gets a blackjack from a well-shuffled standard card deck of 52 cards?

Solution. Let E denote the event that the dealer gets a blackjack. Let F denote the event that you get a blackjack. Then

$$P(E) = \frac{\binom{4}{1}\binom{16}{1}}{\binom{52}{2}} = 0.048265, \quad P(F) = \frac{\binom{4}{1}\binom{16}{1}}{\binom{52}{2}} = 0.048625.$$

Now

$$P(EF) = \frac{\binom{4}{2}\binom{16}{2}}{\binom{52}{4}} = 0.0026595.$$

Thus

$$P(\text{at least one of the two gets a blackjack})$$
$$= P(E) + P(F) - P(EF)$$
$$= 2(0.048265) - 0.0026595 = 0.093871.$$

Therefore, we have

$$P(\text{neither you nor the dealer gets a blackjack})$$
$$= 1 - P(\text{at least one of the two gets a blackjack})$$
$$= 1 - 0.093871 = 0.90613. \qquad \square$$

Example 8. Fifteen new students, of which 12 are boys and 3 are girls, are to be randomly assigned to three classes. (a) What is the probability that each class gets one girl? (b) What is the probability that one class gets all the girls?

Solution. (a) To distribute the 15 students to three classes, there are

$$\frac{15!}{5!5!5!}$$

ways to do it. This gives $n(\Omega)$. Since the three girls are to be assigned to each one of the three classes, the total number ways to assign the 12 boys to three classes with 4 each are

$$\frac{12!}{4!4!4!}.$$

Now the number of ways to assign the 3 girls to three classes are 3!. Applying the multiplication principle, we obtain

$$n(E) = 3!\frac{12!}{4!4!4!},$$

where E is the event that each class gets one girl. Hence

$$P(E) = \frac{n(E)}{n(\Omega)} = \frac{3!\frac{12!}{4!4!4!}}{\frac{15!}{5!5!5!}} = 0.2747.$$

(b) The sample space remains the same as in (a). Now the 12 boys will be assigned into 3 classes of sizes 5, 5, and 2. There are

$$\frac{12!}{5!5!2!}$$

ways to do this. Now there are 3 ways to assign all three girls to one class. Applying the multiplication principle, we obtain

$$n(F) = 3 \times \frac{12!}{5!5!2!},$$

where F is the event that one class gets all the three girls. Hence

$$P(F) = \frac{n(F)}{n(\Omega)} = \frac{3 \times \frac{12!}{5!5!2!}}{\frac{15!}{5!5!5!}} = \frac{6}{91}. \qquad \square$$

Example 9. Assume N cowboys enter a barbecue joint. Each leaves his cowboy hat to the attendant. When they leave the restaurant, the absent-minded attendant randomly gives the hats to the cowboys. What is the probability that no one gets his own hat back?

Solution. Let E_i denote the event that cowboy i gets his own hat back. Then

$$P\left(\bigcup_{i=1}^{N} E_i\right) = P(\text{at least one cowboy gets his own hat}).$$

Let E_0 denote the event that no one gets his own hat back. Then

$$P(E_0) = 1 - P\left(\bigcup_{i=1}^{N} E_i\right).$$

To find $P(\bigcup_{i=1}^{N} E_i)$, we apply the inclusion–exclusion formula. We note

$$P(E_i) = \frac{(N-1)!}{N!} = \frac{1}{N}.$$

There are $\binom{N}{1}$ terms of the like in the first term of the inclusion–exclusion formula. Thus, the first term is

$$\binom{N}{1} \frac{1}{N} = 1.$$

For its second term, with $i \neq j$, we have

$$P(E_i E_j) = \frac{(N-2)!}{N!}$$

and there are $\binom{N}{2}$ of the like. Thus the second term of the formula reads

$$-\binom{N}{2} \frac{(N-2)!}{N!} = -\frac{N!}{2!(N-2)!} \frac{(N-2)!}{N!} = -\frac{1}{2!}.$$

A similar argument is repeated until we reach the last term of the formula. Thus

$$P(E_0) = 1 - \left(1 - \frac{1}{2!} + \cdots + (-1)^{N+1} \frac{1}{N!}\right) = \sum_{i=0}^{N} \frac{(-1)^i}{i!}.$$

Recall

$$e^x = \sum_{i=0}^{\infty} \frac{x^i}{i!}.$$

When N is large, we set x to -1 and reach an approximation

$$P(E_0) \approx e^{-1} = 0.3670.$$

This is an asymptotic approximation. $\qquad\square$

Example 10. (A Game of Poker). The game of poker provides a nice paradigm to fine-tune our expertise in doing probability. A *poker hand* is the outcome of five cards drawn from a standard card deck of 52 cards. Some of the outcomes of such a drawn have well-known names. Let E denote a poker hand. We first determine the size of the sample space $n(\Omega)$, namely, the number of possible outcomes from a random draw. It is easy to see that there are

$$n(\Omega) = \binom{52}{5} = 2,5908,960$$

unique poker hands. In the following, for these hands listed we simply derive the corresponding $n(E)$.

(a) *Royal flush.* Here, a hand has ten, jack, queen, king, ace in a single suit. As an example, the hand has $10\spadesuit, J\spadesuit, Q\spadesuit, K\spadesuit, A\spadesuit$. It is clear that $n(E) = 4$.

(b) *Full house.* Here, a hand has one pair and one triple of cards with equal face values. As an example, the hand has $(8\spadesuit 8\clubsuit)(9\diamondsuit 9\spadesuit 9\clubsuit)$. We first pick the value of the pair (say, 8 in the sample). There are 13 possibilities. We then select suits for the selected value. There are $\binom{4}{2}$ ways. Now we pick the value for the triple. There are only 12 possibilities left (say, 9 in the example). We then select suits for the selected value. There are $\binom{4}{3}$ ways to make the choice. We conclude that

$$n(E) = 13\binom{4}{2} \cdot 12\binom{4}{3} = 3,744.$$

(c) *One pair.* Here, a hand has one pair of equal face values plus three different cards. As an example, the hand has $(8\spadesuit 8\clubsuit)(9\clubsuit 3\spadesuit 4\diamondsuit)$. We first select the value of the pair from the 13 possibilities. One the value is chosen, there are $\binom{4}{2}$ to choose the suits. We then pick 3 values from the remaining 12 possible values. Of the 3 distinct values, any suit from the four possible choices will do. Hence, there are $\binom{12}{3}4^3$ ways to select the three different cards. We conclude that

$$n(E) = \binom{13}{1}\binom{4}{2} \times \binom{12}{3}4^3 = 1,098,240.$$

(d) *Two pair.* Here a hand has two pairs of equal face value plus one other card. As an example, the hand has $(8\spadesuit 8\clubsuit)(9\clubsuit 9\spadesuit)(4\diamondsuit)$. There are $\binom{13}{2}$ ways to choose the two face values and $\binom{4}{2}\binom{4}{2}$ to select the corresponding suits. The total number ways to obtain the two pairs is $\binom{13}{2}\binom{4}{2}^2$. There are remaining 44 cards from which we select one. Hence

$$n(E) = \binom{13}{2}\binom{4}{2}^2 \times 44 = 123,552.$$

The answer to the last case is given in Illustration 3. □

Example 11. In a period from 10/4 to 11/12 of the 2019–2020 season, the NBA team Houston Rockets had 5 runs of wins out of 20 games and a total of 13 wins. Moreover, its opponents had 5 runs of losses. The specifics are given in the following table

Date	score (Rockets/Opponent)	Rockets' Fate
10/4/19	109/96	W
10/8/19	129/134	L
10/10/19	118/111	W
10/16/19	114/128	L
10/18/19	144/133	W
10/24/19	111/117	L
10/26/19	126/123	W
10/28/19	116/112	W
10/30/19	159/158	W
11/1/19	116/123	L

Date	score (Rockets/Opponent)	Rockets' Fate
11/3/19	100/129	L
11/4/19	107/100	W
11/6/19	129/112	W
11/9/19	117/94	W
11/11/19	122/116	W
11/16/19	125/105	W
11/18/19	132/108	W
11/20/19	125/105	W
11/22/19	119/122	L
11/24/19	123/137	L

We want to ask the following question: if the *sequence* of occurrences of wins (W's) and losses (L's) occur purely randomly, what is the probability that the result, denoted by E, shown in above table would have occurred?

Assuming that the $n(\Omega) = \binom{13+7}{13}$ orderings are equally likely. We want to find the probability of having 5 runs of wins.

Let y_1 denote the number of losses before the first run of wins; y_6 be the number of losses after the last run of wins (namely, the 5th run of wins). From the record, we have $(y_1, \ldots, y_6) = (0, 1, 1, 1, 2, 2)$. Let x_i be the number of wins sandwiched in between two sets of contiguous losses. Hence $(x_1, \ldots, x_5) = (1, 1, 1, 3, 7)$.

Thus for the W's we have occupancy problem A:

$$x_1 + \cdots + x_5 = 13, \quad x_i = 1, 2, \ldots \text{ for all } i.$$

From the result of the occupancy problem, we have $n(A) = \binom{13-1}{5-1}$.

Thus for the L's we have occupancy problem B:

$$y_1 + \cdots + y_5 + y_6 = 7, \quad y_i = 1, 2, \ldots \text{ for } i = 2, \ldots, 5; \quad y_1, y_6 = 0, 1, \ldots.$$

We convert the last occupancy problem by setting $z_1 = y_1 + 1$ and $z_6 = y_6 + 1$ and $z_i = y_i$, $i = 2, \ldots, 4$. Making the substitutions yields

$$(z_1 - 1) + z_2 + \cdots + (z_6 - 1) = 7, \quad z_i = 1, 2, \ldots \text{ for all } i$$

or equivalently

$$z_1 + \cdots + z_6 = 9, \quad z_i = 1, 2, \ldots \text{ for all } i$$

Again, we use the result of the occupancy problem, and obtain $n(B) = \binom{9-1}{6-1} = \binom{8}{5} = 56$. By multiplication principle, we conclude $n(E) = n(A) \times n(B) = (495) \times (56) = 27\,720$. Therefore, we find

$$P(E) = \frac{\binom{12}{4}\binom{8}{5}}{\binom{20}{13}} = 0.3576.$$

The generalization of the concept presented in this example is related to nonparametric statistical methods in the context of *runs*. □

Example 12. A football teams has 20 offensive and 20 defensive players. They are to be paired in groups of 2 for determining roommates. Assuming that pairing is done randomly. Let E_i denote the event that there are i offensive–defensive (O–D) roommate pairs, where $i = 0, 1, \ldots, 20$. We are interested in computing $P(E_i)$ for all i.

The example serves three purposes. The first is to illustrate that how do we solve an easy part of the problem and leverage its conclusion to solve a more complicated problem. The second objective is to demonstrate the use of Mathematica in Illustration 4 to handle the somewhat messy numerical problem. The last is to illustrate the verification of our computation of results.

(a) We line up the 20 rooms. The number of ways to randomly assigning 40 players to the 20 rooms (with 2 for each room) is clearly $\binom{40}{2,\ldots,2}$. Since the room designators are not our concern, we actually have

$$n(\Omega) = \binom{40}{2, \ldots, 2} / 20!$$

ways to make the assignments.

When offensive players are to be paired all by themselves, we mimic what we just did and conclude there are

$$\binom{20}{2,\ldots,2}\Big/10!$$

ways to make the assignments. Similarly, the above condition applied for paring the defensive players all by themselves. Using the multiplication principle, it follows that

$$P(E_0) = \left[\binom{20}{2,\ldots,2}\Big/10!\right]^2\Big/\binom{40}{2,\ldots,2}\Big/20!$$

(b) We now leverage the insight obtained from solving part (a) to solve for $P(E_i)$, for $i = 1, \ldots, 10$. For $i = 1, 2, \ldots, 10$, we observe that if there are $2i$ $O-D$ pairs, for the remaining offensive players, there are no $O-D$ pairs. So, we can use the result obtained in part (a) and conclude there are

$$\binom{20-i}{2,\ldots,2}\Big/(10-i)i$$

ways to assign the $O-O$ pairs But there are $\binom{20}{i}$ to choose those $20-i$ offensive $O-O$ players. By the multiplication principle, there are

$$\binom{20-i}{2,\ldots,2}\Big/(10-i)i \times \binom{20}{2i}$$

assignments possible.

For the same $2i$ $O-D$ pairs, the identical reasoning (or by symmetry) gives us the number of way to pair the remaining defensive plays to form $D-D$ pairs. Thus there are

$$\binom{20-i}{2,\ldots,2}\Big/(10-i)i \times \binom{20}{2i}$$

ways to come up with the $D-D$ pairs

Finally, we deal with the pairing of $2i$ $O-D$ players. We have the $2i$ rooms lined up, there are $(2i)!$ ways to assign one offensive player to each room. Once this is done, the other roommate must be a defensive player. Combining all the above, we conclude

$$P(E_{2i}) = (20!)\left[\binom{20-i}{2,\ldots,2}\Big/(10-i)i \times \binom{20}{2i}\right]^2\Big/\binom{40}{2,\ldots,2}\Big/20!$$

for $i = 1, \ldots, 10$.   □

2.5. The Importance of Being Truly Random

In computing the probability that an event E will occur, we use

$$P(E) = \frac{n(E)}{n(\Omega)},$$

where $n(E)$ denotes the number of sample points $\{\omega_i\}$ in E and $n(\Omega)$ denotes the number of sample points $\{\omega_i\}$ in Ω, provided each sample point will have equal probability of occurring, i.e., being random. If we define sample points and events carefully, we can avoid the likelihood of making mistakes. The following examples elaborate about the issue.

Example 13. Consider a simple experiment involving the toss of two fair coins. If we define the outcomes of the experiment by $\{HH, HT, TH, TT\} \equiv \{\omega_1, \omega_2, \omega_3, \omega_4\}$ and if we define E as the event that one head occurs, then $n(E) = 2$ and $n(\Omega) = 4$. Hence $P(E) = \frac{1}{2}$.

On the other hand, if we define the outcomes of the experiment by the number of heads occurring. The outcomes will be $\{0, 1, 2\} = \{(TT), (HT, TH), (HH)\} = \{\omega_1, \omega_2, \omega_3\}$. With $n(\Omega) = 3$ and the same definition for E, then $n(E) = 1$. But clearly, $P(E) \neq \frac{1}{3}$ because the three sample points do not have an equal probability of occurrence. □

A not-so-obvious case on this issue is illustrated in the following example.

Example 14. An airport shuttle bus makes 4 scheduled stops for 15 passengers. Assume that each passenger is equally likely to stop at each one of the four stops. What is the probability that at least one person gets off each stop?

Solution. Some may model this problem as a direct application of the occupancy problem. Let x_i denote the number of passengers get off at stop i. Then the sample space is characterized by the following:

$$x_1 + \cdots + x_4 = 15,$$

$$x_i = 0, 1, \ldots, 15.$$

Applying the result from the occupancy problem, we know that

$$n(\Omega) = \binom{n+r-1}{r-1} = \binom{15+4-1}{4-1} = \binom{18}{3} = 816.$$

The number of cases in which at least one person gets off at each stop is characterized as the follows:

$$x_1 + \cdots + x_4 = 15,$$

$$x_i = 1, \ldots, 15.$$

Applying the result from the occupancy problem, we know that

$$n(E) = \binom{n-1}{r-1} = \binom{15-1}{4-1} = \binom{14}{3} = 364,$$

where E is the event that at least one person gets off at each stop. Thus

$$P(E) = \frac{n(E)}{n(\Omega)} = \frac{364}{816} = 0.44608.$$

But this seemingly nice application of the results from the occupancy problem is incorrect. We illustrate this point as follows. Consider two sample points $\omega_1 \equiv (1, 1, 1, 12)$ and $\omega_2 \equiv (3, 3, 4, 5)$, where $\omega_1, \omega_2 \in E$. Now

$$P(\omega_1) = \binom{15}{1, 1, 1, 12} \left(\frac{1}{4}\right)^{15} = \frac{15!}{12!} \left(\frac{1}{4}\right)^{15} = 2.5425 \times 10^{-6}$$

and

$$P(\omega_2) = \binom{15}{3, 3, 4, 5} \left(\frac{1}{4}\right)^{15} = 0.011746.$$

Clearly, the sample points are not equally likely to occur.

We now solve this problem using the inclusion–exclusion formula. Let $A_i \equiv \{\text{no one gets off at stop } i\}$ and $E = \{\text{at least one gets off at each stop}\}$. Then

$$P(E) = 1 - P(A_1 \cup \cdots \cup A_4).$$

Since $P(A_i) = \left(\frac{3}{4}\right)^{15}$, $P(A_i A_j) = \left(\frac{2}{4}\right)^{15}$ for $i \neq j$ and $P(A_i A_j A_k) = \left(\frac{1}{4}\right)^{15}$ for $i \neq j \neq k$, then

$$P(E) = 1 - \left[\binom{4}{1}\left(\frac{3}{4}\right)^{15} - \binom{4}{2}\left(\frac{2}{4}\right)^{15} + \binom{4}{3}\left(\frac{1}{4}\right)^{15} \right]$$

$$= 1 - 0.05327 = 0.9467. \qquad \square$$

Example 15. We use the combinatorial generating function described in Chapter 1 to do the calculations given in Example 14. The computation of $P(\omega_1)$ and $P(\omega_2)$ is simple. In Illustration 5, we look at the coefficients associated with the terms $x_1 x_2 x_3 x_4^{12}$ and $x_1^3 x_2^3 x_3^4 x_4^5$ in the polynomial expansion of

$$(x_1 + x_2 + x_3 + x_4)^{15}.$$

The two terms pop out immediately.

To find the probability that at least one gets off at each stop, we use the definition of event $A_1 = \{$no one gets off at stop 1$\}$. To accomplish the tracking of this event, we use the polynomial

$$(x_1 + t\, x_2 + t\, x_3 + t\, x_4)^{15}$$

to track the event that no passengers get off at floor 1. Let n_{X_1} denote the number of sample points for which A_1 occurs. By symmetry, this setup applies to the other three stops. Thus we can count the number of sample points $n(A_1 \cup A_2 \cup A_3 \cup A_4) = 4 \times n_{X1}$. Thus

$$P(E) = 1 - \frac{(4 \times n_{X_1})}{4^{15}}.$$

Illustration 5 clearly demonstrates the computations. □

Example 16. Seven balls are tosses randomly into four numbered boxes. Let E denote the event that the first box will contain two balls. Find $P(E)$.

Solution. We see that $n(\Omega) = 4^7$. Then $\binom{7}{2}$ are the number of ways that two balls are chosen to go to box 1. Once that is done, the remaining 5 balls can go to any one of the other 3 boxes. There are 3^5 ways to do the latter. Hence $n(E) = \binom{7}{2} 3^5$. This gives

$$P(E) = \frac{n(E)}{n(\Omega)} = 0.31146.$$

In Illustration 6, we use the combinatorial generating function to solve the same problem. □

Example 17. An box contains 5 red balls and 4 blue balls. We draw three balls randomly from the box and put them in a red urn and put the rest of them in a blue urn. Let E be the event that the number of red balls in the red box plus the number of blue balls in the blue box is equal to five.

Solution. Before we start, we first identify the various possible scenarios under this experiment. They are shown in the following table:

Scenarios	Urn contents after the draws	
Num of red balls drawn from the box	Red urn (R, B)	Blue urn (R, B)
0	$(0, 3)$	$(5, 1)$
1	$(1, 2)$	$(4, 2)$
2	$(2, 1)$	$(3, 3)$
3	$(3, 0)$	$(2, 3)$

where (R, B) denotes the number of red and blue balls in each urn after the three red balls are drawn from the box. From the table, we see only when two red balls are drawn from the box then resulting configurations meet the description of the event E. Hence

$$P(E) = \frac{\binom{5}{2}\binom{4}{1}}{\binom{9}{3}} = 0.47619.$$

An alternative for solving the problem is by using the combinatorial generating function. We set up the needed polynomial as

$$(tx_1 + x_2)^5(x_1 + tx_2)^4.$$

We can envision that the experiment has two independent parts: placing the 5 red balls into two urns — hence the term $(tx_1 + x_2)$ and use a tag "t" to count when a red ball goes into the red urn; placing the 4 blue balls into two urns — hence the term $(x_2 + tx_2)$ and use a tag "t" to count when a blue ball goes into the blue urn. In the above polynomial expansion, event E signals that we must sum all coefficients associated with the term $t^5 x_1^3 x_2^6$. This mean that the needed number of matching in colors is 5, and also the red urn must receive a total of 5 balls and the blue urn 6. The solution by Mathematica is given in Illustration 7. □

Example 18. Consider a zookeeper keeps 4 red birds in a red nest, 5 white birds in a white nest, and 6 blue birds in a blue nest. One day the zookeeper sets all the birds free. Assume that after one day all birds will randomly return to the zoo but perhaps, for some, to nests of different colors. Let E be the event that exactly eight birds returning to the nests of matching colors. Find $P(E)$.

Solution. Following the reasoning given in Example 17, we define the following polynomial:

$$(tx_1 + x_2 + x_3)^4(x_1 + tx_2 + x_3)^5(x_1 + x_2 + tx_3)^6.$$

We now collect all coefficients associated with terms with t^8 in the above polynomial expansion while setting values of all $\{x_i\}$ to ones. In Illustration 8, we obtain $P(E) = 0.0131$. □

2.6. Probability as a Continuous Set Function

A *set function* is a real-valued function defined on subsets of a probability space Ω.

Let $\{E_n\}_{n=1}^\infty$ be sequence of of events

$$E_1 \subset E_2 \subset \cdots \subset E_n \subset \cdots .$$

then we say that the sequence is increasing. If

$$E_1 \supset E_2 \supset \cdots \supset E_n \supset \cdots ,$$

then we say that the sequence is decreasing. In either case, we say the above sequence is a monotone sequence of events.

Example 19. Define the sequence of events $\left\{E_n > \frac{1}{n}\right\}$, where $n = 1, 2, \ldots$. Then, we conclude that the sequence is decreasing. □

Lemma 1. *Let $\{E_n\}_{n=1}^\infty$ be a monotone sequence of events. Then*

$$\lim_{n\to\infty} P(E_n) = P\left(\lim_{n\to\infty} E_n\right).$$

Proof. We consider the case where $\{E_n\}_{n=1}^\infty$ is an increasing sequence of events. The other case can be done in a similar manner. For $n = 1, 2, \ldots$, we define $F_{n+1} = E_{n+1} \backslash E_n$. Then, the sets $\{F_i\}_{i=1}^{\infty}$ are pairwise disjoint and

$$\bigcup_{i=1}^{\infty} F_i = \bigcup_{i=1}^{\infty} E_i.$$

Now

$$P\left(\lim_{n\to\infty} E_n\right) = P\left(\bigcup_{i=1}^{\infty} E_i\right) = P\left(\bigcup_{i=1}^{\infty} F_i\right)$$

$$= \sum_{i=1}^{\infty} P(F_i) \quad \text{by Axiom 3}$$

$$= \lim_{n\to\infty} P\left(\bigcup_{i=1}^{\infty} F_i\right) = \lim_{n\to\infty} P\left(\bigcup_{i=1}^{n} E_i\right)$$

$$= \lim_{n\to\infty} P(E_n) \quad \text{as } \{E_n\} \text{ is increasing.}$$

$\square$

Theorem 1 (Borel–Cantelli Lemmas). *Define the event*

$$E = \bigcap_n \bigcup_{m=n}^{\infty} E_m.$$

We say events $\{E_n\}$ occur infinitely often. Then

(a) $P(E) = 0$ *if* $\sum_n P(E_n) < \infty$;
(b) $P(E) = 1$ *if* $\sum_n P(E_n) = \infty$ *and* $\{E_i\}$ *are independent.*

Proof. (a) For any n, we have

$$E \subseteq \bigcup_{m=n}^{\infty} E_m.$$

By Boole's inequality, we have

$$P(E) \leq \sum_{m=n}^{\infty} P(E_m).$$

Finiteness of the sum implies $\sum_{m=n}^{\infty} P(E_n) \to 0$ as $n \to \infty$. This implies $P(E) = 0$.

(b) If we can show $P(E^c) = 0$, then we are done. Applying DeMorgan's law, we obtain

$$E^c = \left(\bigcap_n \bigcup_{m=n}^{\infty} E_n \right)^c = \bigcup_n \bigcap_{m=n}^{\infty} E_n^c.$$

For $m \geq n$, by mutual independence of $\{E_i\}$ and Lemma 1, we find

$$P\left(\bigcap_{m=n}^{\infty} E_n^c \right) = \lim_{k \to \infty} P\left(\bigcap_{m=n}^{k} E_n^c \right) = \prod_{m=n}^{\infty} P(E_m^c) = \prod_{m=n}^{\infty} (1 - P(E_m)).$$

If the sum $\sum_n P(E_n) = \infty$, due to the fact that

$$\prod_{m=n}^{\infty} (1 - P(E_m)) \leq \prod_{m=n}^{\infty} \exp\left(1 - P(E_m)\right) = \exp\left(-\sum_{m=n}^{\infty} P(E_n) \right) = 0$$

(as $1 - x \leq e^{-x}$ for $x \geq 0$), we conclude that $P(E^c) = 0$ and $P(E) = 1$. $\square$

The Borel–Cantelli lemmas are important tool to show almost sure convergence of a sequence of random variables — a subject to be addressed in Chapter 9.

Example 20. Consider a sequence of independent tosses of a fair coin. Whenever there are r successive heads occur, we call it a success. If we toss the coin an infinite number of times, what is the probability a success is certain to occur?

Solution. Whenever a success occurs, we consider that an experiment has occurred. Let A_n be the event that n experiments have occurred:

$$P(A_n) = \left(1 - \left(\frac{1}{2} \right)^r \right)^n, \quad n = 1, 2, \ldots.$$

We see that A_n is a monotone sequence. By the continuity property of probabilities, we conclude

$$P\left(\bigcap_{n=1}^{\infty} A_i \right) = \lim_{n \to \infty} P(A_n) = 0.$$

Thus

$$1 - P\left(\bigcap_{n=1}^{\infty} A_i \right) = 1.$$

$\square$

2.7. The Subjective Probability

The probability theory that we have studied so far as well as in the remainder of this book is sometimes known as the *axiomatic probability theory*. It is based on a set of basic axioms. The rest is developed mathematically as sequels. However, there are human endeavors where the notion of likelihood persists and the term probability appears quite frequently. As an example, Las Vegas book makers routinely post a large numbers of odds on events of societal interests. What they provide are actually their assessments of probabilities. The probabilities so developed are called subjective probabilities and typically they reflect degrees of personal beliefs about the likelihoods of occurrences of events. While their assessments are personal in nature, the results so obtained must adhere to basic axioms of probability theory otherwise they would deviate from the principles of logical reasoning. An excellent resource covering the use and assessment of subject probabilities is the book *Reading in Decision Analysis* [2].

2.8. Exploring with Mathematica

Illustration 1. Consider a small condo where A represents the set of residents who are bridge players and B the set of residents who are tennis players. Let $A = \{$Bob, Joe, Johnny, James, Eddie, Maria, Jenny, Martin, Wayne$\}$ and $B = \{$Maggie, Rachel, Joe, Eric, George, Eddie, Billy, James, Joy$\}$. We would like to find the set of residents who only play bridge or tennis. So we are looking for the symmetric difference of A and B

$$A \triangle B = (A \backslash B) \cup (B \backslash A) = (AB^c) \cup (A^c B) = (A \cup B) \backslash (A \cap B).$$

We use Mathematica to sort out the result

```
A = {"Bob", "Joe", "Johnny", "James", "Eddie", "Maria", "Jenny", "Martin", "Wayne"};

B = {"Maggie", "Rachel", "Joe", "Eric", "George", "Eddie", "Billy", "James", "Joy"};

Complement[A ∪ B, A ∩ B]

{Billy, Bob, Eric, George, Jenny, Johnny, Joy, Maggie, Maria, Martin, Rachel, Wayne}
```

□

Illustration 2. We show the computation involving the use of the inclusion–exclusion formula with $n = 3$. The problem description is given in Example 4. One of the objectives is to show how data structures are used in Mathematica involving lists.

```
d1 = 5; d2 = 7; d3 = 13;
lista = {{d1, d2, d3}, {d1 * d2, d1 * d3, d2 * d3}, {d1 * d2 * d3}};
f[x_] := Total[If[Divisible[#, x], 1, 0] & /@ Range[1000]] / 1000 // N
listb = Flatten[Take[lista, {1}]]
x = Total[Table[f[listb[[i]]], {i, 1, 3}]];
listb = Flatten[Take[Drop[lista, {1}], {1}]];
y = -Total[Table[f[listb[[i]]], {i, 1, 3}]];
listb = Flatten[Take[Drop[lista, {1, 2}], {1}]];
z = Table[f[listb[[i]]], {i, 1, 1}];
probE = x + y + z
{0.367}                                                              □
```

Illustration 3. We now introduce the use of Mathematica to find the probabilities of poker hands discussed in Example 10 via Mathematica | Alpha, an online query software. We only need to type "=" and type the poker hand of interest. We then hit the enter key. Voila! the answer pops out. The following confirms the result found in part (d).

	number of possible hands	approximate probability	approximate chance
5-card hand	123 552	0.04754	≈ 1 in 21

This handy software would help us to check out our argument leading to the answer. Many times solving probability problems are not trivial. Having such an instrument at our disposal is a plus. □

Illustration 4. We now revisit Example 12 about pairing roommates of a football team. We use Mathematica to show that how would the messy computation be handled.

$$\text{p2i}[i\_] := (\text{Binomial}[20, 2*i])^2 \, ((2*i)\,!) \left(\frac{(20-2*i)\,!}{2^{(10-i)}\,(10-i)\,!}\right)^2 \Big/ \left(\frac{40\,!}{2^{20}\,(20!)}\right);$$

(a) The following gives $P(E_0)$:

`p2i[0] // N`

1.3403×10^{-6}

(b) The following computes $P(E_{2i})$:

`pout = Table[p2i[i], {i, 0, 10}] // N;`

We check out the answers by summing the probabilities:

`Total[pout]`

`1.`

The following graph shows that the most likely paring is 5 offensive plays are paired with 5 defensive plays as it has the largest probability of occurrence.

`ListPlot[pout, Ticks → {Table[{i, i - 1}, {i, 11}], Automatic}]`

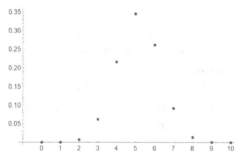

Illustration 5. Numerical computations involve Example 15.

`nw1 = Coefficient[Expand[` $(x_1 + x_2 + x_3 + x_4)^{15}$ `],` $\{x_1\, x_2\, x_3\, x_4^{12}\}$ `];`

`nS = ` 4^{15} `;`

`pw1 = nw1 / nS  // N`

$\{2.54251 \times 10^{-6}\}$

`nw2 = Coefficient[Expand[` $(x_1 + x_2 + x_3 + x_4)^{15}$ `],` $\{x_1^3\, x_2^3\, x_3^4\, x_4^5\}$ `];`

`pw2 = nw2 / nS // N`

$\{0.0117464\}$

`nE1 = Coefficient[Expand[` $(x_1 + t\, x_2 + t\, x_3 + t\, x_4)^{15}$ `],` $\{t^{15}\}$ `];`

`nX1 = nE1 /. {`$x_2 \to 1$`, `$x_3 \to 1$`, `$x_4 \to 1$`};`

`1 - (nX1 * 4) / nS // N`

$\{0.946546\}$

Illustration 6. The following result verifies the results given in Example 16.

nE1 = Coefficient[Expand[(t x₁ + x₂ + x₃ + x₄)⁷], {t²}];

nS = 4⁷; nX1 = nE1 /. {x₁ → 1, x₂ → 1, x₃ → 1, x₄ → 1, t → 1};

pw1 = nX1 / nS // N

{0.311462} ☐

Illustration 7. The solution of Example 17 by the combinatorial generating function is given below:

nS = Coefficient[Expand[(x₁ + x₂)⁹], {x₁³ x₂⁶}]

{84}

nE = Coefficient[Expand[(x₁ t + x₂)⁵ (x₁ + x₂ t)⁴], {t⁵ x₁³ x₂⁶}]

{40}

pE = nE / nS // N

{0.47619} ☐

Illustration 8. The use of the combinatorial generating function in Example 18 for finding $P(E)$.

nE = Coefficient[Expand[(t x₁ + x₂ + x₃ + x₄)⁴ (x₁ + t x₂ + x₃ + x₄)⁵ (x₁ + t x₂ + x₃ + x₄)⁶], {t⁸}];

nE1 = nE /. {x₁ → 1, x₂ → 1, x₃ → 1, x₄ → 1, t → 1}

{14 073 345}

pE = nE1/4¹⁵ // N

{0.0131068}

☐

Problems

1. Sixty percent of the students carry neither an umbrella nor a raincoat to school. Twenty percent carry an umbrella and thirty percent carry a raincoat. If one of the students is selected at random, what is the probability that this student is carrying (a) an umbrella or a raincoat? (b) an umbrella and a raincoat?

2. If six dice are rolled, what is the probability that all show different face?

3. Suppose seven coins are flipped. Computing the probability, we get (a) no heads, (b) one head, (c) two heads, (d) three heads.

4. If 6 balls are thrown at random into 10 boxed, what is the probability no box will contain more than 1 ball?

5. An absent-minded professor has n keys of which one turns on the engine.

   (a) If the professor tries the keys randomly one by one, what is the probability that the professor will find the correct key at the kth trial, where $k < n$?

   (b) If every time one key is tried and finds incorrect, the professor will randomly select one from the n keys once more. What is the probability that the correct key is found at the kth trial?

6. Assume $P(A) = 0.4$, $P(B) = 0.2$, and $P(A \cup B) = 0.5$. Find $P(A \cup B^c)$.

7. Consider two events E and F. Let $P(E) = \frac{3}{4}$ and $P(F) = \frac{1}{3}$. (a) Establish a upper bound and a lower bound for $P(EF)$. (b) Do the same for $P(A \cup B)$.

8. We pick a card from a well-shuffled standard card deck. We do it three times. Once a card is selected with its denomination noted, it is return to the deck and the deck gets reshuffled again. Find the probability that we get at least one diamond?

9. We roll three fair dice. (a) What is the probability that the sum is 14? (b) Use the combinatorial generating function and Mathematica to verify your result.

10. Assume there are 365 days in a year. Assume that the probability a random chosen person is born on any given day of the year is the same. In a room with k people, develop a recursive equation that computes p_k, the probability that no two persons have the same birthday for $k = 1, \ldots, 50$. Use Mathematica to plot p_k.

11. In a room with 30 persons, each chooses two distinct numbers randomly from $\{1, \ldots, 20\}$. Let E be the event that at least two persons will choose the same two numbers. What is $P(E)$?

12. Rice-a-Rodney is a box of cereal for breakfast. In side of each box, it contains a picture card of one of the eleven best soccer players in San Francisco. Assume that each box will have an equal probability of containing the picture of one of the eleven players. Let E be the event that you will get the cards of your two favorite players given that you have bought six boxes. Find $P(E)$.

13. An urn contains 12 black balls and 12 white balls. You pick a pairs of balls from the urn three times without replacement. Let E be the event that at least one pair had chosen two balls of different color? Find $P(E)$.

14. Six tea pots and their caps come in pairs. Each pair has its distinct colors. If pot-cap pairs are randomly matched, what is the probability no pot is matched with the cap of the same color?

15. We toss 6 balls into 3 urns. Assume each is equally likely to land in any urn. Let E be the event that each urn will contain at least one ball. Find $P(E)$.

16. Consider a four-sided fair die with its sides numbered 1, 2, 3, and 4. If we toss this die five times randomly, what is the probability that the sum of the five faces shown will be 12?

17. Zora takes three cookies at random from a cookie jar containing 5 chocolate cookies, 7 blueberry cookies, and 8 strawberry cookies. (a) Let E be the event that Zora gets all the cookies of the same type. Find $P(E)$. (b) Let F be the event that Zora gets three distinct types of cookie. Find $P(F)$.

18. Consider an urn contains ten balls numbered $0, 1, \ldots, 9$. Four balls are drawn randomly and one by one from the urn *with* replacement. (a) What is the probability that the four balls drawn are numbered 1, 2, 3, and 4. (b) What is the probability that the four balls drawn are all different. (c) What is the probability that the four balls drawn are all the same. (d) What is the probability that three balls drawn have the same number but the other is different.

19. An automechanic has an unusual way to charge customers for the service. When the time comes at which a customer must pay the bill, the customer is asked to pick four balls randomly from the urn without replacement. The urn contains 8 balls marked $5, 10 balls marked $10, and 7 balls marked $20. If the balls customer selected have the same amount shown on the balls, the customer pays nothing, otherwise the customer pays the sum of the amounts shown on the three balls. (a) What is the probability that the customer pays nothing? (b) What is the probability that the customer pays $50?

20. Assume that 8 Texans and 4 New Yorkers are to be randomly paired to share 6 double rooms in a Holiday Inn. What is the probability that no Texan will share a room with a New Yorker?

21. An airport shuttle bus makes 7 scheduled stops for 25 passengers. Assume that each passenger is equally likely to stop at each on the 7 stops. Let E be the event that at least one person gets off at each stop. (a) Use the combinatorial generation function and Mathematica to find $P(E)$. (b) Use the inclusion–exclusion formula approach to verify your answer.

22. Consider a standard card deck of 52 cards. We draw cards one after the other without replacement. (a) Let E be the event that an ace will appear before any king, queen, and jack. Find $P(E)$. (b) Let F be the event that two aces will appear before any king, queen, and jacks? Find $P(F)$.

23. There are 10 people in a room. Each randomly chooses the name of the other 9 people. Let E denote the event that at least two people chooses each other's names. Find $P(E)$.

24. Julie has a penny, nickel, dime, and quarter in her pocket. So does Avis. Each takes one coin from the pocket randomly. What is the probability that Julie's coin is worth more than Avis's?

25. Let A and B be two events and $P(A)$ and $P(B)$ denote the probabilities of their respective occurrence. Find the probability that exactly one of the two events will occur in terms of $P(A)$, $P(B)$, and $P(A \cap B)$.

26. Five cards are drawn one by one from a standard card deck of 52 cards. Let the event E denote that there is an ace and either a queen or king, or both. Find $P(E)$.

27. Avis, Bree, Claudia, Donna, and Elisa are all lined up in front of a stage. Assume that their orderings are done completely randomly. (a) What is probability that there is exactly one person between Avis and Bree? (b) What is the probability that there are exactly two persons between Avis and Bree? (c) What is the probability that there are three persons between Avis and Bree?

28. Let W be the event that a bridge hand is void in at least one suit. Find $P(W)$.

29. An urn contains m white balls and n black balls. We draw these balls one at a time until a total of k white balls are drawn, where $k \leq m$. Let E be the event that a total of r balls are drawn. Find $P(E)$.

30. Five cards are drawn one by one from a standard card deck of 52 cards. What is the probability that the hand contains at least one card from each one of the four suits?

31. (a) Consider a well-shuffled standard card deck. (a) If we randomly divide the deck into four piles of 13 cards each, what is the probability that each pile contains exactly one ace? (b) If we pick one card randomly, what is the probability that the card drawn will be either a heart or an ace or both?

32. A producer of a new brand of breakfast cereal offers a baseball card in each cereal box. Ten different baseball cards are distributed equally among the cereal boxes produced. What is the probability of getting the cards of your two favorite teams if you buy five boxes?

33. Consider an elevator in a building with 7 floors and a basement. Assume that four people starts from the basement and each is equally likely to get off at any floor, what is the probability that no two passengers get off at the same floor?

34. Consider a sequence of events $\{A_i\}$. Show by induction that

$$P\left(\bigcap_{i=1}^{n} A_i\right) \geq 1 - \sum_{i=1}^{n} P(A_i^c)$$

35. Suppose we place 14 indistinguishable balls in 7 boxes. What is the probability that there will be at least one ball per box?

36. A and B are playing a sequence of games. For each play, each have an equal chance of winning. The first person who wins five plays will be the winner of the game. (a) Assume that A has already won 4 plays and B has already won 3 plays. Let E_1 denote the probability that A will win the game. What is $P(E_1)$? (b) Assume that A has already won 3 plays and B has already won 2 plays. Let E_2 denote the probability that A will win the game. What is $P(E_2)$?

37. (A generalization of Problem 26) A and B are playing a sequence of games. For each play, each have an equal change of winning. The first person who wins n plays will be the winner of the game. Assume that A has already won i plays and B has already won j plays. Let E denote the probability that A will win the game. (a) Express $P(E)$ in terms of n, i, and j. (b) Verify the results obtained in Problem 26 using the expression for $P(E)$.

38. Suppose that we roll four dice at the same time. (a) Use Mathematica to find the probability $P(E_1) = (\{\text{sum} = 24\})$ (b) $P(E_2) = P(\{\text{sum} = 35\})$.

39. (The Banach's matchbox problem) A smoker carries a matchbox in each one of the two pockets. Originally, each box contains N matches. Every time the smoker needs a match, it is equally likely to be drawn one of the two pockets. (a) Let E be the event that when the last match is drawn from one of the pocket the other pocket contains k matches, where $k = 0, 1, \ldots, N$. (b) For $N = 20$, use Mathematica to compute $P(X = k)$, $k = 0, 1, \ldots, N$, where X is the number of matches left in the "other" pocket. Plot the PMF of X.

40. In a flip of a fair coin twelve times, use Mathematica to compute the probability that there will be more heads than tails.

Remarks and References

The football roommate pairing problem is based on Example 5k of Ross [3]. Example 17 is related to that given in [2]. In [1], there are many practical guides for estimating and using subjective probabilities for real-world applications.

[1] Howard, R. A., J. E. Matheson, and Miller, K. L. (eds), *Reading in Decision Analysis*, Decision Analysis Group, Menlo Park, CA, 1977.

[2] Mood, A. M. and Graybill, F. A. *Introduction to the Theory of Statistics*, 2nd edn., McGraw Hill, 1963.

[3] Ross, S. M., *A First Course in Probability*, 10th edn., Pearson, 2019.

Chapter 3

Conditional Probability

3.1. Introduction

Conditional probability is about information processing. Given we know something, how would it alter our assessment about the likelihood of the occurrence of a given event. If the possession of some information will not change our assessment, then it is logical to assert that the event of interest and the added information are independent. As it turns out, our mathematical definitions of conditional probability and independence coincide with these common beliefs. In this chapter, we will develop the tools needed to facilitate updating information so as to find-tune our assessment of probabilities.

3.2. Conditional Probabilities and Independence

The conditional probability of an event E given that event F has occurred is defined by

$$P(E|F) = \frac{P(EF)}{P(F)} \qquad (3.1)$$

provided that $P(F) > 0$. The symbol "|" means "given that." The idea is best illustrated by the following Venn diagram:

51

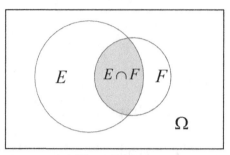

A Venn Diagram

Assume that the sample points $\{\omega_i\}$ in the sample space Ω in the above Venn diagram are equally likely to occur. The probability E occurs is

$$P(E) = \frac{n(E)}{n(\Omega)}. \tag{3.2}$$

Now we know event F has occurred, our sample space is being switched from Ω to the set F. Assume the $E \cap F \neq \emptyset$. We make the substitutions $\Omega \leftarrow F$ and $E \leftarrow E \cap F$ in (3.2). This means (3.2) reverts back to (3.1) due to the change of the sample space involved.

Two events E and F said to be *independent* if

$$P(E|F) = P(E).$$

We see that when the two events are independent, then knowing the occurrence of event F does not change our assessment of $P(E)$. Based on this definition, we see that

$$\frac{P(EF)}{P(F)} = P(E) \quad \text{or} \quad P(EF) = P(E)P(F).$$

The last expression can also be considered as an alternative definition of the independence of the two events E and F. A handy notation to express the independence between events E and F, is $E \perp F$.

Recall that events E and F are disjoint if $E \cap F = \emptyset$. Assume $P(E) > 0$ and $P(F) > 0$. Are these two independent events? Obviously, the answer is no because if we know if one event has occurred, the other cannot possibly occur. To establish this formally: if E and F are disjoint, then $P(A \cap B) = P(\emptyset) = 0$ and it is impossible that $P(A \cap B) = P(A)P(B) > 0$. On the other hand, if they are independent, then $P(A \cap B) = P(A)P(B) > 0$ and it is impossible that $A \cap B = \emptyset$.

Conditional Independence. Consider three events A, B, and C. We say events A and B are conditional independent given C if and only if

$$P(A|BC) = P(A|C). \tag{3.3}$$

In other words, if we know that C occurs, then knowing B has also occurred does not contribute to our assessment of the likelihood of the occurrence of A. An equivalent condition of conditional independence is

$$P(AB|C) = P(A|C)P(B|C). \tag{3.4}$$

Lemma. *Conditions (3.3) and (3.4) are equivalent.*

Proof. Assume that (3.3) holds, then

$$P(AB|C) = \frac{P(ABC)}{P(C)} = \frac{P(ABC)}{P(BC)} \times \frac{P(BC)}{P(C)} = P(A|C)P(B|C),$$

where the last equality is obtained by invoking (3.3). So we have (3.4). Now we assume that (3.4) holds, we expand it to write

$$\frac{P(ABC)}{P(C)} = \frac{P(AC)}{P(C)} \times \frac{P(BC)}{P(C)}.$$

We now multiply the last expression by $P(C)/P(AB)$. The resulting expression is (3.3). □

The most glaring usage of conditional independence is the role it plays in Markov chains. Let $\{A_n\}_{n=1}^{\infty}$ be a sequence of events. If

$$P(A_{n+1}|A_1, \ldots, A_n) = P(A_{n+1}|A_n\} \quad \text{for} \quad n = 1, 2, \ldots,$$

then the above implies that the future is independent of the past given the present. When the above equation holds, we say that $\{A_n\}_{n=1}^{\infty}$ follows a *discrete-time Markov chain*. In this text, we will not address this subject.

Example 1. Consider the rolling of two fair dice. Define $A \equiv \{$the sum of the two is 9$\}$ and $B \equiv \{$the first is 2$\}$. We see that

$$A = \{(3,6), (4,5), (5,4), (6,3)\}.$$

Thus $A \cap B = \varnothing$ and $P(AB) = P(\varnothing) = 0 \neq P(A)P(B) > 0$. The two events are dependent.

Consider a variant. Let $E \equiv \{$the sum is 6$\}$ and $F \equiv \{$the first is a 4$\}$. We see that

$$P(E) = P\{(1,5), (2,4), (3,3), (4,2), (5,1)\} = \frac{5}{36},$$

$$P(F) = \frac{1}{6},$$

$$P(EF) = P\{(4,2)\} = \frac{1}{36}.$$

We see now

$$P(EF) \neq P(E)P(F) = \frac{5}{36} \times \frac{1}{6}.$$

Thus, E and F are not independent.

Consider another variant. Let $G \equiv \{$the sum is 7$\}$ and again $F \equiv \{$the first is a 4$\}$. Then

$$P(G) = P\{(1,6), (2,5), (3,4), (4,3), (5,2), (6,1)\} = \frac{1}{6},$$

$$P(FG) = P\{(4,3)\} = \frac{1}{36}$$

and

$$P(FG) = \frac{1}{36} = P(F) \times P(G) = \frac{1}{6} \times \frac{1}{6}.$$

Hence $F \perp G$.   □

The above example demonstrates that we need to treat the validation of the independence property with care. When in doubt, it is best to check against the definition of independence. We consider the following questions.

Example 2. Given that a throw of three unbiased dice show different faces, what is the probability that (a) the total is eight? and (b) at least one is a six?

Solution. (a) Let A be the event that a throw of three unbiased dice show different faces and B be the event that its total is eight. We find

$$P(A) = \frac{6 \times 5 \times 4}{6^3} = \frac{5}{9}.$$

We note that $A \cap B = \{$the throw shows $(1, 2, 5)$ and $(1, 3, 4)\}$, namely, the 3 faces are different and the sum is 8 while ignoring the order of the appearance of the three numbers. Now the probability of observing $(1, 2, 5)$ is $3!/6^3$ and so is observing $(1, 3, 4)$. Hence

$$P(AB) = 2 \times \frac{3!}{6^3} = \frac{1}{18}$$

and

$$P(B|A) = \frac{P(AB)}{P(A)} = \frac{\frac{1}{18}}{\frac{5}{9}} = \frac{1}{10}.$$

(b) Let C be the event that at least one is a six. We notice the event $A \cap C = \{$of the three numbers shown, one is a 6 and the other two are not 6$\}$. We see that

$$P(AC) = \frac{5 \times 4}{6^3} \times 3 = \frac{5}{18}.$$

Thus

$$P(C|A) = \frac{P(AC)}{P(A)} = \frac{\frac{5}{18}}{\frac{5}{9}} = \frac{1}{2}. \qquad \square$$

The notion of independence poses some interesting questions. We now examine them below.

Question 1. Given $G \perp F$ and $G \perp H$, is $G \perp FH$?

Answer. We would speculate that the answer is yes. As it turns out, the answer is not necessarily. This is done by producing a counter example. As in Example 1, we let $G \equiv \{$the sum is 7$\}$ and again $F \equiv \{$the first is 4$\}$. Also, we let $H \equiv \{$the second is a 3$\}$. Then Example 1 shows that $G \perp F$.

Similarly, we have $G \perp H$. We see

$$P(G|FH) = 1.$$

But

$$P(G) = \frac{1}{6}.$$

This means

$$P(G|FH) \neq P(G)$$

i.e., it is not true that $G \perp FH$.   □

Question 2. Is the independence relation transitive, i.e., does $E \perp F$ and $F \perp G$ imply $E \perp G$?

Answer.   The answer is not necessarily. Here is one such example. Consider a family has two children. The sample space Ω is $\{(b, b), (b, g), (g, b), (g, g)\}$. Assume each sample point is equally likely. Denote

$$E \equiv \{\text{first child is a boy}\},$$
$$F \equiv \{\text{the two children are of different gender}\},$$
$$G \equiv \{\text{first child is a girl}\}.$$

We see that

$$P\{E\} = \frac{1}{2}, \quad P\{F\} = P\{(b, g), (g, b)\} = \frac{2}{4} = \frac{1}{2}$$

and

$$P\{EF\} = P\{(b, g)\} = \frac{1}{4} \quad \Longrightarrow \quad P(EF) = P(E)P(F) = \frac{1}{4}.$$

Thus $E \perp F$. By symmetry, we conclude $F \perp G$. Now

$$P(EG) = P(\varnothing) = 0 \neq P(E)P(F) = \frac{1}{4}.$$

Hence E and G are not independent.   □

Question 3. If E and F are independent, is E and F^c independent?

Answer. Intuitively, we would concur. We give a simple proof to show this is indeed so. We see

$$P(E) = P(E \cap \Omega) = P(E \cap (F \cup F^c))$$
$$= P(EF \cup EF^c) \quad \text{also } EF \cap EF^c = \varnothing.$$

Thus

$$P(E) = P(EF) + P(EF^c) = P(E)P(F) + P(EF^c) \quad \text{as } E \perp F.$$

This implies

$$P(EF^c) = (1 - P(F))P(E) = P(E)P(F^c).$$

Hence E and F^c are independent. □

Independence of Three Events. We say events A, B, and C are *mutually independent* if

(a) $P(ABC) = P(A)P(B)P(C)$, and
(b) $P(AB) = P(A)P(B)$, $P(AC) = P(A)P(C)$, and $P(BC) = P(B)P(C)$.

Condition (b) is known as the *pairwise independence*. It is important to note that for events A, B, and C, pairwise independence does not imply mutual independence. To illustrate, we consider a ball is drawn from randomly from an urn containing four balls numbered $\{1, 2, 3, 4\}$. Define events

$$A \equiv \{1, 2\}, \quad B = \{1, 3\}, \quad C = \{1, 4\}.$$

Then, it is easy to verify that the three events are pairwise independent. But $P(ABC) = P(\{1\}) = \frac{1}{4}$. But $P(A)P(B)P(C) = \left(\frac{1}{2}\right)^3 = \frac{1}{8}$. Thus condition (a) is not met.

To generalize the above to more than three events, we consider events $E_1, E_2, \ldots, E_n$. They are mutually independent if for every subset $E_{i_1}, E_{i_2}, \ldots, E_{i_r}$ from $\{E_i\}_{i=1}^n$, the following condition is satisfied:

$$P(E_{i_1}, E_{i_2}, \ldots, E_{i_r}) = P(E_{i_1})P(E_{i_2}) \cdots P(E_{i_r}).$$

Recall in Chapter 2, we mentioned the term conditional multiplication rule. When viewed in the context of conditional probability, it translates into the following useful result: for events $E_1, E_2, \ldots, E_n$, we have

$$P(E_1 E_2 \cdots E_n) = P(E_1)P(E_2|E_1)P(E_3|E_1 E_2) \cdots P(E_n|E_1 E_2 \cdots E_{n-1})$$
(3.5)

provided $P(A_1 A_2 \cdots A_{n-1}) > 0$. Equation (3.5) is sometimes called the *chain rule* formula. To show the above is true, we see

$$P(E_1) \geq P(E_1 E_2) \geq \cdots \geq P(E_1 E_2 \cdots A_{n-1}) > 0.$$

Thence we can rewrite (the right-hand side of (3.5) as

$$\frac{P(E_1)}{P(\Omega)} \frac{P(E_1 E_2)}{P(E_1)} \frac{P(E_1 E_2 E_3)}{P(E_2 E_3)} \cdots \frac{P(E_1 E_2 \cdots E_n)}{P(E_1 E_2 \cdots E_{n-1})}.$$

The above reduces to the left-hand side of (3.5) as $P(E_1 E_2 \cdots E_n)$. It can be viewed as the *conditional multiplication rule* in probability. It is stated in the form of a "probability tree" shown below:

Each component on the top branch of the tree gives the conditional probability showing at the right side of (3.5). When they are multiplied together, it gives the probability of reaching the right-most node, namely, $P(E_1, E_2, \ldots, E_n)$.

3.3. The Law of Total Probabilities

Consider the sample space Ω is being partitioned into n mutually exclusive and collectively exhaustive subsets $\{A_i\}_{i=1}^n$, i.e.,

$$\Omega = A_1 \cup A_2 \cdots \cup A_n$$

and $A_i A_j = \varnothing$ for $i \neq j$. For any event E, we see

$$E = \Omega \cap E$$
$$= (A_1 \cup A_2 \cdots \cup A_n) \cap E$$
$$= A_1 E \cup A_2 E \cup \cdots \cup A_n E.$$

More importantly, $A_i E \cap A_j E = \varnothing$ for $i \neq j$. This is shown in the following picture.

The above partition leads to the *law of total probability*:

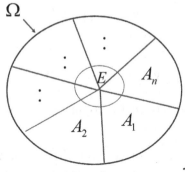

Ω

Partition of $\Omega \cap E$

$$P(E) = P\left(\Omega \cap E\right)$$
$$= P\left(A_1 E \cup A_2 E \cup \cdots \cup A_n E\right)$$
$$= P(A_1 E) + P(A_2 E) + \cdots + P(A_n E)$$
$$= P(E|A_1)P(A_1) + P(E|A_2)P(A_2) + \cdots + P(E|A_n)P(A_n)$$
$$= \sum_{i=1}^{n} P(E|A_n)P(A_n).$$

Example 3. Assume we toss a dice twice. Let X denote the outcome of the first toss and Y denote the outcome of the second toss. Define A be the event that $Y > X$. What is $P(A)$?

Solution. We use the law of total probabilities and condition on the outcome of the first toss. This gives us

$$P(A) = \sum_{i=1}^{6} P(A|X = i)P(X = i) = \sum_{i=1}^{6} P(Y > i|X = i) \times \frac{1}{6}.$$

Because of the two tosses are independent, we have

$$P(Y > i|X_i = i) = \sum_{j=i+1}^{6} P(Y = j) = \frac{6-i}{6}, \quad i = 1, \ldots, 6$$

Hence

$$P(A) = \frac{1}{6} \sum_{i=1}^{6} \frac{6-i}{6} = \frac{5}{12}.$$

It is interesting to observe that by symmetry, we also have

$$P(Y < X) = \frac{5}{12}$$

and

$$P(X = Y) = 1 - P(X > Y) - P(X < Y) = 1 - \frac{5}{12} - \frac{5}{12} = \frac{1}{6}.$$

We can easily verify the above by counting or applying the law of total probabilities. □

Example 4. We draw cards successively from a standard card deck of 52 cards. Define E_i be the event the ith drawn is a diamond. (a) What are $P(E_2)$ and $P(E_3)$?

Solution. First we see

$$P(E_2) = P(E_2 E_1) + P(E_2 E_1^c)$$

$$= P(E_2|E_1)P(E_1) + P(E_2|E_1^c)P(E_1^c)$$

$$= \frac{12}{51}\frac{1}{4} + \frac{13}{51}\frac{3}{4} = \frac{12}{204} + \frac{39}{204} = \frac{1}{4}.$$

Similarly, we have

$$P(E_2^c) = P(E_2^c E_1) + P(E_2^c E_1^c)$$

$$= P(E_2^c|E_1)P(E_1) + P(E_2^2|E_1^c)P(E_1^c)$$

$$= \frac{39}{51}\frac{1}{4} + \frac{38}{51}\frac{3}{4} = \frac{39}{204} + \frac{114}{204} = \frac{3}{4}.$$

Before we draw the third card, we partition the sample space into $\Omega = E_1 E_2 \cup E_1^c E_2 \cup E_1 E_2^c \cup E_1^c E_2^c$. We then apply the law of total probability

$$P(E_3) = P(E_3|E_1 E_2)P(E_1 E_2) + P(E_3|E_1^c E_2)P(E_1^c E_2)$$

$$+ P(E_3|E_1 E_2^c)P(E_1 E_2^c) + P(E_3|E_1^c E_2^c)P(E_1^c E_2^c)$$

$$= \frac{11}{50}\frac{12}{204} + \frac{12}{50}\frac{39}{204} + \frac{12}{50}\frac{39}{204} + \frac{13}{50}\frac{114}{204} = \frac{1}{4}. □$$

Example 5. Consider once more the previous example. We now want to compute $P(E_4)$. We use a different approach avoiding having to trace the trajectory of computation. We introduce the notation of *state*. The state carries the minimal information that is needed to carry on the computation. Before we draw the fourth card, the state of the system is described by the number of diamonds that has already being drawn, i.e., 0, 1, 2 and 3. We use S_i, $i = 0, 1, 2, 3$ to denote the four states. Thus we tally the states of the system as follows:

State of the System	S_0	S_1	S_2	S_3
Number of diamonds left after 3 draws	13	12	11	10
Number of non-diamonds left after 3 draws	36	37	38	39

We now apply the law of total probability

$$P(E_4) = \sum_{i=0}^{3} P(E_4 S_i) = \sum_{i=0}^{3} P(E_4|S_i)P(S_i)$$

$$= \frac{13}{49} \frac{\binom{13}{0}\binom{39}{3}}{\binom{52}{3}} + \frac{12}{49} \frac{\binom{13}{1}\binom{39}{2}}{\binom{52}{3}} + \frac{11}{49} \frac{\binom{13}{2}\binom{39}{1}}{\binom{52}{3}}$$

$$+ \frac{10}{49} \frac{\binom{13}{3}\binom{39}{0}}{\binom{52}{3}} = \frac{1}{4}.$$

This seems to be a lesser messy way to do the problem. □

In Examples 4 and 5, we have demonstrated that the probability that in drawing a card from a standard card deck without replacement is diamond is always 1/4 regardless of the order of their respective drawings. This phenomenon has its root in the notion of exchangeable random variables. We will explore this interesting and important topic in Chapter 6.

3.4. Bayes Theorem

Using the definition of conditional probability and the law of total probability, we obtain the *Bayes formula* below

$$P(A_k|E) = \frac{P(A_k E)}{P(E)} = \frac{P(E|A_k)P(A_k)}{\sum_{i=1}^{n} P(E|A_i)P(A_i)}, \tag{3.6}$$

where we assume that $P(E) > 0$.

In the Bayes formula, on its left side we have $P(A_k|E)$ whereas on its right side we have $P(E|A_k)$. This reversal of the two events makes it attractive in many applications. For example, E could mean that a medical test displays a specific result and A_k denote being in a given disease state. Many times, we many have a better information about $P(E|A_k)$. For an another example, in a casino in Las Vegas, someone is yelling "twelve." There are two related probabilities: $A \equiv P$ (it is from the toss of a pair of dice | "twelve") or $B \equiv P(\text{"twelve"} \mid$ it is from the toss of a pair of dice)? It is obvious that B is easier to compute; whereas we would need more information so that we can apply (3.6) to find A, e.g., number of gambling machines (e.g., roulettes) in the vicinity of the sound that can generate "twelve" and their respective fractions. In summary, the Bayes formula enables us to process the prediction of $P(A_k|E)$. In *Bayesian statistical analysis*, we note that $P(A_k)$ is called the *prior* probability and $P(A_k|E)$ is called the *posterior* probability. The conditional probability $P(E|A_k)$ is called the *likelihood*.

Example 6. Assume that the number of children in a family is 1, 2, 3, and 4, with equal probability. Little Jenny has no sisters. What is the probability she is the only child?

Solution. Let A_k denote the event that the family has k children, $k = 1, 2, 3, 4$. Let E be the event that the family has only one girl. We note that

$$P(EA_1) = P(E|A_1)P(A_1) = \frac{1}{2}\frac{1}{4} = \frac{1}{8},$$

$$P(EA_2) = P(E|A_2)P(A_2) = \frac{2}{2^2}\frac{1}{4} = \frac{1}{8},$$

$$P(EA_3) = P(E|A_3)P(A_3) = \frac{3}{2^3}\frac{1}{4} = \frac{3}{32},$$

$$P(EA_4) = P(E|A_4)P(A_4) = \frac{4}{2^4}\frac{1}{4} = \frac{1}{16}.$$

We see that

$$P(E) = \sum_{k=1}^{4} P(EA_k) = \frac{1}{8} + \frac{1}{8} + \frac{3}{32} + \frac{1}{16} = \frac{13}{32},$$

Hence

$$P(A_1|E) = \frac{P(EA_1)}{P(E)} = \frac{\frac{1}{8}}{\frac{13}{32}} = \frac{4}{13}. \qquad \square$$

Example 7. An insurance company believes that people can be divided into two classes: those who are accident prone and those who are not. The statistics show that an accident prone person will have an accident in one year with probability 0.4, whereas this probability decreases to 0.2 for a person who is not accident prone. Assume that 30 percent of the population is accident prone.

Assume that Johnny is a potential policy holder of the company.

(a) What is the probability that Johnny will have an accident in the first year of policy ownership?
(b) What is the conditional probability that Johnny will have an accident in the second year of policy ownership given that Johnny has had an accident in the first year of the policy ownership?
(c) Consider a variant of part (b). If a policy holder has an accident in the first year, the company would change the assessment of the probability of having an accident in the second year to 0.5. Moreover, if a none-accident prone policy holder does have an accident in a year, then it will be classified as an accident-prone person in the subsequent year. What is the answer for part (b) under this changed scenario?

Solution. (a) Let A_1 denote the event that Johnny will have an accident in the first of the policy ownership. Then

$$P(A_1) = P(A_1|A)P(A) + P(A_1|A^c)P(A^c)$$
$$= (0.4)(0.3) + (0.2)(0.7) = 0.26.$$

(b) Let A_2 denote the event that Johnny will have an accident in the second year of the policy ownership. Then

$$P(A_2|A_1) = P(A_2A|A_1) + P(A_2A^c|A_1)$$
$$= \frac{P(A_2A_1A)}{P(A_1)}\frac{P(AA_1)}{P(AA_1)} + \frac{P(A_2A_1^cA)}{P(A_1)}\frac{P(A_1A^c)}{P(A_1A^c)}$$
$$= P(A_2|A_1A)P(A|A_1) + P(A_2|A_1A^c)P(A^c|A_1). \qquad (3.7)$$

From part (a), we find

$$P(A|A_1) = \frac{P(A_1|A)P(A)}{p(A_1)} = \frac{(0.4)(0.3)}{0.26} = 0.4615,$$

$$P(A^c|A_1) = 1 - 0.4615 = 0.5385.$$

We assume conditional independence and thence $(A_2|A_1A) = P(A_2|A) = 0.4$ and $P(A_2|A_1A^c) = P(A_2|A^c) = 0.2$. Applying (3.7), we find

$$P(A_2|A_1) = (0.4)(0.4615) + (0.2)(0.5385) = 0.2923.$$

(c) Here $P(A_2|A_1A) = 0.5$ and $P(A_2|A_1A^c) = 0.4$. Applying (3.7) once more yields

$$P(A_2|A_1) = (0.5)(0.4615) + (0.4)(0.5385) = 0.44620.$$

Therefore, without the assumption of conditional independence, the probability of having an accident in the second year increases noticeably if Johnny has an accident in the first year. □

Example 8. Suppose that a laboratory test on a blood sample yields two results, positive or negative. Assume that 95% of people with COVID-19 produce a positive result. But 2% of people without COVID-19 will also produce a positive result (*a false positive*). Suppose that 1% of the population actually has COVID-19. What is the probability that a person chosen at random from the population will have COVID-19, given that the person's blood test yields a positive result?

Solution. Let C denote the event the patient has COVID-19. Let TP denote the event that the patient tested positive. We are given with $P(TP|C) = 0.95$ and $P(TP|C^c) = 0.02$. Also $P(C) = 0.01$ and $P(C^c) = 0.99$. We now apply Bayes formula:

$$P(C|TP) = \frac{P(TP|C)P(C)}{P(TP|C)P(C) + P(TP|C^c)P(C^c)}$$

$$= \frac{(0.95)(0.01)}{(0.95)(0.01) + (0.02)(0.99)} = 0.3242.$$

It seems that the probability 0.3242 is lower. This is because a large number of non-COVID people also contributes to test positive results. □

The Definition of Odds. The term odds are often used to denote likelihoods. The odds of an event E is defined by

$$O(E) \equiv \frac{P(E)}{1 - P(E)} = \frac{P(E)}{P(E^c)}.$$

Conversely, the odds $O(E)$ of an event E gives the probability

$$P(E) = \frac{O(E)}{1 + O(E)}.$$

As an example, when $P(E) = \frac{2}{3}$, we say that the odds are 2 to 1 in favor of the event E.

We apply the Bayes theorem to write

$$P(E|F) = \frac{P(F|E)P(E)}{P(F)}.$$ (3.7)

The odds form of Bayes formula is given by

$$O(E|F) = O(E) \times L(F|E),$$ (3.8)

where

$$\text{Prior odds} = O(E) = \frac{P(E)}{P(E^c)}, \quad \text{Posterior odds} = O(E|F) = \frac{P(E|F)}{P(E^c|F)},$$

$$\text{Likelihood ratio} = L(F|E) = \frac{P(F|E)}{P(F|E^c)}.$$

We note $L(F|E)$ is the ratio of two likelihoods — hence the term.

Lemma. *Equations* (3.7) *and* (3.8) *are equivalent.*

Proof. We write (3.7) as

$$P(E|F) = \frac{P(F|E)P(E)}{P(F)}.$$ (3.9)

We divide (3.9) by $P(E^c|F)$. Its left side becomes $O(E|F)$. The right side of (3.9) reads

$$\frac{P(F|E)P(E)}{P(E^c|F)P(F)} = \frac{P(F|E)P(E)}{\frac{P(E^cF)}{P(F)}P(F)} = \frac{P(F|E)P(E)}{P(E^cF)}$$

$$= \frac{P(F|E)P(E)}{P(E^cF)}\frac{P(E^c)}{P(E^c)} = \frac{P(F|E)P(E^c)}{P(E^cF)}\frac{P(E)}{P(E^c)}$$

$$= \frac{P(F|E)}{P(F|E^c)}O(E) = O(E)O(E|F).$$

$\square$

Example 9. Consider the COVID-19 example once more. Assume that a patient having a high fever walked into a health clinic one day. The doctor assesses that the patient has a 30% chance of having contracted COVID-19. Now the patient is to be given a blood test whose reliability shows the same probabilities: $P(TP|C) = 0.95$ and $P(TP|C^c) = 0.02.$

To demonstrate the use odds, we see that the prior odds in favor of COVID-19 is

$$\frac{P(C)}{P(C^c)} = \frac{3}{7}.$$

The posterior odds in favor of COVID-19 is

$$\frac{P(C|TC)}{P(C^c|TC)} = \frac{P(C)}{P(C^c)} \times \frac{P(TP|C)}{P(TP|C^c)} = \frac{3}{7} \times \frac{95}{2} = \frac{285}{14}.$$

In terms of posterior probability, we have equivalently

$$P(C|TC) = \frac{\frac{285}{14}}{1 + \frac{285}{14}} = 0.95318.$$

We see that change of prior odds from 1 to 99 to 30 to 70 has a drastic effect on the conclusion. □

3.5. Artificial Intelligence and Bayes Theorem

Many books on artificial intelligence (AI) starts with a review of Bayes theorem. Having covered some rudimentary issues about the Bayes theorem in the last section, this is a good juncture to take a detour so that a nature connection is made to a topic of growing importance.

Assume we are interested in establishing the belief about the likelihood that an event H will occur. This belief is manifested itself in term of $P(H)$. Let $E_1, \ldots, E_m$ be a sequence of evidences that may affect $P(H)$. They may alter our assessment of $P(H)$ in either direction. Let the reliability of evidence E_i be expressed by the likelihood ratio

$$L(E_i|H) = \frac{P(E_i|H)}{P(E_i|H^c)}.$$

Then the Bayes theorem in the odds form (3.8) states

$$O(H|E_i) = O(H)L(E_i|H). \tag{3.10}$$

Moreover, the combined belief in the hypothesis H is given by

$$O(H|E_1, \ldots, E_n) = O(H)L(E_1, \ldots, E_m|H).$$

If we assume that conditional independence of $\{E_i\}$ given H, then the ratio of the following two terms

$$P(E_1, \ldots, E_m|H) = \prod_{k=1}^{m} P(E_k|H), \quad P(E_1, \ldots, E_m|H^c) = \prod_{k=1}^{m} P(E_k|H^c)$$

implies

$$O(H|E_1, \ldots, E_k) = O(H) \prod_{k=1}^{m} L(E_k|H). \tag{3.11}$$

We see that (3.11) is a generalization of (3.10) under the assumption of conditional independence where the last product term gives the joint likelihood ratio.

Sequential Updating. Let H denote a hypothesis. Assume we have collected evidences $\mathbf{E}_n = \{E_1, \ldots, E_n\}$ and already obtained $P(H \mid \mathbf{E}_n)$. Now that an an additional piece of evidence E_{n+1} becomes available. How can we efficiently update our estimate of $P(H|\mathbf{E}_{n+1})$, where $\mathbf{E}_{n+1} = \{\mathbf{E}_n, E_{n+1}\}$. To reduce computational complication, we assume conditional independence in the sense that

$$P(E_{n+1}|H\mathbf{E}_n) = P(E_{n+1}|H), \quad P(E_{n+1}|H^c\mathbf{E}_n) = P(E_{n+1}|H^c). \tag{3.12}$$

We recall in Example 7 that the notation of conditional independence has already been used in above manner.

Now we observe that

$$
\begin{aligned}
P(H|\mathbf{E}_{n+1}) &= \frac{P(H\mathbf{E}_n E_{n+1})}{P(\mathbf{E}_n E_{n+1})} \times \frac{P(H\mathbf{E}_n)}{P(H\mathbf{E}_n)} \times \frac{P(\mathbf{E}_n)}{P(\mathbf{E}_n)} \\
&= \frac{P(H\mathbf{E}_n E_{n+1})}{P(H\mathbf{E}_n)} \times \frac{P(H\mathbf{E}_n)}{P(\mathbf{E}_n)} \times \frac{P(\mathbf{E}_n)}{P(\mathbf{E}_n E_{n+1})} \\
&= P(E_{n+1}|H\mathbf{E}_n) \times P(H|\mathbf{E}_n) \times \frac{1}{P(E_{n+1}|\mathbf{E}_n)} \\
&= P(H|\mathbf{E}_n) \frac{P(E_{n+1}|H)}{P(E_{n+1}|\mathbf{E}_n)}, \tag{3.13}
\end{aligned}
$$

where the last equality is established using (3.12). We apply (3.13) with $H \longleftarrow H^c$ and write

$$P(H^c|\mathbf{E}_{n+1}) = P(H^c|\mathbf{E}_n) \frac{P(E_{n+1}|H^c)}{P(E_{n+1}|\mathbf{E}_n)}. \tag{3.14}$$

We now divide (3.13) by (3.14). This gives

$$\frac{P(H|\mathbf{E}_{n+1})}{P(H^c|\mathbf{E}_{n+1})} = \frac{P(H|\mathbf{E}_n)}{P(H^c|\mathbf{E}_n)} \frac{P(E_{n+1}|H)}{P(E_{n+1}|H^c)}$$

or in terms of odds, we have

$$O(H|\mathbf{E}_{n+1}) = O(H|\mathbf{E}_n)L(E_{n+1}|H). \tag{3.15}$$

Taking the log of the above yields

$$\log O(H|\mathbf{E}_{n+1}) = \log O(H|\mathbf{E}_n) + \log L(E_{n+1}|H). \tag{3.16}$$

We observe that in (3.15), the posterior probability $O(H \mid \mathbf{E}_n)$ now becomes the prior in the updating process. In (3.16), we note that the log of the likelihood provides the added weight induced by the new evidence E_{n+1}. It enables us to update the log of the posterior probability showing on the left side of (3.16).

Example 10. Let H denote the event that a defendant is guilty of a crime. Based on independent reports from three detectives, an application of Bayes theorem yields

$$P(H|\mathbf{E}_3) = 0.6, \quad P(H^c|\mathbf{E}_3) = 0.4,$$

where $\mathbf{E}_3 = \{E_1, E_2, E_3\}$ denotes the evidences collected by the three detectives. Expressed in odds, we have

$$O(H|\mathbf{E}_3) = \frac{0.6}{0.4} = 1.5. \tag{3.17}$$

Two weeks later, a DNA report indicates

$$P(E_4|H) = 0.7, \quad P(E_4|H^c) = 0.1$$

or

$$L(E_4|H) = \frac{0.7}{0.1} = 7,$$

where E_4 is the evidence collected by the DNA analysis.

We now use (3.15) and find

$$O(H|\mathbf{E}_4) = O(H|\mathbf{E}_3)L(E_4|H) = 1.5 \times 7 = 10.5$$

and

$$P(H|\mathbf{E}_4) = \frac{O(H|\mathbf{E}_4)}{1 + O(H|\mathbf{E}_4)} = \frac{10.5}{1 + 10.5} = 0.913.$$

Thus the DNA report moves the guilty probability from 0.6 to 0.913 because the fourth piece of evidence. □

When Hypotheses are Vector-Valued. We now consider the case when a hypothesis is multi-valued Let the hypothesis be represented by $\mathbf{H} = \{H_1, \ldots, H_m\}$. We also introduce a set of N sources that generate the evidence vector $\mathbf{E} = \{E^1, \ldots, E^N\}$. As an example, in a health-care setting, hypothesis H_1 may represent the state of being well, H_2 having a stroke, and H_3 having a heart attack, etc. Evidence E^1 may represent a report from an ambulance attendant, E^2 from an hospital lab, and E^3 from an consulting specialist.

Given $\mathbf{E} = \{E^1, \ldots, E^N\}$, the overall belief in the hypothesis H_i can be found from (3.6), namely,

$$P(H_i|\mathbf{E}) = \frac{P(\mathbf{E}|H_i)P(H_i)}{P(\mathbf{E})} = \alpha P(\mathbf{E}|H_i)P(H_i),$$

where α is the normalizing constant by requiring the elements of the vector resulting from the right-side multiplication sum to one. Then $\alpha = 1/(P(\mathbf{E}))$.

We now assume conditional independence of $\{E_i^k\}$ with respect to each H_i. Then we can store the likelihoods associated with $\{E_i^k\}$ in an $m \times N$ matrix $M \equiv \{m_{ij}\} = \{P(E^k|H_i)\}$. By conditional independence,

$$P(H_i\mathbf{E}) = \alpha P(H_i) \left(\prod_{k=1}^{N} P(E^k|H_i) \right).$$

Define

$$\Lambda_i = \left(\prod_{k=1}^{N} P(E^k|H_i) \right) \tag{3.17}$$

then we have

$$P(H_i|\mathbf{E}) = \alpha P(H_i)\Lambda_i. \tag{3.18}$$

The computation of (3.17) requires some clarification. For each k, $k = 1, \ldots, N$, we let the vector $\mathbf{m}^k = (m_{1k}, \ldots, m_{mk})$. Then

$$\Lambda_i = \bigotimes_{k=1}^{N} \mathbf{m}^k, \quad i = 1, \ldots, m, \tag{3.19}$$

where we use $\bigotimes$ to denote the element-by-element product of two vectors. Hence Λ_i is a vector of dimension m. Finally, we observe that the computation of the two vectors on the right side of (3.18) is also done by the operator $\bigotimes$. Summing the resulting elements and normalizing each component yields the posterior probability vector $\{P(\mathbf{H}|\mathbf{E})\}$.

Example 11 (Fire Detection). Consider a house with smoke detectors installed in the kitchen, bedroom, and living room. The evidences emitted from any detector k are given by the vector {silent, humming, loud alarm} denoting the three respective locations. As an example, if two detectors (say, the one in the kitchen and bedroom) emit loud alarm, then two identical columns appear as the corresponding likelihoods, namely, the last columns gives the corresponding likelihoods twice. There are four possible hypotheses $H = \{$cigarette smoke, electric fire, appliances fire, no fire$\}$. The likelihoods are summarized in the matrix M, where

$$M = \begin{bmatrix} 0.7 & 0.2 & 0.1 \\ 0.2 & 0.3 & 0.5 \\ 0.3 & 0.2 & 0.6 \\ 1 & 0 & 0 \end{bmatrix}.$$

Assume the owner senses something is wrong and there might be the possibility of some house fire. Assume that owner's assessment of the prior probabilities are $P(H) = \{0.1, 0.1, 0.1, 0.7\}$. In Illustration 3, we will compute the posterior probabilities based on the alternative usages of the fire detectors. □

3.6. More on Problem Solving by Conditioning

Many probabilistic problems can be solved by clever conditioning. One common approach following this train of thoughts is by conditioning on some event(s) that facilities the development of recursion. This is illustrated in the following examples.

Example 12. We now revisit Example 8 of Chapter 2 — the cowboys and their hats problem in which N cowboys enter a barbecue joint each leaves his hat to the attendant. When leaving the restaurant, the attendant gives the hats randomly back to the cowboys. (a) What is the probability that no one gets his own hat back; (b) what is the probability k cowboys get their own hats back?

Solution. (a) Let E denote the event that no cowboys get their own hats back. Let $P_N = P(E)$. Let Billy be one of the cowboys and let B be the event that Billy gets his own hat back. Conditioning on the fate of Billy, we apply the law of total probabilities and write

$$P_N = P(EB) + P(EB^c)$$
$$= P(E|B)P(B) + P(E|B^c)P(B^c).$$

We see that $P(E|B) = 0$ and thus

$$P_N = P(E|B^c)\frac{N-1}{N}.$$

To make the argument concrete, under B^c we assume that Billy gets the hat of Johnny. Billy's own hat is left in the pile of $N-1$ hats. There are two possible ensuing scenarios. Under both scenarios, we artificially "designate" Johnny's hat is that of Billy's). With a probability of $1/(N-1)$, Johnny gets Billy's hat. In this case we already two two no matches. For the $N-2$ remaining hats, the probability of no matches is P_{N-2}. The product of these two terms gives the probability of no matches under the first scenario. On the other hand, if Johnny does not get Billy's hat (while envisioning Johnny's own hat is actually Billy's hat), then the probability of no-matches of $N-1$ hats is given by P_{N-1}. Thus

$$P(E|B^c) = \frac{1}{N-1}P_{N-2} + P_{N-1}.$$

Hence

$$P_N = \left(\frac{1}{N-1}P_{N-2} + P_{N-1}\right)\left(\frac{N-1}{N}\right) = \frac{1}{N}P_{N-2} + \frac{N-1}{N}P_{N-1}$$

and it follow that

$$P_N - P_{N-1} = -\frac{1}{N}(P_{N-1} - P_{N-2}).$$

We convert the above to a first-order difference equation by defining $A_N = P_N - P_{N-1}$. Hence

$$A_N = -\frac{1}{N}A_{N-1}.$$

With $P_0 \equiv P_0$, $P_1 = 0$ and $P_2 = \frac{1}{2}$, we find $A_2 = \frac{1}{2}$, and

$$A_3 = -\frac{1}{3}A_2 = -\frac{1}{3!}, \quad A_4 = -\frac{1}{4}A_3 = \frac{1}{4!}, \quad \cdots .$$

Since $\sum_{k=2}^{N} A_k = P_N - P_1 = P_N$, we obtain

$$P_N = \frac{1}{2!} - \frac{1}{3!} + \frac{1}{4!} - \cdots + (-1)^N \frac{1}{N!}, \quad N = 2, 3, \ldots .$$

(b) Let F_k denote the probability exactly k cowboys get their own hats. For a specific set of k cowboys, the probability that they get their own hats and the others failed to get their own hats is

$$\frac{1}{N} \times \frac{1}{N-1} \times \cdots \times \frac{1}{N-(k-1)} P_{N-k}$$

$$= \frac{1}{N} \times \frac{1}{N-1} \times \cdots \times \frac{1}{N-(k-1)} \left(\frac{1}{2!} - \frac{1}{3!} + \frac{1}{4!} - \cdots \right.$$

$$\left. +(-1)^{N-k} \frac{1}{(N-k)!} \right)$$

$$= \frac{(N-k)!}{N!} \left(\frac{1}{2!} - \frac{1}{3!} + \frac{1}{4!} - \cdots + (-1)^{N-k} \frac{1}{(N-k)!} \right).$$

However, there are $\binom{N}{k}$ ways to choose these k lucky cowboys who get their own hats back. So the probability that any arbitrary k of them get their own hats is

$$P(F_k) = \frac{(N-k)!}{N!} \left(\frac{1}{2!} - \frac{1}{3!} + \frac{1}{4!} - \cdots + (-1)^{N-k} \frac{1}{(N-k)!} \right) \times \frac{N!}{k!(N-k)!}$$

$$= \frac{1}{k!} \left(\frac{1}{2!} - \frac{1}{3!} + \frac{1}{4!} - \cdots + (-1)^{N-k} \frac{1}{(N-k)!} \right).$$

A numerical example is given in Illustration 2. $\qquad\square$

Example 13 (Craps Game). In a game of craps, the "shooter" is the person who rolls a pair of fair dice. On the first roll, if the sum of the dice is 7 or 11, the shooter wins, and if the sum is 2, 3, or 12, he loses. If the sum is 4, 5, 6, 8, 9, or 10, the number becomes shooter's "point". If so, the game continues until if the shooter "makes his point", i.e., his number comes up again before he throw a 7. If 7 occurs first, the shooter loses. The game is depicted in the following picture:

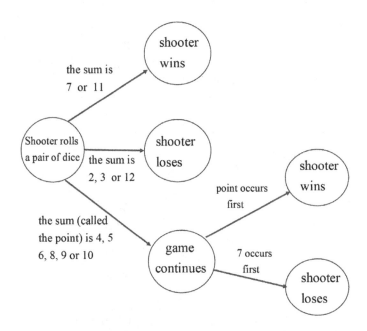

Let the sample space $\Omega = \{(i,j), i = 1, \ldots, 6; j = 1, \ldots, 6\}$ with $n(\Omega) = 36$. Outcomes from a roll of two dice are summarized below:

k	Outcome of a roll of 2 dice	$n(E_k)$	After the 1st roll
2	(1,1)	1	Shooter loses
3	(1,2),(2,1)	2	Shooter loses
4	(1,3),(2,2),(3,1)	3	Point
5	(1,4),(2,3),(3,2),(4,1)	4	Point
6	(1,5),(2,4),(3,3),(4,2),(1,5)	5	Point
7	(1,6),(2,5),(3,4),(4,3),(5,2),(6,1)	6	Shooter wins
8	(2,6),(3,5),(4,4),(5,3),(6,2)	5	Point
9	(3,6),(4,5),(5,4),(6,3)	4	Point
10	(4,6),(5,5),(6,4)	3	Point
11	(5,6),(6,5)	2	Shooter wins
12	(6,6)	1	Shooter loses

We consider the game backward. After the first roll, if the game is to continues, we face the following scenarios:

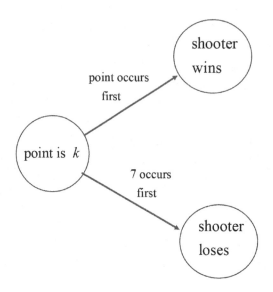

Assume that the shooter's point is $x \in \{4, 5, 6, 8, 9, 10\}$. We partition the sample space Ω into three parts: $\{x\}, \{7\}, \Omega_x$, where $\Omega_x = \Omega \backslash \{x, 7\}$. The shooter wins if $\{x\}$ occurs first; the shooter looses if $\{7\}$ occurs first; otherwise the shooter flips the dice again. Let r_x denote the probability that the game continues at the second stage. Then it is clear that

$$r_x = \frac{n(\Omega) - (n(\{x\} + n\{7\}))}{n(\Omega)} = \frac{n(\Omega_x)}{n(\Omega)}$$

and

$$P(\{x\} \text{ occurs first at the second stage}$$

$$= \frac{n(\{x\})/n(\Omega)}{(n(\Omega) - n(\Omega_n))/n(\Omega)} = \frac{p}{1 - r_x}$$

where $p = P(\text{the shooter's point occurs})$.

To illustrate the above computation, we consider $x = 5$. We have $n(\{5\}) = 4$ and $n(\{7\}) = 6$. Thus $p = \frac{4}{36}$ and $r_5 = \frac{36-10}{36} = \frac{26}{36}$. This means $P(\text{the shooter wins at the second stage given the point is 5}) = \frac{4/36}{10/36} = 4/10$. The results of the computation for the second stage are summarized below:

x	4	5	6	8	9	10
$n(\{x\})$	3	4	5	5	4	3
p	$\frac{3}{36}$	$\frac{4}{36}$	$\frac{5}{36}$	$\frac{5}{36}$	$\frac{4}{36}$	$\frac{3}{36}$
$\frac{p}{1-r_x}$	$\frac{3}{9}$	$\frac{4}{10}$	$\frac{5}{11}$	$\frac{5}{11}$	$\frac{4}{10}$	$\frac{3}{9}$

Let A denote the event that the shooter wins. We go back to the first stage and condition on the outcome of the first roll. This gives

$$P(A) = \sum_{i=2}^{12} P(A|\{i\})P(\{i\})$$

$$= P(\{7\}) + P(\{11\}) + \sum_{i=4,5,6,8,9,10} P(A|\{i\})P(\{i\})$$

$$= \frac{6}{36} + \frac{2}{36} + \frac{3}{9}\frac{3}{36} + \frac{4}{10}\frac{4}{36} + \frac{5}{11}\frac{5}{36} + \frac{5}{11}\frac{5}{36} + \frac{4}{10}\frac{4}{36} + \frac{3}{9}\frac{3}{36}$$

$$= \frac{244}{495} = 0.4929.$$

A Short Cut. Define event $B_k = (\{k\}$ occurs before $\{7\}$ in the second stage). Our earlier exposition implies that $P(B_k) = n(\{x\}/(n(\Omega) - n(\Omega_x))$. Hence $P(B_4) = \frac{3}{3+6}$, $P(B_5) = \frac{4}{4+6}$, $P(B_6) = \frac{5}{5+6}$, $P(B_8) = \frac{5}{11}$, $P(B_9) = \frac{4}{4+6}$, and $P(B_{10}) = \frac{3}{3+6}$. It follows that

$$P(A) = P(\{7\}) + P(\{11\}) + \sum_{i=4,5,6,8,9,10} P(B_i|\{i\})P(\{i\})$$

$$= \frac{6}{36} + \frac{2}{36} + \frac{3}{9}\frac{3}{36} + \frac{4}{10}\frac{4}{36} + \frac{5}{11}\frac{5}{36} + \frac{5}{11}\frac{5}{36} + \frac{4}{10}\frac{4}{36} + \frac{3}{9}\frac{3}{36}$$

$$= \frac{244}{495} = 0.4929. \qquad \square$$

Example 14 (Random Walk with Absorbing Barriers). Assume that there are two players, A and B, who are engaged in a game of successive coin flips. In a single flip, the probability of getting heads is p. Initially, A has i dollars and B has $N - i$ dollars. If a given flip yields heads, then A wins one dollar from B. Otherwise, A loses one dollars to B. The game ends when either person's wealth reaches 0. For all practical purposes, we focus on the wealth level of A. From the vintage point of player A, the random walk has two absorbing barriers: 0 and N. Once player A's wealth reaches either barrier, the game ends. A sample path is shown in the following graph.

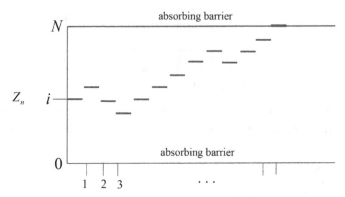

A sample path of a random walk with two absorbing barriers

Let X_n denote the outcome of the nth flip. Then

$$X_n = \begin{cases} +1 & \text{with probability } p, \\ -1 & \text{with probability } q. \end{cases}$$

Let Z_n be player A's wealth level after the nth flip. Then

$$Z_n = Z_0 + \sum_{k=1}^{n} X_n,$$

where $Z_0 = i$.

Let E denote the event that A wins the game. Define

$$P_i = P(E).$$

Conditioning on the outcomes of the first flip, the law of total probability implies

$$P(E) = P(E|H)P(H) + P(E|T)P(T).$$

Given that the player A's initial wealth level is i, if the first flip lands on heads then A's wealth level moves up to $i+1$; else down to $i-1$. Consequently, using the fact that $P_i = P(E)$, the above equation can be stated as

$$P_i = P_{i+1}p + P_{i-1}q, \quad i = 1, 2, \ldots, N - 1$$

with the two boundary conditions $P_0 = 0$ and $P_N = 1$.

We convert the above to a system of a first-order linear equations using the identity $p + q = 1$. Since

$$(p + q)P_i = P_{i+1}p + P_{i-1}q.$$

We have

$$p(P_{i+1} - P_i) = q(P_i - P_{i-1}), \quad i = 1, \ldots, N - 1.$$

Define $W_i = P_{i+1} - P_i$ and $a = \frac{q}{p}$. Then, we have

$$W_i = aW_{i-1}, \quad i = 1, \ldots, N - 1.$$

Using one of the boundary condition $P_0 = 0$, with $W_1 = aW_0 = a(P_1 - P_0) = aP_1$, we have $W_i = a^i P_1$, $i = 2, \ldots, N - 1$. Note

$$\sum_{k=1}^{i-1} W_k = (P_i - P_{i-1}) + (P_{i-1} - P_{i-2}) + \cdots + (P_2 - P_1) = P_i - P_1$$

$$= (a + a^2 + \cdots + a^{i-1})P_1.$$

Assume $a \neq 1$, then

$$P_i = (1 + a + a^2 + \cdots + a^{i-1})P_1, \quad i = 1, \ldots, N,$$

$$= \left(\frac{1 - a^i}{1 - a}\right)P_1.$$

Now we use the other boundary condition $P_N = 1$, we find

$$P_N = \left(\frac{1 - a^N}{1 - a}\right)P_1 = 1 \quad \Longrightarrow \quad P_1 = \frac{1 - a}{1 - a^N}$$

and

$$P_i = \left(\frac{1 - a^i}{1 - a}\right)\left(\frac{1 - a}{1 - a^N}\right) = \frac{1 - a^i}{1 - a^N} = \frac{1 - \left(\frac{q}{p}\right)^i}{1 - \left(\frac{q}{p}\right)^N}, \quad i = 1, \ldots, N.$$

When $a = 1$, or equivalently $p = 1$, we have

$$P_i = iP_1, \quad i = 1, \ldots, N.$$

Thus $P_1 = \frac{1}{N}$ and

$$P_i = \frac{i}{N}, \quad i = 1, \ldots, N.$$

To summarize, we conclude that, for $i = 0, 1, \ldots, M$, we have

$$
P_i = \begin{cases} \dfrac{1 - \left(\frac{q}{p}\right)^i}{1 - \left(\frac{q}{p}\right)^N} & \text{if } p \neq \frac{1}{2}, \\[4mm] \dfrac{i}{N} & \text{if } p = \frac{1}{2}. \end{cases}
$$

□

Example 15. A fair coin is flipped n times, where $n = 4, \ldots, 10$. Let E_n be the event that we obtain at least four heads in a row. We want to find $P(E_n)$, $n - 4, \ldots, 10$. To solve the problem, we consider a preliminary experiment in which the coin is flipped repeatedly until we get a tail for the first time before the first three flips, or we stop at the fourth flip. Let $\{F_i\}_{i=1}^4$ denote the sample points $(\{T\}, \{HT\}, \{HHT\}, \{HHHT\})$. The sample space associated with this experiment is $\Omega = \{\{F_i\}_{i=1}^4 \cup F_4^c\}$, where $F_4^c = \{HHHH\}$. We see that

$$
P(F_i) = \left(\frac{1}{2}\right)^i \quad i = 1, 2, 3, 4,
$$

$$
P(F_4^c) = \left(\frac{1}{2}\right)^4.
$$

Also we see that

$$
P(\Omega) = \sum_{i=1}^4 \left(\frac{1}{2}\right)^i + \left(\frac{1}{2}\right)^4 = 1.
$$

and the five events form a partition of Ω.

Let P_n denote the probability of not observing four heads in a row in n flips, i.e., $P(E_n) = 1 - P_n$. Applying the law of total probability, we obtain

$$
P_n = \sum_{i=1}^4 P(F_i) P_{n-i} + P(F_4^c)(0) = \sum_{i=1}^4 \left(\frac{1}{2}\right)^i P_{k-i}, \quad n = 4, \ldots, 10
$$

where $P(F_i) = 1$, $i = 1, \ldots, 4$. The key observation in this example is that once a tail is observed, the problem starts afresh. However the number of flips remaining is to be adjusted accordingly.

In Illustration 4, we obtain the numerical results $\{P(E_n)\}$, for $n = 4, \ldots, 10$. For a small n, it is easy to use enumeration to find the answer.

As a example, for $n = 5$, there are 3 cases for which E_5 occurs. They are $\{HHHHT, HHHHH, THHHH\}$. Hence

$$P(E_5) = \frac{3}{2^5} = 0.09375.$$

This is exactly what is show in Illustration 4. □

3.7. Exploring with *Mathematica*

Illustration 1 Application of the results obtained from random walk (cf., Example 14). A pharmaceutical company is testing the efficacy of treating COVID-19 patients by two competing vaccines, called A and B. The company use pairs of patients for administering A and B to each pair randomly. Let P_i denote the probability that Vaccine i will cure a COVID-19 patient, where $i = 1$ and 2 corresponds to A and B, respectively. We introduce two indicator random variables

$$W_i = \begin{cases} 1 & \text{if the patient in the } i\text{th pair using Vaccine } A \text{ is cured,} \\ 0 & \text{otherwise,} \end{cases}$$

$$Y_i = \begin{cases} 1 & \text{if the patient in the } i\text{th pair using Vaccine } B \text{ is cured,} \\ 0 & \text{otherwise.} \end{cases}$$

We toss away the tied cases and keep track of the cumulative differences

$$Z_n = \sum_{i=1}^{n} X_n,$$

where

$$X_n = W_n - Y_n = \begin{cases} +1 & \text{with probability } p, \\ -1 & \text{with probability } q = 1 - p \end{cases}$$

and

$$p = P\{A \text{ Moves up by one} | \text{either } A \text{ moves up by one or } A$$
$$\text{moves down by one}\}$$
$$= \frac{P_1(1 - P_2)}{P_1(1 - P_2) + (1 - P_1)P_2}.$$

Thus, the drug testing is framed entirely in the context of a random walk.

Assume the design parameter for the pharmaceutical company is M. As soon as Z_n reaches M, then the company declares that Vaccine A is better than Vaccine B; on the other hand, if Z_n reaches $-M$, the company will assert the converse. If follows that for testing the hypothesis that one is better than the other, we set $Z_0 = M$ and the two absorbing barriers at $2M$ and 0. Under the revised Y-*axes* label, we are able to immediately employ the formula derived in Example 11. When Z_n reaches $2M$, then the company declares that Vaccine A is better than Vaccine B. On the other hand, when Z_n reaches 0 first, then the opposite verdict is rendered. This is depicted in the following graph.

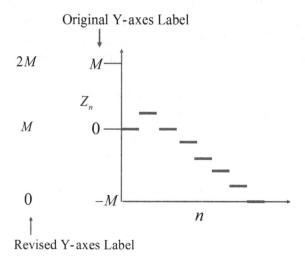

We see that the realization shown in the above picture displaying $Z_n = 0$. Using the formula given in Example 12, the error probability is given by

$$\text{Error Probability} = P(\text{Vaccine}\,B \text{ performs better than Vaccine } B | P_1 > P_2)$$

$$= P(Z_n = 0 \,|\, P_1 > P_2 \quad \text{and} \quad Z_0 = M)$$

$$= 1 - \frac{1 - \left(\frac{q}{p}\right)^M}{1 - \left(\frac{q}{p}\right)^{2M}}$$

A numerical example. Assume we have $P_1 = 0.6$ and $P_2 = 0.4$. Then, the following computation shows that when $M = 11$, the probability of making the wrong assertion reaches zero.

$$P1 = 0.6; \; P2 = 0.4; \; p = \frac{P1 * (1 - P2)}{P1 * (1 - P2) + (1 - P1) * P2}; \; q = 1 - p; \; a = \frac{q}{p};$$

$$fn[M\_] := 1 - (1 - a^M) / (1 - a^{2*M});$$

$$rs = Table[\{M, fn[M]\}, \{M, 5, 15\}];$$

ListPlot[rs, AxesLabel → {"M", "Prob"}, LabelStyle → {Black, Medium},
PlotStyle → {Black, Thick}, AxesStyle → {Black, Medium}]

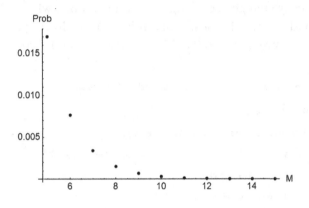

Illustration 2. Consider Example 12 once more. Assume $N = 10$, in the following, we compute $P(F_k)$, the probability that k cowboys get their own hats back, for $k = 1, \ldots, 10$.

$$P1 = 0.6; \; P2 = 0.4; \; p = \frac{P1 * (1 - P2)}{P1 * (1 - P2) + (1 - P1) * P2}; \; q = 1 - p; \; a = \frac{q}{p};$$

$$fn[M\_] := 1 - (1 - a^M) / (1 - a^{2*M});$$

$$rs = Table[\{M, fn[M]\}, \{M, 5, 15\}];$$

ListPlot[rs, AxesLabel → {"M", "Prob"}, LabelStyle → {Black, Medium},
PlotStyle → {Black, Thick}, AxesStyle → {Black, Medium}]

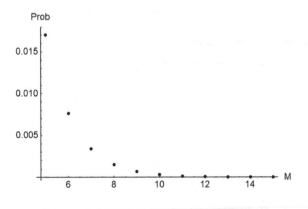

The above graph shows that with 10 in the restaurant, it is unlikely that more than four cowboys will get their own hats back in a random returning of the hats.   □

Illustration 3. We now give a numerical demonstration of Example 11. Based on the data given in the example, depending on the number and the types of warning these detectors produce, the posterior probabilities of the various hypotheses occur change accordingly. As an example, when all three detectors produce loud alarm, the posterior probability of having either an appliance fire or kitchen fire go up to $0.3655 + 0.6316 = 0.9971$ (from a prior probability 0.2).

Case 1: if two detectors are making loud sound and one is silent

```
PH = {0.1, 0.1, 0.1, 0.7};

m1 = {0.7, 0.2, 0.3, 1}; m3 = {0.1, 0.5, 0.6, 0};

BG1 = m1 * m3 * m3;                        (* this applies (3.19) *)

PosteriorH = PH * BG1 / Total[ PH * BG1]   (* this uses  (3.18)  *)
{0.0424242, 0.30303, 0.654545, 0.}
```

Case 2: if all three detectors are making loud sound

```
BG2 = m3 * m3 * m3;

PosteriorH = PH * BG2 / Total[ PH * BG2]
{0.00292398, 0.365497, 0.631579, 0.}
```

Case 3: If two detectors are silent and one is making loud sound

```
BG3 = m1 * m1 * m3;

PosteriorH = PH * BG3 / Total[ PH * BG3]
{0.398374, 0.162602, 0.439024, 0.}
```

Case 4: If all three detectors are silent

```
BG4 = m1 * m1 * m1;

PosteriorH = PH * BG4 / Total[ PH * BG4]
{0.0464896, 0.0010843, 0.00365953, 0.948767}
```

□

Illustration 4. The following is the computation involving Example 15.

```
x = {1, 1, 1, 1}; y = Table[2^-i, {i, 1, 4}]; newX = x;
xhead = x.y; pE[4] = N[1 - xhead];

Do[{newX = Drop[Prepend[newX, xhead], -1]; xhead = newX.y;
   pE[i + 4] = N[1 - xhead]}, {i, 6}]

Print[Table[pE[i], {i, 4, 10}]]
   {0.0625, 0.09375, 0.125, 0.15625, 0.1875, 0.216797, 0.245117}
```

□

Problems

1. Given that a throw of three unbiased dice shows different faces, what is the probability that (a) at least one is a six; (b) the total is eight?

2. An urn contains 5 white and 10 black balls. Let Y be the number observed by rolling a fair die. We then choose Y balls from the urn. (a) Let E be the event the all the chosen balls are white. Find $P(E)$. (b) Find $P(E| Y = 3)$.

3. There are two identical tables. Each table has two drawers. Table I has a white balls in each drawer. Table II has a white ball in one of the drawer and a black ball in the other drawer. We choose a table randomly and open one of its drawer randomly. We find a white ball. What is the probability that we find a white ball in the other drawer?

4. It is expected that 60% of the people in a small town will be getting the flu in the next month. If you are not inoculated then the probability of you getting the flu is 80%. But if you are inoculated then the probability drops to 15%. What is the probability that a person is inoculated?

5. An auto insurance company classifies its clients into three groups: low risks, average risks, and high risks. Its past history suggested that the probabilities that low, average, and high risk clients will involved in an accident in an one-year period are 0.05, 0.15, and 0.30, respectively. Assume that 20% of the drivers are low risks, 50% average risks, and 30% are high risks. (a) If Johnny does not have any accidents in one year, what is probability that Johnny is a high-risk driver? (b) If Johnny has an accidents in a year, what is probability that Johnny is a low-risk driver?

6. In a jury pool, there are 4 college educated men and 6 college educated women, and 6 non-college educated men. How many non-college educated women must be present if gender and college-education are to be independent when all jurors are selected at random?

7. (a) If the odds of event E is to occur is $\frac{2}{5}$, what is the probability that event E will occurs? (b) If the odds of event F is 7, what is the probability that event F will occur?

8. A fair coin is tossed three times. Let E be the event that all tosses land heads. (a) What is the odds of the event E? (b) What is the posterior odds of the event E given at least one of the three coins landed heads?

9. You choose at random a letter from the word "chance" and a letter from the word "choice." What is the probability that the two chosen letters are the same?

10. There are 15 golf balls in a box, of which 9 are new. Three of them are randomly chosen and played with. We then returned them to the box. Later, another 3 balls are randomly chosen from the box. Find the probability that none of these three chosen balls have ever been used.

11. Suppose that events A, B, and C are independent with probabilities of $\frac{1}{5}, \frac{1}{4}$, and $\frac{1}{3}$ of occurring, respectively. (a) Let $E = \{A \cap B \cap C\}$. Find $P(E)$. (b) Let $F = \{A \text{ or } B \text{ or } C\}$. Find $P(F)$. (c) Let $G = \{\text{exactly one of the three events occurs}\}$. Find $P(G)$.

12. Five cards are dealt from a standard deck of 52. Find (a) The probability that the third card is an ace. (b) The probability that the third card is an ace given that the last two cards are not aces. (c) The probability that all card are of the same suit. (d) The probability of two or more aces.

13. Suppose an urn has seven balls numbered $1, 2, \ldots, 7$. A person draws a ball, returns it, draws another, returns it, and so on until he gets a ball which has been drawn before and then stops. Of course, each draw constitutes a random drawing. Find the probability that its takes exactly four drawings to accomplish this objective.

14. Consider two coins named Coins 1 and 2. Coin 1 shows heads with probability $\frac{1}{10}$ when spun. Coin 2 shows heads with probability $\frac{1}{2}$ when spun. You pick one of the two coins at random and spin it twice. Find (a) $P(\text{heads on the first spin})$, (b) $P(\text{heads on the second spin})$, (c) $P(\text{heads on both spins})$, and (d) $P(\text{the coin chosen is Coin 2 given that heads on both spins})$.

15. Assume that approximately 1% of women aged 40–50 years have breast cancer. Assume that a woman with breast cancer has a 90% chance of a positive test from a mammogram, while a woman without has a 10% chance of a false-positive result. What is the probability that a woman has breast cancer given that she just has a positive test?

16. *The Monty Hall Problem.* The problem is named after the host of the TV show *Let's Make a Deal,* in which contestants were often placed in situations such as the following: Three curtains are numbered $1, 2,$ and $3.$ A car is randomly parks behind one curtain. Behind the other two curtains, each has a donkey. You pick a curtain, say Curtain 1. To build some suspense the host opens up one of the two remaining curtains, say Curtain 3, to reveal a donkey. (a) Should you switch curtains and Curtain 2 if you are given the chance? (2) What is the probability that there is a donkey behind Curtain 3 if the host opens Curtain 3?

17. Abgail, Briggette, and Concetta are experienced target shooters. The probability that Abgail hits the target in a single shot is $\frac{1}{3}$. For the other two shooters the probabilities are $\frac{1}{5}$ and $\frac{1}{4}$, respectively. In a given contest, the three shooters fire at the target independently. Two shots hit the target and one misses. What is the probability that Abgail misses the target?

18. In a given city, we assume that 50% of its households subscribe to cable TVs, 70% have internet services, and 40% have both. What is the probability that a randomly selected household does not subscribe to cable TV given that the household has internet service?

19. A jar contains five white balls and five black balls. You roll a fair die. You then draw at random as many balls from the jar as the face shown on the die being rolled. (a) What is the probability that each of the balls drawn is white? (b) What is the probability that the face shown on the die is five given that the each of the balls drawn is white?

20. Assume that the number of children in a family is 1, 2, 3, with equal probability. Little Bobby has no sisters. What is the probability he is an only child? (Assume that the probability that a child is a boy or a girl is equal.)

21. Assume that the number of children in a family is 1, 2, 3, with equal probability. Little Bobby has no brothers. What is the probability he is an only child? (Assume that the probability that a child is a boy or a girl is equal.) Let E_i be the event that a family has i children, $i = 1, 2, 3.$

22. A family has j children with probability p_j, where $p_1 = 0.1$, $p_2 = 0.25$, $p_3 = 0.35$, $p_4 = 0.3$. A child from this family is randomly chosen. Given that this child is the oldest child in the family, find the probability that the family has (a) only one child, and (b) 4 children.

23. Assume that the fraction young people who is a smoker in a population is 25%. Assume also that the probability that their fathers are smoker is 35%. If a young person's father is a smoker then that person's being a smoker increases to 40%. What is the probability that a young person is a smoker given this person's father is a non smoker?

24. Let the sample space $\Omega = \{w_1, w_2, w_3, w_4\}$. Assume each sample point is equally likely to occur. Define

$$A = \{w_1, w_2\}, \quad B = \{w_2, w_3\}, \quad C = \{w_1, w_3\}.$$

(a) Are are events A, B, and C pairwise independent events? (b) (a) Are are events A, B, and C mutually independent events?

25. An oil wildcatter in West Texas asks SlumberJack to do a seismic test in an area he thinks has potential. It is generally believed that the probability of oil being present in the area is 35%. If the seismic test shows a positive result, then the chances of having oil underneath increases to 85%. On the other hand, the test would show false positive 20% of the time. The wildcatter prefers to use the odds form of Bayes rule to update the probability that there is oil present when the test shows a positive results. What is the needed conditional probability?

26. (a) Let E and F be two events. Prove

$$P(EF) \geq P(E) + P(F) - 1.$$

(b) Generalize the above to n events $E_1, \ldots, E_n$ to the following

$$P(E_1 \cap E_2 \cap \cdots \cap E_n) \geq P(E_1) + P(E_2) + \cdots + P(E_n) - (n-1).$$

27. We toss a fair die repeatedly. Let X_n denote the sum of all the points observed after n tosses. Let A be the event that $\{X_n = n\}$ and $p_n = P(A)$. Find a recursive equation for p_n. Use *Mathematica* to compute p_n for $n \leq 20$ and plot $\{p_n, n \leq 20\}$.

28. Consider a sequence of independent coin tosses. Let p denote the probability that a toss lands heads and $q = 1 - p$ be the probability that a toss lands in tails. Let E_n be the event that we observe a run of r more successes in a total of n tosses. Define $p_n = P(E_n)$. Find $\{p_n\}$.

Let $r = 3$, $p = 0.6$, use *Mathematica* to compute $\{p_n, n \leq 20\}$ and plot the results so obtained.

29. Consider a coin that comes up heads with probability p and tails with probability $q = 1 - p$. Let P_n be the probability that after n independent tosses, there have been an even number of heads. (a) Derive a recursive equation relating P_n to P_{n-1}. (b) Prove by induction that

$$P_n = \frac{1 + (1 - 2p)^n}{2} \quad n = 1, 2, \ldots (*).$$

30. Consider a coin, when it is tossed, has a probability p of landing heads, and has a probability of $q = 1 - p$ of landing tails. Let E be the event there is a run of r heads in a row before there is a run of s tails in a row. (a) Find $P(E)$ in terms of p, q, r, and s. (b) For $p = 0.6$, $s = 3$, use *Mathematica* to compute $P(E)$ for $r = 1, 2, \ldots, 7$.

31. A coin has probability θ of landing on its edge. Professors Abel, Braz, and Clarence made the assumption that θ is 10^{-10}, 10^{-11}, and 10^{-12}, respectively. You assign prior probabilities of $\frac{3}{8}$ to each of Professor Abel's and Braz's hypothesis, and probability $\frac{1}{4}$ to Professor Clarence's. The coin is tossed and lands on its edge. Find your posterior probabilities for the three professors' hypothesis.

32. A fair coin is tosses twice. Find the conditional probability that both tosses show heads given at least one of them is a head.

33. An run contains r red balls and b blue balls. A ball is chosen at random from the urn, its color is noted. It is returned together with d more balls of the same color. (a) What is the probability that the first ball drawn is blue? (b) What is the probability that the second ball drawn is blue?

Remarks and References

The notion of conditional independence plays an important role in artificial intelligence. For example, see Jordan [1] and Pearl [2, 3]. The section on artificial intelligence and the Bayes theorem is influenced by Pearl [4, Chapter 1]. Example 14 on random walk and its application shown in Illustration 1 are done in the spirit of Ross [4, Chapter 3, Example 4m].

[1] Jordan, M. J. (ed)., *Learning in Graphical Models*, MIT Press, 1999.
[2] Pearl, J. Causality: Models, *Reasoning, and Inference*, 2nd edn., Cambridge University Press, 2009.
[3] Pearl, J., *Probabilistic Reasoning in Intelligent Systems: Networks of Plausible Inference*, revised 2nd printing, Morgan Kaufmann Publishing, 1988.
[4] Ross, S. M., *A First Course in Probability*, 10th edn., Pearson, 2019.

Chapter 4

Random Variables

4.1. Introduction

A numerically valued *function* X of ω with domain Ω:

$$\omega \in \Omega: \quad \omega \to X(\omega)$$

is called a *random variable*. This is a generic definition of a random variable. Its implication will become clear as we look at some examples. Roughly speaking, a random variable is a *quantification* of a random event.

When the range of the function X is the set of *countable* number of possible values. Then X is called a *discrete* random variable. For a discrete random variable X, we call

$$p_X(x) = P(X = x)$$

the *probability mass function* (PMF).

Following the axioms of probability, we require

$$p_X(x) \geq 0 \quad \text{for all } x$$

and

$$\sum_x p_X(x) = 1.$$

Example 1. Assume $X(\omega) \in \{1.52, 0.247, 7.879\}$ for all $\omega \in \Omega$. Since the cardinality of Ω is 3, X is a discrete random variable. Note that, in this example, X does not assume integer values. □

Example 2. Consider the toss of two fair coins. The sample space $\Omega = \{\omega_1, \omega_2, \omega_3, \omega_4\} = \{HH, HT, TH, TT\}$. Each sample point ω_i occurs with probability $\frac{1}{4}$. If we define X as the number of times heads occur. Then $X \in \{0, 1, 2\}$ and the probability mass function of X is given by

$$p_X(0) = P(X = 0) = P(\omega_4) = \frac{1}{4},$$

$$p_X(1) = P(X = 1) = P(\omega_2, \omega_3) = \frac{1}{2},$$

$$p_X(2) = P(X = 2) = P(\omega_1) = \frac{1}{4}.$$

We define Y as the event that heads occur at least once. Then $Y \in \{0, 1\}$, and $Y = 1$ denotes the event that heads occurs at least once. Then we have

$$p_Y(0) = P(Y = 0) = P(\omega_4) = \frac{1}{4},$$

$$p_Y(1) = P(Y = 1) = P(\omega_1, \omega_2, \omega_3) = \frac{3}{4}.$$

□

Example 3. Consider the toss of a coin with probability of p of coming up heads and $q = 1 - p$ of yielding tails. Let X be the number of tosses needed in order to obtain a head for the first time. Then $X = \{1, 2, \ldots\}$ and the probability mass function is given by

$$P(X = i) = q^{i-1}p, \quad i = 1, 2, \ldots.$$

If random variable Z is defined as the number of tosses to obtain two heads in succession. Then $Z = \{2, 3, 4, \ldots\} = \{HH, THH, \{HTHH, TTHH\}, \ldots\}$ and the probability mass function is given by

$$p_Z(2) = P(Z = 2) = p^2,$$

$$p_{Z(}(3) = P(Z = 3) = qp^2,$$

$$p_Z(4) = P(Z = 4) = (pq + q^2)p^2,$$

$$p_Z(i) = P(Z = i) = \cdots, \quad i = 5, 6, \ldots.$$

□

When the range of the function X is the set of real numbers, then X is called a *continuous* random variable. For a continuous random variable X, we call $f_X(x)$ the *probability density function* (PDF). The notion of density can be interpreted as the likelihood that random variable X assumes values in the 'neighborhood' of x. In other words, we have

$$P\left(X \in \left\{x - \frac{\varepsilon}{2}, \ x + \frac{\varepsilon}{2}\right\}\right) \approx f_X(x) \cdot \varepsilon,$$

where $\varepsilon > 0$. Thus $f_X(x)$ can be interpreted as the *rate* of likelihood.

Given the density f_X, we can find the probability that X will assume values in the interval U using

$$P(X \in U) = \int_U f_X(x)dx.$$

Following the axioms of probability, we require

$$f_X(x) \geq 0 \quad \text{for all } x$$

and

$$\int_{-\infty}^{\infty} f_X(x) = 1.$$

For a continuous random variable X, it is important to remember that $f_X(x) = 0$ for each x.

Example 4. Assume that the number of pounds of coffee bean sold, denoted by random variable X, on a single day at a given store follows the probability density function:

$$f_X(x) = \begin{cases} 0 & \text{if } 0 \leq x < 60, \\ \frac{1}{60} & \text{if } 60 \leq x < 120, \\ 0 & \text{if } x \geq 120. \end{cases}$$

The density shows that it is equally likely X will be between 60 and 120 pounds. If we want to know the probability that X is greater than 100, then we see

$$P(X > 100) = \int_{100}^{120} f_X(x)dx = \int_{100}^{120} \frac{1}{60}dx = \frac{1}{3}. \qquad \square$$

Linking Events and Random Variables: In the previous chapters, we talked about events. We can express events in terms of random variables via indicator random variables. Let A be an event. We define indicator random variable

$$I_A = \begin{cases} 1 & \text{if } A \text{ occurs,} \\ 0 & \text{otherwise.} \end{cases}$$

Thus I_A is a random variable assuming two specific values 1 and 0, where event $\{I_A = 1\}$ occurs with probability $P(A)$. In §4.4, we will elaborate on the notion of the expectation $E(X)$ of a random variable X. It follows that for the indicator random variable I_A, we have $E(I_A) = P(A)$. Following this link, all results developed for random variables work equally well for events.

Example 5 (The Inclusion–Exclusion Formula (of Chapter 2) Revisited). Define indicator random variable

$$X_i = \begin{cases} 1 & \text{if } E_i \text{ occurs,} \\ 0 & \text{otherwise.} \end{cases}$$

We want to express the event $A = \bigcup_{i=1}^n E_i$ in terms of random variables $\{X_i\}_{i=1}^n$. By DeMorgan's law, we see that $A^c = \bigcap_{i=1}^n E_i^c = \{\text{all } E_i \text{ fail to occur}\} = \{\text{all } X_i = 0\}$. Define

$$Y = \prod_{i=1}^n (1 - X_i).$$

Then $\{A^c\} \iff \{Y = 1\}$. We now define indicator random variable

$$Y = \begin{cases} 1 & \text{if } A^c \text{ occurs,} \\ 0 & \text{otherwise.} \end{cases}$$

This implies that $P(A^c) = E(Y)$ and

$$P(A) = 1 - P(A^c) = 1 - E(Y)$$

$$= 1 - E\left(\prod_{i=1}^n (1 - X_i)\right)$$

$$= E(X_1 + \cdots + X_n) - E\left(\sum_{i<j}(X_i X_j)\right)$$

$$+ \cdots + (-1)^{n+1} E(X_1 + \cdots + X_n)$$

Applying the definitions of the indicator random variables $\{X_i\}_{i=1}^n$ leads immediately to the formula. $\qquad\square$

4.2. Distribution Functions

The *distribution function* of random variable X, also known as the cumulative distribution function (CDF), is defined by

$$F_X(x) = P(X \le x), \quad x \in (-\infty, \infty).$$

When X is a discrete random variable, the above equation becomes

$$F_X(x) = \sum_{k \le x} p_X(k).$$

Similarly, when X is a continuous random variable, we have

$$F_X(x) = \int_{-\infty}^{x} f_X(x)dx.$$

In terms of operations about a random variable, the above equation indicates that when working with a discrete random variable, we simply use the needed summation; whereas for a continuous random variable, we just use the respective integration. In Section 4.5, we introduce the Riemann–Stieltjes integral that unifies the two in a single representation.

The distribution function of X contains all the information about the random variable. Moreover, both versions share the following properties:

(i) F_X is non-decreasing, i.e., if $u \le v$, then $F_X(u) \le F_X(v)$.
(ii) $\lim_{x \to -\infty} F_X(x) = 0$.
(iii) $\lim_{x \to +\infty} F_X(x) = 1$.
(iv) F_X is right continuous with left limits (a *càdlàg* function, a French acronym for *continu à droite, limité à gauche*)

$$\lim_{x \downarrow x_0} F_X(x) = F(x_0).$$

The following graph displays a salient feature of a distribution function. It is the distribution function of random variable X which is not strictly discrete nor continuous. We see that the figure shows that when $X < x_0$ and $X > x_0$ it is continuous. But when $X = x_0$, it has a jump. As we see in the following that $P(X = x_0)$ is equal to the magnitude of the jump. Again, the graph motivates the need to introduce the notion of Riemann–Stieltjes integral given in Section 4.5. In real-world applications, such occurrences are not unusual. As an example, we let X denote the waiting time to see a barber. If 30% of time, a customer wants to get a hair cut without having to wait; otherwise the customer will have to wait for a random length of time. Then $P(X = 0) = 0.3$ and X is a continuous random variable when $X > 0$.

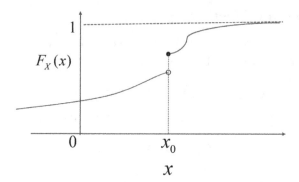

Consequences from the definitions of distribution function:

(i) $P(a < X \le b) = F_X(b) - F_X(a)$,

(ii) $P(X < b) = F_X(b^-) = \lim_{n \to \infty} F_X\left\{b - \frac{1}{n}\right\}$,

(iii) $P(X = a) = F_X(a) - F_X(a^-)$,

where we use b^- denote $b - \varepsilon$ and $\varepsilon > 0$ is an arbitrary small number.

Example 6. Consider the distribution function shown below:

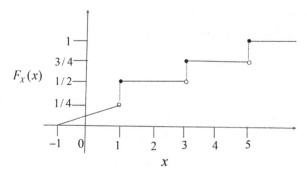

Since the distribution function $F_X(x)$ tells us about everything about the random variable X, we want to construct the probability distribution about X. The figure shows that there are three jumps indicating

$$P(X = i) = \frac{1}{4}, \quad i = 1, 3, 5.$$

For $-1 \le X < 1$, X is continuous with the density

$$f_X(x) = \frac{1}{8}(x + 1), \quad -1 \le x < 1.$$

We see

$$\int_{-1}^{1} f_X(x)dx = \int_{-1}^{1} \frac{1}{8}(x + 1)dx = \frac{1}{4}.$$

Thus all the probabilities sum to 1. In this example, X is neither continuous nor discrete. □

4.3. Functions of a Random Variable

Let X be a random variable and g be a function of X. For example, X represents the number of bagels sold on a given day and $g(X)$ represents the profit from selling X bagels on a given day. Knowing the PMF of X, then $Y \equiv g(X)$ is a discrete random variable. We see that

$$p_Y(y) = \sum_{\{x_i : g(x_i) = y\}} p_X(x_i).$$

When X is a continuous random variable, we have

$$F_Y(y) = P(Y \le y) = \int_{\{x : g(x) \le y\}} f_X(x) dx.$$

We differentiate the above with respect to y to obtain $f_Y(y)$.

Example 7. Consider random variable X with

$$p_X(x) = \begin{cases} \frac{1}{4} & \text{if } x = 1, \\ \frac{1}{2} & \text{if } x = 0, \\ \frac{1}{4} & \text{if } x = -1. \end{cases}$$

Assume $Y = g(X) = X^2$. Then

$$p_Y(y) = \begin{cases} \frac{1}{2} & y = 0, \\ \frac{1}{2} & y = 1. \end{cases}$$ □

Example 8. Consider a continuous random variable X with

$$f_X(x) = \begin{cases} 1 & 0 \le x \le 1, \\ 0 & \text{otherwise.} \end{cases}$$

A simple calculation shows that $F_X(x) = x$ for $0 \le x \le 1$. Let $Y = g(X) = e^X$. We would like to find the density f_Y. We see

$$\begin{aligned} F_Y(y) &= P(Y \le y) \\ &= P(e^X \le y) \\ &= P(X \le \ln y) \\ &= F_X(\ln y) \\ &= \ln y, \quad 1 \le y \le e. \end{aligned}$$

Differentiating the above with respect to y gives

$$f_Y(y) = \begin{cases} 0 & y < 0, \\ \frac{1}{y} & 1 \le y \le e, \\ 0 & y > e. \end{cases}$$

□

4.4.   Expectation of a Random Variable

One of the most prominent feature of a random variable is its mean, also known its expectation, defined by

$$\mu_X \equiv E(X) = \sum_x x p_X(x) \quad \text{when } X \text{ is discrete}$$

$$= \int_{-\infty}^{\infty} x f_X(x) dx \quad \text{when } X \text{ is continuous.}$$

The mean of X measures the central tendency of the random variable.

Consider the special case for a discrete nonnegative integer-valued random variable X where $p_X(x) \ge 0$ for $x = 0, 1, 2, \ldots$ and $p_X(x) = 0$, otherwise. Then we see

$$E(X) = \sum_{x=0}^{\infty} x p_X(x) = \sum_{x=1}^{\infty} x p_X(x)$$

$$= \sum_{x=1}^{\infty} \sum_{k=1}^{x} 1 \cdot p_X(x) = \sum_{k=1}^{\infty} \sum_{x=k}^{\infty} p_X(x)$$

$$= \sum_{k=1}^{\infty} P(X \ge k) = \sum_{k=1}^{\infty} p_X^{\ge}(x) = \sum_{k=0}^{\infty} P(X > k), \qquad (4.1)$$

where $p_X^{\ge}(k) = P(X \ge k)$ denotes the complementary cumulative distribution of X. We will illustrate the use of this relation in the sequel.

We now derive an analogous identity when X is a continuous nonnegative random variable. We see

$$E(X) = \int_0^{\infty} x f_X(x) dx = \int_0^{\infty} \left(\int_0^x dy \right) f_X(x) dx$$

$$= \int_0^{\infty} \int_0^x f_X(x) dy dx = \int_0^{\infty} \int_y^{\infty} f_X(x) dx dy$$

$$= \int_0^{\infty} P(X > y) dy = f_X^{\ge}(x). \qquad (4.2)$$

We now consider the expectation of the function g of a random variable X. One way to compute $E[g(X)]$ is to let $Y = g(X)$. Once we know the distribution of Y, we can then compute $E(Y)$ accordingly. Another way to apply the following identity:

$$E[g(X)] = \begin{cases} \sum_x g(x)p_X(x) & \text{when } X \text{ is discrete,} \\ \int g(x)f_X(x)dx & \text{when } X \text{ is continuous.} \end{cases} \qquad (4.3)$$

We give the proof of the above for the case when X is a discrete random variable. Let $Y = g(X)$. Then

$$E[g(X)] = \sum_{x_i} g(x_i)p_X(x_i)$$

$$= \sum_j y_j \sum_{i:g(x_i)=y_j} p_X(x_i)$$

$$= \sum_j y_j P(Y = y_j)$$

$$= E(Y).$$

Example 9. Let X be a random variable and $g(X) = a + bX$, where a and b are two given constants. For example, Y can be the temperature in Manhattan at noon on a given day measured in Fahrenheit and X is that measured in Celsius. Thus $Y = 32 + 1.8X$, and $a = 32$ and $b = 1.8$. For this example, we have

$$E[Y] = E[g(X)] = E[a + bX] = a + bE[X],$$

where the last equality is obtained by noting a and b are two given constants. $\qquad \square$

A generalization of the above is the case when

$$Y = a_0 + a_1 X_i + \cdots + a_n X_n,$$

where $\{X_i\}$ are random variables and $\{a_i\}$ are given constants, then

$$E[Y] = a_0 + a_1 E[X_1] + \cdots + a_n E[X_n].$$

An *important* observation is that nothing is said whether $\{X_i\}$ are mutually independent. As we will see in the sequel, this observation will enable us to solve many seemingly complex problems easily. The proof of this assertion will be given in Chapter 8 where we discuss jointly distributed random variables.

Example 10 (Pricing a Single-Period Call Option). We are at time 0. A call option gives an option holder the *right* to purchase a stock at a prespecified price (called the strike price) K at time 1. Assume that the price of the stock at time 0, denoted by S_0, is known. The price of the stock at time 1, denoted by S_1, is unknown to the option holder at time 0. Then the payoff to the option holder is the random variable $g(S_1) = (S_1 - K)^+$, where

$$x^+ = \begin{cases} x & \text{if } x > 0, \\ 0 & \text{otherwise.} \end{cases}$$

To be specific, we assume that $S_0 = 90$, $K = 90$, and

$$S_1 = \begin{cases} 100 & \text{with probability } \frac{1}{2}, \\ 80 & \text{with probability } \frac{1}{2}. \end{cases}$$

Then the expected payoff to the option holder at time 0 is

$$E[g(S_1)] = E\left[(S_1 - K)^+\right]$$

$$= (100 - 90)^+ \times \frac{1}{2} + (80 - 90)^+ \times \frac{1}{2}$$

$$= 10 \times \frac{1}{2} + 0 \times \frac{1}{2} = 5. \qquad \square$$

Example 11 (The Single-Period Inventory Problem). Let X denote the number of bagels sold at Bernstein bagel on a given day. Assume that

$$p_X(x) = P(X = x), \quad x = 0, 1, 2, \ldots.$$

Let S be the number of bagels baked per day to meet the demand for the day. Let b denote the profit per each bagel sold, and l be the loss per unit left unsold. Let

$$g(X|S) = \text{the net gain if the demand for the day is } X$$

$$\text{and Bernstein makes } S \text{ bagels for the day to meet its demand.}$$

Then we have

$$g(X|S) = \begin{cases} bX - l(S - X) & \text{if } X \leq S, \\ bS & \text{if } X > S. \end{cases}$$

Hence the expected net gain for the day is

$$E[g(X|S)] = \sum_{x=0}^{S}(xb - l(s - x))p_X(x) + bS \sum_{x=S+1}^{\infty} p_X(x).$$

□

4.5. Random Variables that are Neither Discrete nor Continuous

In applications, there are random variables that are neither discrete nor continuous. In an earlier example, we used X to denote the waiting time to get a hair cut. For another example, we let Y denote the number of SUV's sold in a car dealership. On a bad day, we may have $Y = 0$. When $Y > 0$, Y may follow a particular probability mass function. In Example 6, we have already displayed the distribution function of this type of random variable. One way to handle the situation is to split the domain of the distribution in two parts: the continuous and discrete parts and treat each part separately. But this is awkward. An alternative is through the use of Riemann–Stieltjes integral.

Let the range of a random variable X be defined over the interval $[a, b]$. Over the interval, we divide it into subintervals such that $a = x_0 < x_1 < \cdots < x_n = b$. Let F denote the distribution function of the random variable X and g be a function defined over the same interval. In the context of probability theory, the Riemann–Stieltjes integral of g with respect to F from a to b is defined by

$$\int_a^b g(x)dF((x) = \lim_{\|\triangle\| \to 0} \sum_{k=1}^{n} g(\xi_k)[F(x_j) - F(x_{k-1}),$$

where $x_{k-1} < \xi_k \leq x_k$, $k = 1, \ldots, n$ and $\|\triangle\| = \max\{x_1 - x_0, \ldots, x_n - x_{n-1}\}$. When X is a discrete random variable, it follows that

$$E[g(X)] = \int_a^b g(x)dF(x) = \sum_{k=1}^{n} g(x_i)p_X(x_i),$$

whereas when X is continuous, we have

$$E[g(X)] = \int_a^b g(x)dF(x) = \int_a^b g(x)f(x)dx.$$

4.6.   Various Transforms for Applications in Probability

Studying subjects in probability can be made easier if we can make use of various forms of transforms. One prominent case is the handing of sum of independent random variables. This is known as the convolution of random variables. When transforms are in use, they convert coevolution operations into multiplications of functions. When we want to compute moments of random variables, transforms of the probability density or mass functions enable us to find them with relative ease. For examining asymptotic behaviors of random variables, they can provide easier means for analysis.

We review four transforms in the following. They all involve taking the expectation of function of random variable.

(1) **The Geometric Transform:** When X is a discrete non-negative integer-valued random variable, we define

$$p_X^g(z) \equiv E[z^X] = \sum_{i=0}^{\infty} z^i p_X(i).$$

The above is commonly known as the *probability generating function* of X, also known as the Z-transform of the probability mass function of X.

(2) **The Laplace Transform:** To simplify the exposition, we only consider the case when X is a continuous *non-negative* random variable. We define

$$f_X^e(s) \equiv E[s^{-sX}] = \int_0^{\infty} e^{-sx} f_X(x) dx.$$

The above is the Laplace transform of the probability density of X.

(3) **The Moment Generating Function:** Let X be a random variable. The moment generation function of X is defined by

$$M_X(t) = E[e^{tX}] = \int e^{tx} f_X(x) dx = \sum e^{tx} p_X(x).$$

(4) **The Characteristic Function:** Let X be a random variable X. The characteristic function of X is defined by

$$\varphi_X(t) \equiv E\left(e^{itX}\right) = E\left[\cos(tX) + i\sin(tX)\right],$$

where $i = \sqrt{-1}$.

Of the four types of transforms stated above, the first three are special cases of the characteristic function. Their connections are shown below:

$$\varphi_X\left(\frac{\log z}{i}\right) = E\big[e^{i(\frac{\log z}{i})X}\big] = E(z^X) = p_X^g(z),$$

$$\varphi_X(is) = E\big[e^{i(is)X}\big] = E(e^{-sX}) = f_X^e(s),$$

$$\varphi_X(-it) = E\big[e^{i(-it)X}\big] = E(e^{tX}) = M_X(t).$$

If we know the characteristic function of a random variable, then other transforms can be readily obtained. We note that the characteristic function is actually an application of Fourier transform to the distribution function of random variable X. We address issues relating to their uses when they are encountered in the sequel.

4.7. Higher Moments of a Random Variable

Let X be a random variable. Define the nth moments of X by

$$E(X^n) = \begin{cases} \displaystyle\sum_x x^n p_X(x) & \text{when } X \text{ is discrete,} \\ \displaystyle\int x^n f_X(x)dx & \text{when } X \text{ is continuous.} \end{cases}$$

Having defined the notion of moments, the nth *central moment* of X is defined by

$$\mu_n(X) \equiv E[(X - \mu_X)^n] = \sum_x (x - \mu_X)^2 p_X(x) \quad \text{when } X \text{ is discrete,}$$

$$= \int_{-\infty}^{\infty} (x - \mu_X)^2 f_X(x)dx \quad \text{when } X \text{ is continuous.}$$

When $n = 2$, we have the classical definition of random variable

$$\mu_2(X) = \text{Var}(X) = E\left[(X - \mu_X)^2\right]$$
$$= E\left[X^2 - 2\mu_X X + \mu_X^2\right]$$
$$= E[X^2] - \mu_X^2.$$

The last expression is a common way to compute $\text{Var}(X)$.

For $n = 3$, the following expression is called *skewness* of X:

$$\frac{\mu_3(X)}{\text{Var}(X)^{3/2}} = \frac{E\left[(X - \mu_X)^3\right]}{\text{Var}(X)^{3/2}}$$

and with $n = 4$, we define the *kurtosis* of X

$$\frac{\mu_4(X)}{\text{Var}(X)^2} = \frac{E\left[(X - \mu_X)^4\right]}{\text{Var}(X)^2}.$$

As mentioned earlier, the first moment measures the central tendency of the random variable X. The second moment measures its variability. The third moment measures its skewness, and the fourth moment measures its kurtosis. If a random variable is symmetric about its mean, then its skewness is 0. A positively skewed distribution is a type of distribution in which most values are clustered around the left tail of the distribution while the right tail is longer. A negatively skewed distribution has the opposite narrative. If the distribution of X has a large kurtosis, then it tends to have more outliers.

Cumulant Generating Function: Nowadays computing moments of random variables are typically done on computers. The use of the cumulant generation function $c_n(X)$ is more convenient. Let

$$\Phi_X(s) = \log[\varphi(s)] \iff \varphi(s) = e^{\Phi_X(s)}.$$

The nth cumulant $c_n(X)$ can be found from

$$c_n(X) = \frac{1}{i^n} \frac{\partial^n \Phi_X(s)}{\partial s^n}\bigg|_{s=0}.$$

Once cumulants are found, we compute the four measures of X as follows:

$$c_1(x) = E(X), \quad c_2(X) = \mu_2(X),$$
$$c_3(X) = \mu_3(X), \quad c_4(X) = \mu_4(X) - 3\mu_2^2(X).$$

It follows that

$$\text{skewness}(X) = \frac{c_3(X)}{c_2(X)^{3/2}},$$

$$\text{excess kurtosis}(X) = \frac{c_4(X)}{c_2(X)^2} = \frac{\mu_4(X)}{c_2(X)^2} - 3 = \text{kurtosis}(X) - 3.$$

Examples demonstrate the use of cumulant generating function for producing higher moments of random variables are given in Example 12, Problem 26, and in Illustration 12 of Chapter 7.

Example 12. Let X be the number of defective cell phones found daily during an outgoing quality control inspection station. The empirically observed frequency distribution of X is given below:

x	Relative frequency
0	0.29
1	0.23
2	0.21
3	0.17
4	0.10

Using the above as the PMF $p_X(x)$, we use the function "CumulantGeneratingFuntion" of Mathematica to compute the mean, variance, skewness, and kurtosis of X. In Illustration 2, we find the following statistics

$$E(X) = 1.56, \quad \text{Var}(X) = 1.7664,$$
$$\text{skewness}(X) = 0.3564, \quad \text{excess kurtosis}(X) = -1.0695,$$

or kurtosis$(X) = 1.9305$. $\square$

Example 13. If X is a random variable and $Y = a + bX$ where a and b are two given constants, then

$$\begin{aligned}
\text{Var}(Y) &= \text{Var}(a + bX) \\
&= E\left[((a + bX) - (a + b\mu_X))^2\right] \\
&= E[(bX - b\mu_X)^2] \\
&= b^2 \text{Var}(X).
\end{aligned}$$

$\square$

As expected, the constant term a exerts no impact on the variability of the random variable X.

Example 14 (A Coupon Collection Problem). Assume there are N distinct types of coupons. Each type of will be selected with equal probability, i.e.,

$$P(\text{Type } i \text{ is chosen}) = \frac{1}{N}, \quad i = 1, 2, \ldots, N.$$

Assume that each selection is independent of previous selections.

Let T be the number of coupons needed to get a complete collection of all types of coupons, i.e., each one of the N types is in your collection. We want to find the probability distribution of T.

Let $A_j = \{$Type j is *not* in the collection given that you have already acquired n types of coupons$\}$. Then we see that

$$P(T > n) = P\left(\bigcup_{j=1}^{N} A_j\right)$$

$$= \sum_{j} P(A_j) - \sum\sum_{j_1 < j_2} P(A_{j_1} A_{j_2})$$

$$+ \cdots + (-1)^{k+1} \sum\sum\sum_{j_1 < \cdots < j_k} P(A_{j_1}, \ldots, A_{j_k})$$

$$+ \cdots + (-1)^{N+1} P(A_{j_1} A_{j_2} \cdots A_N)$$

We see that

$$P(A_{j_1}) = \left(\frac{N-1}{N}\right)^n$$

and there are $\binom{N}{1} = N$ types of terms like the above. There are

$$P(A_{j_1} A_{j_2}) = \left(\frac{N-2}{N}\right)^n$$

and there are $\binom{N}{2}$ terms like the above. In general, we have

$$P(A_{j_1}, \ldots, A_{j_k}) = \left(\frac{N-k}{N}\right)^n$$

and there are $\binom{N}{k}$ terms like the above. Finally, we have

$$P(A_{j_1} A_{j_2}, \ldots, A_{j_{N-1}}) = \left(\frac{N-(N-1)}{N}\right)^n = \left(\frac{1}{N}\right)^n$$

and there are $\binom{N}{N-1}$ terms like the above.

By applying the inclusion–exclusion formula, we find

$$P(T > n) = \binom{N}{1}\left(\frac{N-1}{N}\right)^n - \binom{N}{2}\left(\frac{N-2}{N}\right)^n + \binom{N}{3}\left(\frac{N-3}{N}\right)^n$$

$$- \cdots + (-1)^{N+1} \binom{N}{N-1} \left(\frac{1}{N}\right)^n$$

$$= \sum_{i=1}^{N-1} (-1)^{i+1} \binom{N}{i} \left(\frac{N-i}{N}\right)^n.$$

Now we see that

$$P(T > n - 1) = P(T = n) + P(T > n).$$

Thus

$$P(T = n) = P(T > n - 1) - P(T > n).$$

In Illustrations 3 and 4, we present two applications using the results given in this example. $\qquad\square$

4.8. Exploring with Mathematica

Illustration 1. In the following, we display a plot of the distribution function of random variable X given in Example 2. The graph clearly shows that X is a càglàd function in that its sample path is right continuous and with left limits.

```
Plot[Piecewise[{{0, x < 0}, {1/4, 0 ≤ x < 1}, {3/4, 1 ≤ x < 2}, {1, 2 ≤ x}}], {x, -3.5, 3.5},
  AxesLabel → {"x", "Fx(x)"},
  Epilog → {
    PointSize[0.025],
    Point[#] & /@ {{0, 0.25}, {1, 0.75}, {2, 1}, {0, 0}, {1, 0.25}, {2, 0.75}},
    White,
    PointSize[0.015],
    Point[#] & /@ {{0, 0}, {1, 0.25}, {2, 0.75}},
    Black,
    Thick,
    Dashed,
    Line[{{0, 0}, {0, 0.25}}],
    Line[{{1, 0.25}, {1, 0.75}}],
    Line[{{2, 0.75}, {2, 1}}]
  }
]
```

Illustration 2. Computation of moments using the cumulant generating function.

```
cgf = CumulantGeneratingFunction[p, i t]
```

$$\text{Log}\left[0.29 + 0.23\ e^{i\ t} + 0.21\ e^{2\ i\ t} + 0.17\ e^{3\ i\ t} + 0.1\ e^{4\ i\ t}\right]$$

```
c = Re[Table[ 1/(i^k) Limit[D[cgf, {t, k}], t → 0], {k, 1, 4}] // N]
```

$\{1.56, 1.7664, 0.836832, -3.33701\}$

```
skewnessX = c[[3]] / c[[2]]^(3/2)
```

0.356455

```
ExcesskurtosisX = c[[4]] / c[[2]]^2
```

-1.0695

```
KurtosisX = 3 + ExcesskurtosisX
```

1.9305 □

Illustration 3. We now give an application of the result given in Example 14 (the Coupon Collection problem). Houston Texans has 52 players in its roster. Each cereal box contains the picture of one of its players. A person collects a complete set of coupons gets a pair of tickets to their fancy Sky Boxes in one of their home games. Let T denote the number of cereal boxes needed to be purchased in order to acquire a complete set of coupons.

```
BN = 52;
fn[i_, n_] = Binomial[BN - i, i] * ((BN - i) / BN)^n * (-1)^(i+1);
pn[n_] := Total[Table[fn[i, n - 1], {i, 1, (BN - 1)}]] - Total[Table[fn[i, n], {i, 1, (BN - 1)}]];
rs = Table[{n, pn[n]}, {n, 52, 600}] // N;
ListPlot[rs, AxesLabel → {"k", "Prob"}, LabelStyle → {Black, Medium},
  PlotStyle → {Black, Thick}, AxesStyle → {Black, Medium}]
```

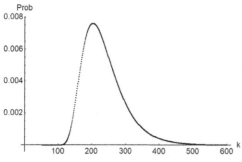

□

Illustration 4. Consider the coupon collection problem given in Example 14 one more time. Recall that there are r distinct types of coupons. Now we are interested in finding the probability that k $(k \leq r)$ distinct types are included in a collection of n coupons. To be specific, let D_n denote the number of distinct types in a collection of n coupons. Then we want to compute $P(D_n = k)$.

For the moment, we consider k specific types. Let $A =$ {each one of these k types are there in the collection of n}. Then we see that

$$P(A) = \left(\frac{k}{r}\right)^n.$$

As we are considering the k specific types, we use the results found in Example 14 to find that out of a collection of n coupons the probability that each one of these k types is represented in the collection. We denote the later event be B. The solution for the probability $P(B|A)$ has been derived in Example 14. It is

$$P(B|A) = 1 - \sum_{i=1}^{k-1} \binom{k}{i} \left(\frac{k-i}{k}\right)^n (-1)^{i+1}.$$

Since there are $\binom{r}{k}$ possible ways to select the specific set of k types, it follows that

$$P(D_n = k) = \binom{r}{k} P(AB) = \binom{r}{k} P(B|A) \times P(A).$$

Consider the case where $r = 7$ and $n = 20$. The following gives the probabilities $P(D_{20} = k)$ for $= 1, \ldots, 7$.

```
fn[i_, n_] := Binomial[k, i] * ((k - i) / k)" * (-1)^{i+1};

PDn[k_] := Module[{n = 20, r = 7, pba}, pa = (k / r)";
    pba = pa * (1 - Total[Table[fn[i, n], {i, 1, (k - 1)}]]);
    pDnk = Binomial[r, k] * pba // N];

result = Table[{k, PDn[k]}, {k, 0, 7}];

ListPlot[result, AxesLabel → {"k", "Prob"}, LabelStyle -> {Black, Medium},
    PlotStyle → {Black, Thick}, AxesStyle → {Black, Medium}]
```

Problems

1. We toss two fair dice. Let X denote the sum of the two faces shown. Derive the PMF $p_X(x)$.

2. The distribution function of random variable X is given below:

$$F_X(x) = \begin{cases} 0 & \text{if } x < 0, \\ \frac{1}{4} & \text{if } 0 \le x < 1, \\ \frac{3}{4} & \text{if } 1 \le x < 2, \\ 1 & x \ge 2. \end{cases}$$

(a) Compute $E(X)$, (b) $\text{Var}(X)$, and (c) $P(0.5 < X < 1)$.

3. Let X be a positive integer-valued random variable. Let the function $p(\cdot)$ be defined as

$$p(x) = \frac{4 \cdot 3^{x-1}}{7^x}, \quad x = 1, 2, \dots.$$

(a) Verify that $p(\cdot)$ a legitimate PMF $p_X(x)$? (b) Find the CDF $F_X(x)$. (c) Find $P(X > k)$. (d) Use (4.1) to find $E(X)$. (e) Use Mathematica to compute $P(X$ is even), where X is the random variable that assumes the PMF $p_X(x)$.

4. Let X be a random variable for which

$$P(X > x) = \frac{1}{1+x}, \quad x = 0, 1, 2, \ldots.$$

(a) Is X a legitimate random variable? If so, find its PMF $p_X(x)$; if not, why not? (b) Compute $E(X)$ if you can; otherwise justify your assertion.

5. The distribution function of a random variable X is given below:

$$F_X(x) = \begin{cases} 0 & \text{if } x < 0, \\ \frac{x}{4} & \text{if } 0 \leq x < 1, \\ \frac{1}{2} & \text{if } 1 \leq x < 2, \\ \frac{1}{12}x + \frac{1}{2} & \text{if } 2 \leq x < 3, \\ 1 & \text{if } x \geq 3. \end{cases}$$

Compute the following quantities: (a) $P(1 \leq X < 3)$, (b) $P(X > \frac{3}{2})$, and (c) $P(2 < X \leq 7)$.

6. Consider the game of rolling a fair dice. You pay $5 to roll the dice once. If you get 1 and 6 you get $10 back; otherwise you get nothing. Let X be a random variable denoting the outcome of the play. (a) What is $E(X)$? (2) What is $\text{Var}(X)$?

7. Let X be a discrete random variable whose PMF is shown below:

$$p_X(x) = \begin{cases} 0.25 & \text{if } x = 2, \\ 0.35 & \text{if } x = 5, \\ 0.40 & \text{if } x = 9. \end{cases}$$

(a) Find $E[e^X]$. (b) Find $E\left[\frac{1}{X-4}\right]$.

8. Consider a random variable X whose distribution function is shown below

$$F_X(x) = \begin{cases} 0 & \text{if } x \leq 1, \\ \frac{2x}{5} & \text{if } 0 < x \leq 1, \\ \frac{x^2+1}{5} & \text{if } 1 < x \leq 2, \\ 1 & \text{if } x > 2. \end{cases}$$

(a) Find $E(X)$. (b) Find $\text{Var}(X)$. (c) Find $P(0.5 < X < 1.85)$.

9. Consider a random variable X follows the following PDF:

$$f_X(x) = \begin{cases} \alpha x^2 & 0 < x < 1, \\ 0 & \text{otherwise.} \end{cases}$$

(a) Find the constant α. (b) Find the CDF $F_X(x)$. (c) Find $E(X)$. (d) Find $\text{Var}(X)$.

10. Let X be a discrete positive integer-valued random variable with PMF shown below:

$$p_X(x) = pq^{x-1}, \quad i = 1, 2, \ldots,$$

where $0 < p < 1$, and $q = 1 - p$.
(a) Find the probability generating function of X. (b) Find the characteristic function of X. (c) Use the result from (b) to verify the probability generating function of X obtained from part (a).

11. Numerical Experiment with PGF. In Example 10, we find the probability generating function of X:

$$p_X^g(z) = \frac{pz}{1 - zq}.$$

For the case when $p = 0.6$ and $q = 0.4$. Use Mathematica's function "Series" to expand $p_X^g(z)$ as a polynomial in z. Use the function "CoefficientList" to collect the coefficients of associated with the terms $z^0, z^1, \ldots, z^6$ in the expansion. These coefficients will give $p_X(x)$, $x = 0, \ldots, 6$. Find these $p_X(x)$'s.

12. Consider a continuous positive random variable X whose PDF is given by

$$f_X(x) = \begin{cases} \lambda e^{-\lambda x} & \text{if } x > 0, \\ 0 & \text{otherwise.} \end{cases}$$

(a) Find the Laplace transform of X. (b) Find the characteristic function of X. (c) Use the result from (b) to verify the Laplace transform of X obtained in Part (a).

13. A standard card deck contains 52 cards of which four are aces. Cards are randomly selected, one at a time, until an ace is obtained (in which case the game ends). Assume that each drawn card is returned to the deck and the deck is reshuffled to maintain its randomness. (a) What is the probability exactly n draws are needed? (b) What is the mean number of draws needed? (c) What is probability that at least k draws are needed?

14. Let X denote the number of cartons of milk demanded at a local Dime–Nickel Store on a day. Assume X follows the following PMF:

$$p_X(x) = \frac{2x + 3}{63}, \quad x = 0, 1, \ldots, 6.$$

(a) Find $E(X)$. (b) The store receives only five cartons per day. What is the expected amount of lost sales per day? (c) What is the standard deviation of the daily lost sales?

15. Assume that random variable X follows the following density

$$f_X(x) = 3e^{-3x}, \quad x > 0.$$

Define $Y = 1 - e^{-3X}$. What is the distribution function of Y?

16. Let $F_X(x)$ denote the cumulative distribution function of a random variable X. Assume that F_X is defined by

$$F_X(x) = \begin{cases} 0 & x < -2, \\ \frac{1}{5}(x + 2) & -2 \le x < 0, \\ \frac{1}{2} & 0 \le x < 4, \\ \frac{1}{12}(x + 6) & 4 \le x < 6, \\ 1 & x \ge 6. \end{cases}$$

Find the following probabilities: (a) $P(X = -1)$; (b) $P(X = 0)$; (c) $P(X = 1)$; (d) $P(X > 2)$; (e) $P(X = 4)$; (f) $P(X > 6)$.

17. Let $F_X(x)$ be the CDF of a random variable X. Assume that $F_X(x)$ is given below:

$$F_X(x) = \begin{cases} 0 & \text{if } x < -1, \\ \frac{1}{4}(x+1) & \text{if } -1 \le x < 0, \\ \frac{1}{2} & \text{if } 0 \le x < 1, \\ \frac{1}{12}(x+7) & \text{if } 1 \le x < 2, \\ 1 & \text{if } x \ge 2. \end{cases}$$

Find (a) $P(X < 1)$; (b) $P(X = 1)$; (c) $P(1 \le X < 2)$; (d) $P(X > \frac{1}{2})$; (e) $P(X = \frac{3}{2})$; (f) $P(1 < X \le 6)$; (g) $P(X = 3)$.

18. Let X be a continuous random variable with the following PDF:

$$f_X(x) = \begin{cases} \frac{2}{x^2} & \text{if } 1 < x < 2, \\ 0 & \text{otherwise.} \end{cases}$$

Let $Y = X^2$. (a) Let $F_Y(y)$ denote the CDF of Y. Find $F_Y(y)$. (b) Let $f_Y(y)$ denote the PDF of Y. Find $f_Y(y)$. (c) Find $E(Y)$.

19. Consider a random variable X whose characteristic functions is $\varphi_X(u) = \frac{1}{2e^{-iu}-1}$. (a) What is the probability generating function $p_X^g(x)$ of this random variable? (b) Use Mathematica to do the series expansion of the probability generating function. By examining the coefficients of the expansion to determine the PMF of X.

20. To determine whether people have CV19, 100 people are to have antibody tests. However, rather than testing each individual separately, it has been decided first to place the people into groups of 10. The samples of the 10 people in each group will be pooled and analyzed together. If the test is negative, one test will suffice for the ten people, whereas if the test is positive, each of the 10 people will also be individually tested and, in all, 11 tests will be made on this group. Assume that the probability that a person has the disease is 0.1 for all people, independently of one another, and compute the expected number of tests necessary for each group. (Note that we are assuming

that the pooled test will be positive if at least one person in the pool has COVID-19.)

21. Consider the game of roulette. After each random spin, assume that red shows up with probability $\frac{18}{38}$. Assume you bet \$10 on each spin. If red shows up, you win \$10 and quit the game; otherwise, you lose your bet. In the latter case, you bet on two more spins then quit regardless of the outcomes. Let X denote your net gain when you leave the game. (a) Find $P(X > 0)$. (b) Find $E(X)$. (c) Is this a good betting strategy?

22. Four school buses carrying 200 students to their basket arena. The buses carry, respectively, 50, 35, 45, and 70 students. One of the students is randomly selected (out of the 200 students). Let X denote the number of students who were on the bus carrying the randomly selected students. One of the four bus drivers is also randomly selected. Let Y denote the number of students on his bus. (a) Which of $E(X)$ or $E(Y)$ do you think is larger? Why? (b) Find $E(X)$ and $E(Y)$.

23. Let X be a continuous random variable following the PDF:

$$f_X(x) = ce^{-|x|}, \quad -\infty < x < \infty.$$

(a) What is value of c so that $f_X(x)$ is a legitimate PDF? (b) Let $Y = aX + b$, where $a, b > 0$. Find the PDF $f_Y(y)$.

24. Let X be a integer-valued positive random variable with PMF given below:

$$p_x(x) = \frac{1}{2^x}, \quad x = 1, 2, \ldots.$$

Define a random variable Y by $Y = 1$ if X is even, and -1 if X is odd. Find (a) the PMF of Y; (b) the CDF of Y; and (c) $E(Y)$.

25. Consider a game involving the toss of two fair dice. If the sum of the two tosses is nine or higher, you receive \$9. If the sum is less than or equal to four, you lose \$4. What is mean, variance, skewness, and kurtosis of your payoff from the game?

26. Consider Problem 25 once more. (a) State the characteristic function $\varphi_X(t)$ of X. (b) State the cumulant generating function $\Phi_X(t)$ of X. (c) Use $\Phi_X(t)$ to compute the mean, variance, skewness, and excess kurtosis of X.

27. Let X be a random variable with PDF

$$f_X(x) = 2x, \quad 0 < x < 1$$

Let $Y = (3X - 1)^2$. Find the PDF $f_Y(y)$ of random variable Y.

28. Let X be a random variable with PDF

$$f_X(x) = \begin{cases} ae^{-ax} & x > 0 \\ 0 & \text{otherwise} \end{cases}$$

where $a > 0$. (a) Find the characteristic function $\varphi_X(t)$. (b) Find the moment generating function $M_X(t)$. (c) Find the Laplace transform of $f_X^e(t)$. (d) Use the answer obtained from part (a) to verify your results found in (b), (c), and (d).

29. Let X be a nonnegative random variable with PMF $p_X(x)$ given by

$$p_X(x) = \begin{cases} \frac{1}{x(x+1)} & x = 1, 2, \ldots \\ 0 & x = 0 \end{cases}$$

(a) Verify that X is a legitimate discrete random variable. (b) Find the PGF $p_X^g(z)$ of X.

30. Let $p_X^g(z)$ denote the probability generating function of the discrete nonnegative random variable X. Thus

$$p_X^g(z) = E(z^X) = \sum_{i=0}^{\infty} z^i p_X(i)$$

(a) Write the expression of the above after differentiating with respect to z^r times, i.e., what is the following

$$\frac{d^r}{dz^r} p_X^g(z)?$$

(b) Dividing the result obtained from part (a) by $r!$ and setting $z = 0$, what would you find? (c) What observation can be drawn from part (b)? (d) In $\frac{d^r}{dz^r} p_X^g(z)$, if you set $z = 1$, what will you get? (e) If you know the PGF of random variable X, how will you find $E(X)$ and $\text{Var}(X)$?

31. Let X be a nonnegative continuous random variable. Recall (4.2), we have

$$E(X) = \int_0^\infty P(X > y)dy$$

Show that

$$E(X^2) = \int_0^\infty 2yP(X > y)dy$$

Remarks and References

The derivations of the expressions relating to the cumulant generating function can be found in Stuart and Ord [3]. The set of examples on coupon collections is inspired by Ross [2, Chapter 4]. In Kao [1, Chapter 4], its solution is done by a discrete-time phase-type distribution.

[1] Kao, E. P., *An Introduction to Stochastic Processes*, Duxbury, 1998 and Dover, 2019.
[2] Ross, S. M., *A First Course in Probability*, 10th edition, Pearson, 2019.
[3] Stuart, A. and Ord, K., *Kendall's Advanced Theory of Statistics*, Hodder Arnold, 1994.

Chapter 5

Discrete Random Variables

5.1. Introduction

Any random variable X having a PMF $p_X(x)$ that satisfies

$$p_X(x) \geq 0 \quad \text{for all} \quad x$$

where x assumes at most a countable number of values, and

$$\sum_x p_X(x) = 1,$$

is a discrete random variable. The properties of many discrete random variables are well known. More importantly, their structural properties are well connected and subject to nice intuitive interpretations. These insights will enhance our understanding of their roles played in real-world applications.

5.2. Bernoulli and Binomial Random Variables

A Bernoulli random variable X is defined by

$$p_X(1) = p,$$
$$p_X(0) = 1 - p \equiv q,$$

where $0 < p < 1$. It is the most primitive discrete random variable. As we have seen before, the outcome of flipping a fair coin is a Bernoulli random variable with parameter $p = 1/2$. We use $X \sim bern(p)$ to denote that X

117

follows a Bernoulli PMS with parameter p. It is easy to see that

$$E[X] = p. \quad \text{and} \quad E[X^2] = 1^2 p + 0^2 q = p.$$

Thus

$$Var[X] = E[X^2] - E^2[X] = p - p^2 = pq.$$

Consider we run a sequence of independent Bernoulli trials with X_n denoting the outcome of the nth trial. Let $X_n = 1$ denote the event that the nth trial results a success, and 0, otherwise. Define

$$Z_n = \sum_{i=1}^{n} X_n.$$

Then Z_n is the number of successes out of n trials and it is called a *Binomial* random variable. We see that Z_n is characterized by parameters (n, p).

The probability mass function of Z_n is given by

$$p_{Z_n}(x) = \binom{n}{x} p^x q^{n-x}, \quad x = 0, 1, \ldots, n.$$

We use $Z_n \sim \text{bino}(n, p)$ express that Z_n follows a binomial PMF with parameters (n, p). To show the rationale underlying the PMF, we see that for a specific sequence of x successes and $n - x$ failures, the probability of its realization is given by $p^x q^{n-x}$. But there are $\binom{n}{x}$ ways the x successes can occur in any n possible positions and hence the multiplier $\binom{n}{x}$.

The mean of Z_n can be obtained by applying its basic definition for a discrete random variable:

$$E[Z_n] = \sum_{x=0}^{n} x \times \binom{n}{x} p^x q^{n-x}$$

$$= np \sum_{x=1}^{n} \frac{(n-1)!}{(x-1)!(n-x)!} p^{x-1} q^{n-x}$$

$$= np,$$

where the last equality is obtained by a change of the indexes. A simple way to find the result is by applying the basic identity about the expectation of the sum of a sequence of random variables. Thus we find

$$E[Z_n] = \sum_{i=1}^{n} E[X_i] = np.$$

For the variance, we recall that in the previous chapter, we asserted that when the random variables $\{X_n\}$ are independent, the sum of the variances is additive. Thus the valance of Z_n is given by

$$\text{Var}[Z_n] = \sum_{i=1}^{n} Var[X_i] = npq,$$

where we recall that for the Bernoulli random variable X_i, we have $\text{Var}[X_i] = pq$.

Example 1. Consider a carnival game of Chuck-a-Luck in which three fair dice are rolled. A player chooses one number from the numbers $1, \ldots, 6$ beforehand. If the player's number does not come up, the player loses one dollar. If the chosen number shows up i times out of the three rolls, the player wins i dollars, where $i = 1, 2$, and 3. Let X denote the number of times the player's chosen number matches with the outcomes of the three rolls. □

We see that $X \sim \text{bino}(n, p)$, where $n = 3$ and $p = 1/6$. Thus

$$p_X(i) = \binom{n}{i} \left(\frac{1}{6}\right)^i \left(\frac{5}{6}\right)^{3-i}, \quad i = 0, 1, 2, 3.$$

The gain to the house per wager is

$$g(i) = \begin{cases} +1 & \text{with probability } p_X(0), \\ -i & \text{with probability } p_X(i). \end{cases}$$

Thus the expected gain to the house is

$$E[g(X)] = 1 \times \binom{3}{0} \left(\frac{1}{6}\right)^0 \left(\frac{5}{6}\right)^3 - 1 \times \binom{3}{1} \left(\frac{1}{6}\right) \left(\frac{5}{6}\right)^2$$

$$- 2 \times \binom{3}{2} \left(\frac{1}{6}\right)^2 \left(\frac{5}{6}\right)^1 - 3 \times \binom{3}{3} \left(\frac{1}{6}\right)^3 \left(\frac{5}{6}\right)^0$$

$$= 0.0787. \qquad \square$$

Example 2. In men's tennis, the winner is the first to win 3 out of the 5 sets. If Rafael Nadal wins a set against his opponent with probability $3/4$, what is the probability that he will win the match?

We note that to win the match, Nadal must win the last set. It follows that

$P\{\text{Nadal wins the match}\}$

$$= \left(\frac{3}{4}\right)^3 + \binom{3}{2}\left(\frac{3}{4}\right)^2\left(\frac{1}{4}\right) \times \frac{3}{4} + \binom{4}{2}\left(\frac{3}{4}\right)^2\left(\frac{1}{4}\right)^2 \times \frac{3}{4} = \frac{459}{512}$$

$$= 0.8965.$$
$\square$

5.3. Hypergeometric Random Variables

Consider an urn contains N balls of which m are blue and $N - m$ are red. We randomly choose n balls from the urn without replacement. Let random variable X denote the number of blue balls chosen. Then it is clear that

$$P(X = i) = \frac{\binom{m}{i}\binom{N-m}{n-i}}{\binom{N}{n}}, \quad i = 0, 1, \ldots, n \wedge m.$$

A hypergeometric random variable is characterized by three parameters: n, m, and N. We use $X \sim hyper(n, m, N)$ to denote it. It is instructive to write the above as

$$X = X_1 + \cdots + X_n,$$

where

$$X_i = \begin{cases} 1 & \text{if the } i\text{th sample is a blue ball,} \\ 0 & \text{otherwise.} \end{cases}$$

What is the difference between those $\{X_i\}$ defined in the context of a Binomial random variable and those defined in the current context? If we consider getting a blue ball a success, then under the hypergeometric paradigm the probability of a success is changing during the sampling process. Stated simply, here $\{X_i\}$ are dependent random variables whereas the $\{X_i\}$ under the Binomial scenario are mutually independent.

Even though $\{X_i\}$ are *dependent* variables, we assert that

$$P(X_i = 1) = \frac{m}{N}.$$

(e.g., see the "nth draw is Diamond" problem in a host of examples in Chapter 3; the fact that $P(X_i = 1)$ remains the same regardless of the value of i is of importance and will be elaborated in Chapter 8 when we address the issue on exchangeable random variables).

Since the expectation of the sum of random variables is additive, we conclude that

$$E[X] = \sum_{n=1}^{n} E(X_i) = \frac{nm}{N}.$$

For the variance, a direct derivation involves algebraic manipulations. The following result can be established by a simple approach that is sketched in Section 6 of Chapter 8:

$$Var(X) = npq \left(\frac{N-n}{N-1} \right).$$

We see that when the term in the pair of braces is approximately one (this occurs when n is small relative to N), then the variance can be approximated by that of the binomial random variable.

Example 3. Consider two urns. Urn 1 contains 5 white and 6 black balls. Urn 2 contains 8 white balls and 10 black balls. At the first stage, we randomly take two balls from urn 1 and put them in urn 2. At stage two, we randomly select three balls from urn 2. Let W denote the number of white balls drawn at stage 2. Find $E(W)$.

Let X denote the outcome observed at the stage 1, namely, the number of white ball removed from urn 1 to urn 2. We see that $X \sim hyper(2, 5, 11)$. Let Y denote the outcome observed at the second stage conditional on X, namely, the number of white balls drawn from urn 2. Then we find the PMF of W by conditioning on X using

$$p_W(w) = P(W = w) = \sum_{x=0}^{2} P(W = w | X = x) p_X(x).$$

The conditional PMF $P(W = w | X = x)$ follow $hyper(3, 8+x, 20)$, where $x = 0, 1, 2$. Once the PMF is obtained, finding $E(W)$ is immediate. In Illustration 1, we show the computation and find $E(W) = 1.3364$. □

Example 4 (Estimating Population Size of Rare Species). Assume we are interested in estimating the population size of pandas in Sichuan Province of China. We first catch m pandas and tag them. We then set them free in the wild. After six months, we recapture n pandas in the wild. We let X denote the number of tagged Panda in the recapture of size n.

Assume each panda (tagged or otherwise) will have equal probability of being recaptured. Then we have

$$P(X = i) = \frac{\binom{m}{i}\binom{N-m}{n-i}}{\binom{N}{n}}, \quad i = 0, 1, \ldots, n \wedge m.$$

In the above expression, we see that N is unknown and i is given as it is the outcome of the recapture. Based on the observed i, we want to estimate the size of N.

One way to estimate N is the use of the method of maximum likelihood. Let

$$g_i(N) \equiv P(X = i | N)$$

we want to find

$$N^* = \arg\max_N g_i(N)$$

namely, we want to find the N such that the probability of observing i is the largest. A numerical example is given in Illustration 2. □

Example 5 (Dispute about an Election Outcome). In a close election between two candidates A and B in a small town called Tiny Point, candidate A has 467 votes, and B 452 votes. Thus candidate A wins by a margin of 15 votes. However someone proclaims that there are 39 fraudulent votes and asks the judge to toss away these so-called fraudulent votes. Assume that these votes are equally likely to go to either candidate. Let E denote the event that the election result will be reversed. What is the probability that event E occurs? □

The key to the solution of this problem lies with the observation that out of the 467 votes for candidate A how many votes needed to remove to candidate B so that a reversal of the election result is possible? Let x be the number of fraudulent votes for A and y be the number of the fraudulent votes for B. Then we have $x + y = 39$. To overturn the election result, we need to have $x - y \geq 15$, or $x - (39 - x) \geq 15$, or $x \geq 27$. Let X be the number of votes to be randomly chosen from the group of 467 votes for candidate A out of a sample of size 39 and population of size 869. Then we have $P(E) = P(X \geq 27)$, where $X \sim hyper(39, 467, 869)$. we have

$$P(E) = \sum_{x=27}^{39} \frac{\binom{467}{x}\binom{452}{39-x}}{\binom{869}{39}}.$$

In Illustration 3, we find that $P(E) = 0.01365$

If we use the binomial distribution to approximate $P(E)$, then

$$P(E) = \sum_{x=27}^{39} \binom{39}{x}\left(\frac{1}{2}\right)^{39} = 0.01185.$$

□

5.4. Poisson Random Variables

A Poisson random variable X has only one parameter λ. The PMF of X is given by

$$p_X(x) = P(X = x) = e^{-\lambda}\frac{\lambda^x}{x!}, \quad \text{where} \quad x = 0, 1, 2, \ldots.$$

We use $X \sim pois(\lambda)$ to denote it. Among many of its usage, the Poisson distribution has been used extensively in actuarial industries to model occurrences of rare events and telecommunication to model network traffics.

We see that $p_x(x) > 0$ for all $x = 0, 1, 2, \ldots$ and

$$\sum_{x=0}^{\infty} p_X(x) = \sum_{x=0}^{\infty} e^{-\lambda}\frac{\lambda^x}{x!} = e^{-\lambda}\sum_{x=0}^{\infty}\frac{\lambda^x}{x!} = 1.$$

Moreover,

$$E(X) = \sum_{x=0}^{\infty} x \cdot e^{-\lambda}\frac{\lambda^x}{x!}$$

$$= \lambda e^{-\lambda}\sum_{x=1}^{\infty}\frac{\lambda^{x-1}}{(x-1)!}$$

$$= \lambda e^{-\lambda}e^{\lambda} = \lambda.$$

To find variance of X, we could first obtain its second moment and then use $\text{Var}(X) = E(X^2) - E^2(X)$. As an alternative, we compute its *second product moment* $E(X(X-1))$ and go from there. We see

$$E(X(X-1)) = \sum_{x=2}^{\infty} x(x-1)e^{-\lambda}\frac{\lambda^x}{x!} = \lambda^2 e^{-\lambda}\sum_{x=2}^{\infty}\frac{\lambda^{x-2}}{(x-2)!} = \lambda^2,$$

where the last equality is obtained by a change of indices. Now

$$Var(X) = E(X(X-1)) + E(X) - E^2(X)$$
$$= \lambda^2 + \lambda - \lambda^2 = \lambda.$$

One reason that a Poisson random variable seems to be plausible in modeling occurrences of rare events can be argued along the following story line. Assume we want to predict the number of traffic accidents occurring on a day in a given locality. Let n be the number of cars on the road on the given day. Each car has a probability p of incurring an accident. Define

$$X_i = \begin{cases} 1 & \text{if the } i\text{th car has an accident,} \\ 0 & \text{otherwise.} \end{cases}$$

Assume $\{X_i\}$ are independent random variables. Then, we see that $X = X_1 + \cdots + X_n$ and $X \sim bino(n,p)$ with

$$P(X = x) = \binom{n}{x}p^i(1-p)^{n-x}$$

and $E(X) = np$. Let $\lambda = np$. Then $p = \lambda/n$. We now write

$$P(X = x) = \frac{n!}{(n-x)!x!}\left(\frac{\lambda}{n}\right)^x\left(1-\frac{\lambda}{n}\right)^{n-x}$$

$$= \frac{n(n-1)\cdots(n-x+1)}{n^x}\frac{\lambda^x\left(1-\frac{\lambda}{n}\right)^n}{x!\left(1-\frac{\lambda}{n}\right)^x}.$$

Now when n is large and λ is small, we have

$$\left(1-\frac{\lambda}{n}\right)^n \approx e^{-\lambda},$$

$$(1-\frac{\lambda}{n})^x \approx 1,$$

$$\frac{n(n-1)\cdots(n-x+1)}{n^x} = \frac{n}{n}\cdot\frac{n-1}{n}\cdots\frac{n-x+1}{n} \approx 1.$$

It follows that

$$P(X = i) \approx e^{-\lambda} \frac{\lambda^x}{x!}.$$

Thus for approximating the occurrences of rare events, it seems that Poisson is indeed a plausible choice.

An Useful Ways to Interpret the Poisson Parameter λ. In the aforementioned traffic accident example, we have $E(X) = \lambda$. We can interpret λ is the average *rate* of accidents per day. If we change the interval of interest to t and let $X(t)$ denote the number of accidents in an interval of length t. Then the average rate in an interval of length t is λt and

$$P(X(t) = x) = e^{-\lambda t} \frac{(\lambda t)^x}{x!}, \qquad x = 0, 1, 2, \dots,$$

where λ represents the average rate of accident occurrence per unit time and t is the length of the interval involved. We shall have more to say about the random variable $X(t)$ in the next chapter when we encounter the exponential random variable.

Conditions Under Which We Expect a Poisson Distribution. Let $X(t)$ denote the number of events occurring in an interval of length t. The Poisson paradigm presumes the following conditions prevail:

(i) $X(0) = 0$;
(ii) $P(X(h) = 1) = \lambda h + o(h)$;
(iii) $P(X(h) \geq 2) = o(h)$;
(iv) the probabilities of two events occurring in two non-overlapping intervals are independent.

Here the function $o(h)$ is defined as

$$\lim_{h \to 0} \frac{o(h))}{h} = 0.$$

The second condition says that in a small interval, the probability of one event will occur is proportional to the length of the interval h. The third condition implies that in the small interval h, at most one event will occur. The other two conditions are self-explanatory.

It is notable that the above "behavior-type" conditions will lead to a system of differential equations whose solution is

$$P(X(t) = x) = e^{-\lambda t}\frac{(\lambda t)^x}{x!}, \quad x = 0, 1, 2, \ldots$$

Whenever we plan to use a Poisson random variable to model a real-world scenario, we need to check whether these requirements are reasonably met.

Example 6. Suppose hurricanes occur along the Gulf Coast at a rate of 2.5 per month during a hurricane season.

(a) Find the probability that at least 2 hurricanes occur in a month.
(b) Find the probability that the are more than 3 hurricanes occur in a three month period.
(c) Find the probability distribution of the time, starting from now, until the occurrence of the next hurricane.

Solution. We use a Poisson random variable to model the occurrence of hurricanes.

(a) With $\lambda = 2.5$, it follows that $\lambda t = 2.5 \times (1) = 2.5$ and

$$P(X(1) \leq 2) = \sum_{x=0}^{2} e^{-2.5}\frac{(2.5)^x}{x!} = 0.5438.$$

(b) Since $t = 3$, we have $\lambda t = 2.5 \times 3 = 7.5$ and

$$P(X(3) > 3) = 1 - P(X(3) \leq 3) = 1 - \sum_{x=0}^{3} e^{-7.5}\frac{(7.5)^x}{x!} = 0.9941.$$

(c) Let Y denote the time in months from now till the occurrence of the next hurricane. We see that the event $\{Y > t\}$ if and only if $\{X(t) = 0\}$. Thus

$$P(Y > t) = P(X(t) = 0) = e^{2.5t}. \qquad \square$$

Example 7 (Probabilistic Decomposition of a Poisson Arrival Stream). Customers arrive at a post office at a rate of 10 per hours. There are 55 percent of customers are females and the rest are males. Let X be the number of female customers arriving during a 8-hour period, Y be the number of male customers arriving in the same period, and $p = 0.55$.

Then X follows a Poisson distribution with parameter $\lambda t = \lambda pt = (10) \times (0.55) \times (8) = 44$ and Y follows a Poisson distribution with parameter $\lambda t = \lambda(1 - p)t = (10) \times (0.45) \times (8) = t\lambda = 36$.

In this example, we see that a Poisson arrival stream is decomposed into two independent sub-Poisson arrivals streams. The proof for this decomposition can be found in many references, e.g., see [2, p. 54]. □

5.5. Geometric and Negative Binomial Random Variables

Consider a sequence of independent Bernoulli trials with $p > 0$ being the probability that a single trial results in a success. Let X denote the number of trials needed in order to obtain a success for the first time. X is called a geometric random variable with

$$p_X(x) = q^{x-1}p, \quad x = 1, 2, \ldots.$$

We use $X \sim g = geom(p)$ to denote it. Using the above PMF, we compute

$$P(X > x) = \sum_{k=x+1}^{\infty} p_X(k) = p_X(x+1) + p_X(x+2) + \cdots$$
$$= q^x p(1 + q + q^2 + \cdots)$$
$$= q^x, \qquad x = 0, 1, 2, \ldots.$$

By a result shown in Section 4.4, we find the mean of X as follows:

$$E(X) = \sum_{x=0}^{\infty} P(X > x) = 1 + q + q^2 + \cdots = \frac{1}{p}.$$

An alternative means to find $E(X)$ is obtained by conditioning. Let S denote obtaining a success and F a failure in a single trial. We condition on the outcome of the first trial by writing

$$E(X) = E(X|S)P(S) + E(X|F\}P(F)$$
$$= 1 \times p + (1 + E(X))q.$$

In the above, the last term is due to the fact if the first trial is a failure, then the game starts afresh but one trial has already occurred. Hence the expected trials need to obtain the first success from the start become $1 + E(X)$. Solving the last equation, we obtain

$$(1 - q)E(X) = 1 \quad \Longrightarrow \quad E(X) = \frac{1}{p}.$$

To find the variance, we use a similar approach to find $E(X^2)$ first. Thus

$$\begin{aligned}
E(X^2) &= E(X^2|S)P(S) + E(X^2|F)P(F) \\
&= 1^2 \times p + E[(1 + X)^2]q \\
&= p + E(1 + 2X + X^2)q \\
&= p + q + 2E(X)q + E(X^2)q \\
&= 1 + \frac{2q}{p} + E(X^2)q \\
&= \frac{p + 2(1 - p)}{p} + E(X^2)q.
\end{aligned}$$

Thus

$$E(X^2) = \frac{2 - p}{p^2} \quad \text{and} \quad Var(X) = \frac{2 - p}{p^2} - \frac{1}{p^2} = \frac{1 - p}{p^2} = \frac{q}{p^2}.$$

Memoryless Property of a Geometric Random Variable. Any random variable X is called *memoryless* if and only if

$$P(X > x + y | X > x) = P(X > y) \quad \text{for all } x, y > 0.$$

Let denote X the age of an item. If X is memoryless, then the property implies that the lifetime of the item does not depend on how long it has lasted. We show that the geometric random variable is memoryless:

$$P(X > x + y | X > x) = \frac{P(X > x + y)}{P(X > x)} = \frac{q^{x+y}}{q^x} = q^y = P(X > y).$$

It can be shown that the geometric random variable is the only discrete random variable that is memoryless.

Negative Binomial Distributions. The negative binomial random variable is related to the geometric random variable in the following way. Let

X_i be a geometric random variable, where $X_i \sim geom(p)$, and $\{X_i\}$ are mutually independent. We define

$$Z_n = X_1 + \cdots + X_n.$$

Then Z_n represents the total number of trials needed in a sequence of independent Bernoulli trials to obtain n successes for the first time.

The probability mass function for Z_n can be obtained by a probabilistic argument. For the case where $Z_n = x$, the xth trial must be a success. There are $x - 1$ remaining trials of which $n - 1$ must be successes. Thence we have

$$P(Z_n = x) = \binom{x-1}{n-1} p^{n-1} q^{x-n} \times p = \binom{x-1}{n-1} p^n q^{x-n}, \quad x = n, n+1, \ldots.$$

The negative binomial random variable is characterized by the parameters n and p. We denote it by $X \sim negb(n, p)$.

The structural relation between Z_n and $\{X_i\}$ enables to write the mean and variance of Z_n immediately:

$$E(Z_n) = \frac{n}{p} \quad \text{and} \quad \text{Var}(Z_n) = \frac{nq}{p^2}.$$

We remark in passing that there is another version of the negative binomial distribution in which the random variable is defined by Y denoting the number of failures has occurred before observing n successes for the first time. Thus we have $Z_n = Y + n$, where $Y = 0, 1, 2, \ldots$. In Mathematica implementation of *negb*, the probabilities are related to those of Y.

Example 8. In 2017 World Series, Houston Astros faced LA Dodgers. The winner of the first four games out of a series of seven games became the World Champion. Assume that the probability of winning a single game by Astros when facing Dodgers is 0.53. Compute the probability that Astros will win the Series in exactly six games.

Let X denote the probability that the number of games needed by Astros to win the World Series. Then we see that $X \sim negb\,(4, 0.53)$. Thus

$$P(X = 6) = \binom{6-1}{4-1} p^4 q^{6-4} = \binom{5}{3} p^4 q^2 = \binom{5}{3} (0.53)^4 (0.47)^2 = 0.17430.$$

An alternative approach is to note for $X = 6$, Astros must win 3 games out of 5. This occurs with probability $\binom{5}{3}(0.53)^3(0.47)^2$ and the 6th game Astros must win. Hence the result follows. $\qquad \square$

Example 9. Consider the experiment involving of drawing a card repeatedly from a standard card deck of 52 cards. Each draw represents a random draw from a reshuffled deck. The experiment stops as soon a diamond has been drawn for 5 times.

We see that Z_5 a negative binomial random variable with parameters $n = 5$ and $p = 1/4$. It follows that

$$P(Z_5 = x) = \binom{x-1}{4} p^5 q^{x-5}, \quad x = 5, 6, \ldots.$$

In Illustration 4, we plot the PDF. $\square$

5.6.   Probability Generating Functions

In the last chapter, we introduced the notion of a probability generating function (PGF) $p_X^g(z)$ for a nonnegative integer-valued random variable X. Recall that

$$p_X^g(z) = E(z^X) = \sum_{i=0}^{\infty} z^i p_X(i).$$

This is a special case of the Z- transform as it applies to the function $p_X(\cdot)$. We will demonstrate its various uses in working with random variable X.

Before we start, we observe that factorial moments can be obtained by differentiating with respect to z and then setting the resulting expression with $z = 1$. For example, we have

$$E(X) = \frac{d}{dz} p_X^g(z) \Big|_{z=1}.$$

To show this is so, we see

$$\frac{d}{dz} p_X^g(z) = \frac{d}{dz} \left(p_X(0) + z p_X(1) + z^2 p_X(2) + z^3 p(x(3) + \cdots \right)$$

$$= 1 \times p_X(1) + 2z p_X(2)) + 3z^2 p_X(3) + \cdots.$$

By setting $z = 1$, we find the resulting expression is $E(X)$. Similarly, we will find

$$E(X(X-1)) = \frac{d^2}{dz^2} p_X^g(z) \Big|_{z=1} = E(X^2) - E(X).$$

Higher factorial moments can be obtained in a similar manner.

An important advantage of using the probability generating function is the ease in which we can do *convolution*. When we are considering the sum of independent random variables, say $Z = X_1 + X_2$, we may use conditioning to find $P(Z = z)$. Specifically, we have

$$P(Z = z) = \sum_{x=0}^{\infty} P(X_1 + X_2 = z | X_1 = x) P(X_1 = x)$$

$$= \sum_{x=0}^{\infty} P(X_2 = z - x) P(X_1 = x)$$

$$= \sum_{x=0}^{\infty} p_{X_2}(z - x) p_{X_1}(x).$$

The above operation is known as a convolution.

Let $p_{X_1}^g(z)$, $p_{X_2}^g(z)$, and $p_Z^g(z)$ be the probability generating functions for X_1, X_2, and Z, respectively. Using the theory of Z transform, it can be shown that

$$p_Z^g(z) = p_{X_1}^g(z) p_{X_2}^g(z).$$

Working with a product relation is easier than working with convolution. Once we obtain the generating function for Z, we invert it into polynomial in z. The coefficients of the resulting polynomial yield $P(Z = k)$ for $k = 0, 1, 2, \ldots$. Generalization the above to n independent random variables is immediate. An important result to remember is that there is a one-to-one correspondence between the probability generating function and the probability distribution.

Another useful application of the PGF is the case when the PGF of a random variable X is known through some operations (e.g., convolution). Since $p_X^g(z)$ is a polynomial in z, we can expand the polynomial in z. Its coefficients give $p_X(x)$ for all x. When necessary, numerical procedures can be applied to retrieve $p_X(x)$. This is demonstrated in Illustration 4.

Example 10. Consider a Poisson random variable X with

$$p_X(x) = e^{-\lambda} \frac{\lambda^x}{x!}, \quad x = 0, 1, 2, \ldots.$$

Its probability generating function is given by

$$p_X^g(z) = \sum_{x=0}^{\infty} e^{-\lambda} \frac{\lambda^x}{x!} \times z^x = e^{-\lambda} \sum_{x=0}^{\infty} \frac{(\lambda z)^x}{x!} = e^{-\lambda(1-z)}.$$

To find its mean, we note

$$\frac{d}{dx}p_X^g(z) = \lambda e^{-\lambda(1-z)}.$$

Thus

$$\frac{d}{dx}p_X^g(z)\bigg|_{z=1} = \lambda.$$

The second factorial moment is find with

$$\frac{d^2}{dx^2}p_X^g(z) = \lambda^2 e^{-\lambda(1-z)}$$

and

$$E(X(X-1)) = \frac{d^2}{dx^2}p_X^g(z)\bigg|_{z=1} = \lambda^2.$$

Hence

$$Var(X) = E(X(X-1)) + E(X) - E^2(X) = \lambda^2 + \lambda - \lambda^2 = \lambda. \qquad \square$$

Example 11. We use this example to demonstrate the connection between the characteristic function and probability generation function of a random variable. Consider the Poisson random variable X stated in the previous example. We derive its characteristic function in a manner similar to that of the previous example:

$$\varphi_X(t) = E[e^{itX}] = \sum_{x=0}^{\infty} e^{-\lambda}\frac{\lambda^x}{x!} \times e^{itx} = e^{-\lambda}\sum_{x=0}^{\infty}\frac{(\lambda e^{it})^x}{x!} = e^{-\lambda}e^{\lambda e^{it}}$$

$$= \exp\left(\lambda(e^{it}-1)\right).$$

Now

$$\varphi_X\left(\frac{\log z}{i}\right) = \exp(\lambda(e^{i(\frac{\log z}{i})}-1)) = e^{-\lambda(1-z)} = p_X^g(z).$$

Since we now have the characteristic function, the MGF of X can be found readily:

$$M_X(t) = \varphi_X(-it) = \exp(\lambda(e^{i(-it)}-1)) = \exp(\lambda(e^t-1)). \qquad \square$$

Example 12 (Merging of Two Independent Poisson Arrival Streams). Let X_1 be the number of male arrivals to a post office on a given day. Assume X_1 follows a Poisson distribution with rate λ_1. Let X_2 be the number of female arrivals to a post office on a given day. Assume X_2 follows a Poisson distribution with rate λ_2. Assume X_1 and X_2 are independent Let $Z = X_1 + X_2$. Then we see that

$$p_Z(z) = e^{-(\lambda_1 + \lambda_2)(1-z)}.$$

We recognize that Z assumes the probability generating function of a Poisson distribution with parameter $\lambda_1 + \lambda_2$. Thus Z is Poisson with its rates being additive. $\qquad\qquad\square$

Example 13 (Negative Binomial Random Variable). For $X_i \sim \text{geom}(p)$, we recall

$$p_{X_i}(x) = q^{x-1}p, \quad x = 1, 2, \ldots.$$

Its PGF is given by

$$p_{X_i}^g(z) = \sum_{x=1}^{\infty} q^{x-1}p \times z^x = \frac{p}{q}\sum_{x=1}^{\infty}(qz)^x = \frac{p}{q}\frac{qz}{1-qz} = \frac{pz}{1-qz}.$$

Let $Z_n = X_1 + \cdots + X_n$ and assume that $\{X_i\}$ are mutually independent. Recall that Z_n represents the number of Bernoulli trials needed in order to obtain n successes for the first time. It follows that the probability generating function of Z_n is given by

$$p_{Z_n}^g(z) = \left(\frac{pz}{1-qz}\right)^n$$

$$= p^n z^n \left(1 + (-qz)\right)^{-n}.$$

Applying Newton's generalized binomial theorem, the above can be written as follows:

$$p_{Z_n}^g(z) = p^n z^n \sum_{j=0}^{\infty} \binom{-n}{j}(-qz)^j$$

$$= \sum_{j=0}^{\infty} \frac{n(n+1)\cdots(n+j-1)}{j!} p^n q^j z^{n+j}$$

$$= \sum_{j=0}^{\infty} \binom{n+j-1}{n-1} p^n q^j z^{n+j}.$$

Let $x = n + j$ and we do a change of the indexing variable. This gives

$$p_{Z_n}^g(z) = \sum_{x=n}^{\infty} \binom{x-1}{n-1} p^n q^{x-n} z^x = \sum_{x=n}^{\infty} P(Z_n = x) z^n.$$

Hence the coefficients of the above polynomial in z yield the desired probabilities that $Z_n = x$ for $x = n, n+1, \ldots$. □

Example 14. Assume a car dealership sells two types of cars: SUVs and sedans. Let the daily sales of each type be denoted by X and Z_2, respectively. Assume that $X \sim \text{pois}(3)$, $Z_2 \sim \text{negb}(2, 0.5)$ and $X \perp Z_2$. Let $W = X + Z_2$. Then W is the convolution of X and Z_2.

Let $p_1^g(z)$ denote the PGF of X and $p_2^g(z)$ denote the PGF of Z_2. Then the PGF of W is given by $p_1^g(z) \times p_2^g(z)$. In Illustration 5, we display the two individual PMS's. We then use Taylor series expansion to extract the coefficients so as to produce the PMF of W. □

Example 15. Consider a sequence of independent coin toss. Each toss lands heads (H) with probability p, tails (T) with probability $q = 1 - p$. Let X denote the number of tosses needed to obtain a run of r H's in succession. We want to find $E(X)$, $\text{Var}(X)$, and $p_X(x)$. □

We note that every time a T is observed the process starts a fresh. Prior to observing the T, there can be $0, 1, \ldots, r - 1$ H's occurring in succession. Whenever r H's occurring in succession, the game is over. Thus the following $r + 1$ events form a partition of all the scenarios leading to a verdict — namely, the first r events lead to a system restart and the last event leads to a game cessation. The following table depicts the situations involved:

Y	Possible scenarios leading to a verdict	Sequels from the left-side scenarios
0	$\{\underbrace{T}\}$	$1 + X$
1	$(\underbrace{H}_{1\ H}, T\}$	$2 + X$
⋮	⋮	⋮
$r - 1$	$(\underbrace{H, \ldots, H}_{r-1\ H's}, T)$	$r + X$
r	$(\underbrace{H, \ldots, H}_{r\ H's}\}$	r

Let Y be the scenario index shown in the above table. To verify that the sum of the probabilities leading to a verdict adds up to one, we see

$$\sum_{i=0}^{r} P(Y = i) = q + pq + p^2q + \cdots + p^{r-1}q + p^r$$

$$= q(1 + p + \cdots + p^{r-1}) + p^r = q\frac{1 - p^r}{1 - p} + p^r = 1.$$

Conditioning on Y, we obtain the PGF for X:

$$p_X^g(z) = E(z^X) = \sum_{i=0}^{r} E(z^X | Y = i)P(Y = i)$$

$$= E(z^{1+X})q + E(z^{2+X})pq + E(z^{3+X})p^2q + \cdots + E(z^{r+X})p^{r-1}q$$
$$+ z^r p^r$$
$$= E(z^X)qz \left(1 + (pz)^2 + \cdots + (pz)^{r-1}\right) + z^r p^r$$
$$= E(z^X)qz\frac{1 - (pz)^r}{1 - pz} + z^r p^r.$$

Therefore, we find

$$\left(1 - qz\frac{1 - (pz)^r}{1 - pz}\right) E(z^X) = z^r p^r$$

or the PGF of X is given by

$$p_X(z) = \frac{(pz)^r (1 - pz)}{1 - z + (1 - p)p^r z^{r+1}}.$$

In Illustration 6, we present the relevant numerical results. $\qquad\square$

5.7. Summary

In the following graph, we briefly summarize the relations between the various random variables considered in this chapter. In a way, they are all relatives to one another.

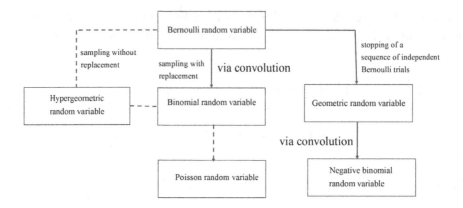

When there are solid links, they indicate direct lineages. When there are dotted lines, they are remote relatives. For example, binomial and negative binomial random variables are just the sums of independently and identically distributed random variables of their respective "parent" random variables.

For a hypergeometric random variable, it is the sum of dependent Bernoulli-type random variables, namely, it deals with sampling without replacement. When the sample size is large, the impact of dependency diminishes and it can be approximated by the binomial.

When the probability of obtaining a success in a single Bernoulli trial is small, then the resulting binomial random variable can be approximated by a Poisson distribution. This has contributed its popularity in actuarial sciences, epidemiology, and telecommunication engineering, among others.

5.8. Exploring with Mathematica

Illustration 1. Numerical solution of Example 3 involving hypergeometric random variables.

```
p1 = Table[Table[PDF[HypergeometricDistribution[3, y, 20], x], {x, 0, 3}], {y, 8, 10}]
```

$$\left\{\left\{\frac{11}{57}, \frac{44}{95}, \frac{28}{95}, \frac{14}{285}\right\}, \left\{\frac{11}{76}, \frac{33}{76}, \frac{33}{95}, \frac{7}{95}\right\}, \left\{\frac{2}{19}, \frac{15}{38}, \frac{15}{38}, \frac{2}{19}\right\}\right\}$$

```
q = Table[PDF[HypergeometricDistribution[2, 5, 11], x], {x, 0, 2}];

y = Total[q * p1];

x = {0, 1, 2, 3};

EW = Total[x * y] // N
1.33636
```

□

Illustration 2 (Estimating Population Size of Rare Specifies). We revisit Example 4 with a numerical example for estimating the sizes of Panda population. Assume that $m = 50$, $n = 40$, and $i = 4$. We plot $g_4(N)$ below:

```
x = Binomial[50, 4] // N;

giN[BN_] := x * Binomial[(BN - 50), 36] / Binomial[BN, 40];

Plot[giN[BN], {BN, 300, 800}, PlotStyle → {Black, Thick}, AxesLabel → {"N", "Prob"},
  AxesStyle → {Black, Medium}, LabelStyle → {Black, Medium}]
```

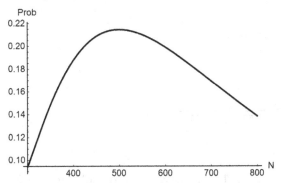

Thus the maximum likelihood estimate of N is around 500 pandas.

Another method to estimate the population size is to use the method of moment. The expectation of X is

$$E(X) = \frac{nm}{N}.$$

We postulate that the observed result corresponds to the expected value. This means $E(X) = i = 4$. With $n = 40$, and $m = 50$. This gives

$$N = \frac{40 \times 50}{4} = 500.$$

Thus the two methods yield the same answer. $\square$

Illustration 3. Finding the complementary cumulative probability for Example 8.

```
1 - CDF[HypergeometricDistribution[39, 467, 919], 26] // N
```

```
0.0136502
```

$\square$

Illustration 4. Using $p = 1/4$ and $q = 3/4$, in the following graph, we plot the PMF of Z_5 defined in Example 9.

```
fn[x_] := Binomial[x - 1, 4] * (1 / 4)^5 * (3 / 4)^(x-5);

Plot[fn[x], {x, 5, 100}, AxesLabel → {"x", "Prob"}, LabelStyle → {Black,
    Medium}, PlotStyle → {Black, Thick}, AxesStyle → {Black, Medium}]
```

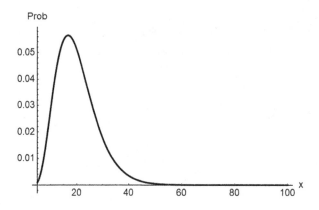

Illustration 5. Based on Example 14, we have $X \sim \text{pois}(\lambda)$, where $\lambda = 3$ and $Z_n \sim \text{negb}(n, p)$, where $n = 2$ and $p = \frac{1}{2}$. The following are the plots of their respective PMFs:

```
p1 = Table[PDF[NegativeBinomialDistribution[2, 1/2], k], {k, 0, 18}]  // N;

negb = Table[{x + 1, p1[[x]]}, {x, 1, 19}];

p2 = Flatten[Table[PDF[PoissonDistribution[3], {x}] , {x, 0, 19}]]  // N

p2a = Flatten[p2]

pois = Table[{(x - 1), p2[[x]]}, {x, 1, 20}]

po = ListPlot[pois, AxesLabel → {"x", "p_X(x)"}, LabelStyle → {Black, Medium},
    PlotStyle → {Black, Thick}, AxesStyle → {Black, Medium}];

nb = ListPlot[negb, AxesLabel → {"x", "p_{z_2}(x)"}, LabelStyle → {Black, Medium},
    PlotStyle → {Black, Thick}, AxesStyle → {Black, Medium}];

GraphicsRow[{po, nb}]
```

Let $f_1(z)$ denote the PGF of X and $f_2(z)$ the PGF of Z_2. Using the results given in the chapter, we know

$$f_1(z) = e^{-3(1-z)}, \quad f_2(z) = \frac{1}{4}z^2 \left(1 - \frac{z}{2}\right)^{-2}.$$

By differentiating the PGFs and a few extra steps, the means and variances of X and Z_2 are obtained below:

```
f1[z_] := Exp[-3 (1 - z)];

EX = f1'[1]

3

VarX = f1''[1] + EX - EX²

3

f2[z_] := 1/4 z² (1 - z/2)^-2;

EZ2 = f2'[1]

4

VarZ2 = f2''[1] + EZ2 - EZ2²

4
```

We recall $E(X) = \text{Var}(X) = 3$ and $E(Z_2) = n/p = 2/(1/2) = 4$ and $\text{Var}(Z_2) = nq/p^2 = 2/(1/2) = 4$. They verify the above results.

In the following, we use Taylor series expansion to extract its coefficients. These coefficients gives the PMF $\{p_W(w)\}$. We display the plot of the PMF below.

```
g[z_] := f1[z] * f2[z];

pw = CoefficientList[Series[g[z], {z, 0, 40}], z];  // N

ListPlot[pw, AxesLabel → {"w", "{p(w)"}, AxesStyle → {Black, Medium},
  LabelStyle → {Black, Medium}, PlotStyle → {Black, Thick}]
```

□

Illustration 6. For Example 15, with $r = 1$, we specialize the PGF to that of a geometric random variable with parameter $p = 0.6$. The following show that $E(X)$ and $\text{Var}(X)$ are indeed in agreement with those of geom(0.6)

```
p = 0.6; r = 1;
```

$$g[z\_] := \frac{(p\,z)^r\,(1 - p\,z)}{1 - z + (1 - p)\,p^r\,z^{r+1}};$$

Check against geometric(0.6):

```
EX = g'[1]
```

```
1.66667
```

```
VarX = g''[1] + EX - EX²
```

```
1.11111
```

We then look at the case when $r = 5$ and compute $E(X)$ and $\text{Var}(X)$. Moreover, we produce a plot of the PMF of X below:

```
Consider the case when r = 5:
```

```
r = 5;
```

```
EX = g'[1]
```

```
29.6502
```

```
VarX1 = g''[1] + EX - EX²
```

```
676.233
```

```
pw = CoefficientList[Series[g[z], {z, 0, 100}], z]; // N
```

```
ListPlot[pw, AxesLabel → {"x", "P(X=x)"}, AxesStyle → {Black, Medium},
  LabelStyle → {Black, Medium}, PlotStyle → {Black, Thick}]
```

Problems

1. Suppose we roll two fair dice 12 times. Let D denote the number of times a double 6 appears. (a) Find the exact probability that $P(D = 1)$. (b) Find the approximate probability that $P(D = 1)$. (c) Let C denote the number of rolls needed in order to observe a double 6 appears for the first time. State the probability mass function $p_C(x)$.

2. In Texas Hold'em poker, in each play, a player is dealt two cards from a well-reshuffled standard card deck. (a) What is the probability that in each play, the player gets a least one ace? (b) Let X be the number of plays needed in order to get at least one ace. State the probability mass function $p_X(x)$. (c) What is $E(X)$?

3. The final match of the soccer at World Cup is a draw at the end of extra time. The teams then proceed to penalty kicks. Each team attempts five shots. Suppose each shot results in a goal with probability 0.7, independent of the other shots. What is the probability that the match is still tied after one round of five shots by each team?

4. A fair coin is tossed n times. Let Y denote the number of heads observed. (a) What is the probability mass function of Y?(b) Let X denote the number of heads obtained minus the number of tails obtained. What is the PMF of X? (c) For $n = 10$, use Mathematica to construct the PMF of X.

5. Cars pass a point on the highway at a Poisson rate of one per minute. Assume that five percent of cars on the road are Fords.

   (a) What is the probability that at least one Ford passed by during an hour?
   (b) If 50 cars have passed by in an hours, what is the probability that five of them were Fords?
   (c) Given that ten Fords have passed by in an hour, what is the probability that a total of 50 cars have passed by in that hour?

6. In the final of the World Series baseball, two teams play a series consisting of at most seven games until one of the two teams has won four games. Two unevenly matched teams are played against each other. The probability that the weaker teams will win any given game is equal to 0.45. Assuming that the results of the various games are independent of each other. (a) Compute the probability of the weaker team winning the final. (b) Compute the expected number of games to be played to end the series. (c) Use Mathematica to verify the results obtained in (a) and (b).

7. In a close election between two candidates Pete and Mary in a small town, the winning margin of Pete is 1,422 to 1,405 votes. However, 101 votes are found to be illegal and have to be thrown out. It is not clear how the illegal votes are divided between the two candidates. Assuming that the illegal votes are not biased in any particular way and the count is otherwise reliable. What is the probability that the removal of the illegal votes will change the result of the election?

8. A standard card deck contains 52 cards of which 4 are aces. Cards are randomly selected, one at a time, until an ace is obtained (in which case the game ends). Assume that each drawn card is returned to the deck and the deck is reshuffled to maintain it randomness. (a) What is the probability exactly n draws are needed? (b) What is the mean number of draws needed? (c) What is probability that at least k draws are needed?

9. Let X be a Poisson random variable with mean λ. (a) Find the characteristic function $\varphi_X(t)$ of X. (b) Use $\varphi_X(t)$ to find the probability generating function p_X^g of X. (c) Use the definition of $p_X^g(z)$ to verify your result obtained in Part (b).

10. Consider the following PMF for random variable X:

$$p_X(x) = \binom{n+x-1}{x} p^n q^x, \quad x = 0, 1, \ldots,$$

where n is a positive integer, $0 < p < 1$, $q = 1 - p$, and $|s| < q^{-1}$. (a) Verify that $p_X(x)$ is a legitimate PMF. (b) Find the probability generating function $p_X^g(x)$ of X. (c) Use the PGF of X to find $E(X)$ and $E(X^2)$. (d) What interpretation can be attached to the random variable X?

11. Consider a sequence of Bernoulli trials with the probability of a success being p, and failure q. Let X denote the number of trials needed so that we observe a total of an even number of successes for the first time. (a) Find the probability generating function $p_X^g(z)$ of X. (b) Find the PMF $p_X(x)$ of X.

13. Consider the repeated tossing a fair die. Let S_n denote the sum of the face values after n tosses. Then

$$S_n = \sum_{n=1}^{n} X_i,$$

where X_i is the face value obtained at the tth toss. (a) Find the probability generating function $p_{S_n}^g(z)$ of S_n. (b) Compute the PMS of S_n for $n = 5$ and plot it by Mathematica.

14. An urn contains m black balls and n white balls. You pick one ball at a time from the urn without replacement and stop as soon as you obtain k black balls. Let X denote the number of picks needed to accomplish your task. Find the PMF $p_X(x)$. (a) Find the PMF $p_x(x)$. (b) For $m = 5$, $n = 7$, and $k = 3$, use Mathematica to compute $E(X)$.

15. Consider a sequence of independent tosses of a biased coin with probability of 0.65 landing heads. Let E be the event that there are fifteen heads before five tails. (a) Find $P(E)$. (b) Use Mathematica to verify the result obtained in part (a).

16. Johnny and Josh are playing a game about rolling a fair die. For Johnny to win, he needs to obtain an one or two. For Josh to win, he needs to obtain four, five, or six. The first player to attain the designated number will be declared as the winner. In the event both attain the designated number in a single roll, then Johnny will also be declared as the winner. (a) What is the probability that Johnny will be the winner of the game? (b) What is the expected length of the game?

17. Consider a standard card deck of 52 cards. You randomly draw one card at a time from the deck with replacement until you find an ace. You repeat the game until all the four distinct aces are found and then the game stops. Every time a card is return to the deck, a reshuffling occurs to ensure its randomness. Let X denote the number of cards drawn till the game stops. (a) Find $E(X)$. (b) Find $\text{Var}(X)$.

18. Clairvoyant Jimmy declares that he has the unusual skill to perform the following feat: in an urn with 25 black balls and 25 white balls, with his eyes blind-folded, if he is asked to pick 25 balls from the urn, he will be able to pick at least 80% of the black balls from the urn. What is the probability for this to occur if drawings are done randomly?

19. An urn contains n balls numbered $1, 2, \ldots, n$. We pick k balls randomly from the urn without replacement. Let X be the ball marked with the largest number drawn from the urn. (a) Find the PMF of X. (b) Find the CDF of X. (c) Use Mathematica to compute $E(X)$ and $\text{Var}(X)$ for the case where $n = 10$ and $k = 5$.

20. Assume that the number of fires occurring in a city follows a Poisson distribution with mean of three per week. (a) What is the probability that in a given week, there will be at most three fires per week? (b) For a given year, what is the probability that there are at most 10 weeks with at least three fires per week?

21. Consider we roll a fair die repeatedly until we obtain an one for the first time. Let X be the sum of all the faces shown when the game stops. (a) Find the probability generating function $p_X^g(z)$. (b) Find $E(X)$ and $Var(X)$. (b) Invert the probability generating function to obtain the PMF of X and plot it PMF.

22. Let X be a random variable with the following PMF:

$$p_X(x) = \begin{cases} \frac{1}{8} & \text{if } x = -2, -1, 1, 2, \\ \frac{1}{4} & \text{if } x = 0, 3, \\ 0 & \text{otherwise.} \end{cases}$$

(a) Find $E(X)$. (b) Find the probability mass function $p_Y(y)$ of Y, where $Y = (X - 1)^2$. (c) Find $E(Y)$.

23. Assume that the probability of finding oil in any hole in a small town is 0.1. The result at any one hole is independent of what happens at any other hole. Mickey Mouse is one drilling company located in the town. Mickey Mouse can only afford to drill one hole at a time. If they hit oil, they stop drilling. Mickey Mouse only can afford to drill seven dry holes due to limited capital. Let X denote the number of holes drilled till Mickey Mouse stops. What is the probability mass function $p_X(x)$ of X? (a) Assume that the cost of drilling a hole is \$1,000,000. Let Y denote the total cost to Mickey Mouse for this venture. Compute $E(Y)$ and $Sdv(Y)$.

24. Consider a population of m seals along the coast of northern California of which n of them have been captured and, tagged, and then released. Let X denote the number of seals needed to recapture so as to obtain k tagged seals. Find the probability mass function $p_X(x)$ of X.

25. Let X be a nonnegative integer-valued random variable. Assume that X has the following probability generating function:

$$p_X^g(z) = \log\left(\frac{1}{1 - qz}\right).$$

(a) Find the PMF $f_X(x)$ of X. (b) Find $E(X)$ and $Var(X)$.

26. A pair of dice is rolled n times, where n is chosen so that the chance of getting at least one double six in the n rolls is close to $\frac{1}{2}$. (a) Find this n. (b) What, approximately, is the chance that you actually get two or more double sixes in this many rolls?

27. We roll 10 dice together. If some turn up six, they are removed. We then roll the remaining dice once more. If some turn up six, they are removed. We repeat the process until there are no sixes left. Let N denote the number of rolls needed so that there are no sixes left. (a) Find the PMF for N. (b) Let T denote the number of individual die rolls. Find the PMF for T. (c) Let L denote the number of dice rolled on the last roll. Find the PMF for L.

28. Use Mathematica to plot PMFs for the three random variables: N, T, and L.

29. Wayne is an avid basketball player. His friend Jimmy wants to find out Wayne's free throw skill. Let θ be the probability that a free throw is a success. The three hypotheses on θ are: 0.25, 0.50, and 0.75. Before actually checking out Wayne's skill, Jimmy's prior assessments for the three levels of success are 0.2, 0.6, and 0.2, respectively. Here comes the test. Wayne made 7 successful shots out of 10. What are the posterior probabilities on θ?

30. Let θ be the rate of success in treating a specific disease. The standard treatment medicine in the market has a success rate of 0.35. Now a new medicine has been developed. It has undergone clinical trials involving 10 patients. As it turns out, the trials yield 7 successes. The prior probabilities for θ follow a discrete uniform distribution on integers $\{0, 0.01, 0.02, \ldots, 1.0\}$. What are the posterior probability that the new medicine is superior than the existing one?

31. Suppose $X \sim bino(n, p)$. Prove by induction on n that

$$P(X \text{ is even}) = \frac{1}{2}(1 + (1 - 2p)^n).$$

32. Suppose $X \sim bino(n, p)$. Let Y be a random variable such that $Y = X \cdot 1_{\{x>0\}}$. Then Y is known as a truncated binomial random variable. Find the PMF of Y.

33. Suppose $X \sim pois(n, p)$. Let Y be a random variable such that $Y = X \cdot 1_{\{x>0\}}$. Then Y is known as a truncated Poisson random variable. Find the PMF of Y.

Remarks and References

The use of *PGF* to model probabilistic systems is given in Abate and Whitt [1]. It is also covered in Howard [3], or Kao [4]. Example 3 is based on Problem 7.23 of Ross [6]. Newton's generalized binomial formula used in Example 13 can be found in Chung [2, p. 131]. Another way to do the conversion is by inverting the corresponding Z transform. Problems 27 and 28 are based on Pitman [5]. Problem 30 stems from Tijms [7].

[1] Abate, J. and W. Whitt, Numerical Inversion of Probabilistic Generating Functions, *Operations Research Letters*, 12(4), 245–51, 1992.
[2] Chung, K. L. *Elementary Probabiity Theory with Stochastic Processes*, Springer-Verlag, 1974.
[3] Howard, R. A. *Dynamic Probabilistic Systems: Volume 1: Markov Models*, Wiley, 1971.
[4] Kao, E. P. *An Introduction to Stochastic Processes*, Duxbury, 1998, and Dover, 2019.
[5] Pitman, J. *Probability*, Springer-Verlag, 1993.
[6] Ross, S. M., *A First Course in Probability*, 10th edition, Pearson, 2019.
[7] Tijms, H. *Probability: A Lively Introduction*, Cambridge University Press, 2018.

Chapter 6

Continuous Random Variables

6.1. Introduction and Transformation of Random Variables

For a continuous random variable X, we recall that $f_X(x)$ is the probability density function (PDF). The density enables us to compute the probability that X assumes values in an interval of length U using

$$P(X \in U) = \int_U f_X(x)dx.$$

For continuous random variables, the probability of observing any specific value is 0. It is equivalent to state $X \le a$ or $X < a$. Namely, the use of the equality sign is optional. For convenience, it is sometimes omitted.

Many continuous random variables are obtained through a transformation $Y = g(X)$. Knowing the distribution function $F_X(x)$ of X, we see

$$
\begin{aligned}
P(Y \le y) &= P(g(X) \le y) && \text{by substitution} \\
&= P(X \le g^{-1}(y) && \text{assume } g(y) \text{ is } \textit{increasing} \\
&= F_X(g^{-1}(y)) && \text{by definition of } F_X.
\end{aligned}
$$

The following graph clearly depicts the event $\{g(X) \le y\}$ $\Longleftrightarrow$ $\{X \le g^{-1}(y)\}$:

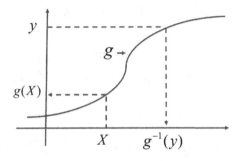

We now differentiate the last expression with respect to y. An application of the chain rule yields

$$f_Y(y) = f_X\left(g^{-1}(y)\right) \frac{d}{dy} g^{-1}(y).$$

When g is decreasing, the converse holds, i.e., $\{g(X) \le y\} \iff \{X \ge g^{-1}(y)\}$. Thus

$$P(Y \le y) = P(X \ge g^{-1}(y))$$
$$= 1 - F_X(g^{-1}(y)).$$

Hence, when differentiating with respect to y, we end up with

$$f_Y(y) = f_X(g^{-1}(x)) \left| \frac{d}{dy} g^{-1}(y) \right|$$

as the derivative of $g^{-1}(y)$ is now negative and it cancels out with the minus sign in front of F_X. Thus the last expression is applicable when g is monotone.

In making the transformation, we may apply the above result directly — particularly when $f_X(x)$ is already known. This is demonstrated in Example 1. Or we may go through the aforementioned process and derive the result accordingly as illustrated in Examples 1 and 2.

Example 1. Assume random variable X follows the following density

$$f_X(x) = \begin{cases} 4x^3 & \text{if } 0 < x < 1, \\ 0 & \text{otherwise.} \end{cases}$$

Let $Y = g(X) = 1 - 3X^2$. We would like to find the density for Y. We see that the range of Y over which the density f_Y is positive is $(-2, 1)$. Since

$$X = \sqrt{\frac{1-Y}{3}} = g^{-1}(Y) \equiv h(Y).$$

We apply the result for making the transformation and write

$$f_Y(y) = f_X(h(y)) \left| \frac{d}{dy} h(y) \right|$$

$$= 4 \left(\sqrt{\frac{1-y}{3}} \right)^3 \left| \frac{d}{dy} \left(\frac{1-y}{3} \right)^{\frac{1}{2}} \right|$$

$$= 4 \left(\frac{1-y}{3} \right)^{\frac{3}{2}} \left(\frac{1}{2} \right) \left(\frac{1-y}{3} \right)^{-\frac{1}{2}} \left(-\frac{1}{3} \right)$$

$$= 4 \left(\frac{1-y}{3} \right) \frac{1}{6}$$

$$= \frac{2}{9} (1-y), \qquad -2 < y < 1.$$

We now use an *alternative* approach to find the density. We note

$$P(Y \leq y) = P\left(1 - 3X^2 \leq y\right)$$

$$= P(-3X^2 \leq y - 1)$$

$$= P\left(X \geq \left(\frac{1-y}{3} \right)^{\frac{1}{2}} \right)$$

$$= 1 - P\left(X \leq \left(\frac{1-y}{3} \right)^{\frac{1}{2}} \right)$$

$$= 1 - F_X \left(\left(\frac{1-y}{3} \right)^{\frac{1}{2}} \right).$$

We see that

$$F_X(x) = \int_0^x 4z^3 dz = x^4, \quad 0 \leq x \leq 1.$$

Therefore, upon substitution we obtain

$$P(Y \leq y) = 1 - \left(\frac{1-y}{3} \right)^2 = F_Y(y).$$

Differentiating the above with respect to y yields

$$f_Y(y) = \frac{2}{9} (1-y), \quad -2 < y < 1. \qquad \square$$

Example 2. Assume $Y = X^2$ where X is a continuous random variable with density $f_X(x)$. Then we know $Y \geq 0$ and

$$
\begin{aligned}
F_Y(y) &= P(Y \leq y) \\
&= P(X^2 \leq y) \\
&= P(-\sqrt{y} \leq X \leq \sqrt{y}) \\
&= F_X(\sqrt{y}) - F_X(-\sqrt{y})
\end{aligned}
$$

We apply the chain rule and conclude

$$
f_Y(y) = \frac{1}{2\sqrt{y}} \left(f_X(\sqrt{y}) + f_X(-\sqrt{y}) \right)
$$

Notice that for each y value, there are two x values that correspond to it. Hence, the likelihood of y is doubled. $\qquad\square$

6.2. Laplace Transforms and Characteristic Functions

To ease up our effort in working with continuous random variables, we first cover the basics of Laplace transforms (LT) and characteristic functions (CF). We will find that with these two useful tools along with the help of Mathematica, studying more elaborate issues in probability would become a pleasing experience. At the absence of these helps, we could easily getting bogged down by calculus related tasks as opposed to focusing on probability at a conceptual level.

In Chapter 4, we introduced the notion of an LT $f_X^e(s)$ for a continuous random variable $X > 0$. To obtain the various moments of X, we see that

$$
\begin{aligned}
\frac{d^n}{ds^n} f_X^e(s) &= \frac{d^n}{ds^n} E(e^{-sX}) = E\left(\frac{d^n}{ds^n} e^{-sX} \right) \\
&= E\left((-1)^n X^n e^{-sX} \right) = (-1)^n E(X^n e^{-sX}).
\end{aligned}
$$

When we evaluate the above expression with $s = 0$, we find

$$
E(X^n) = (-1)^n \left. \frac{d^n}{ds^n} f_X^e(s) \right|_{s=0}.
$$

One of the advantages in using LTs is the ease with which we are working the sum of *independently* distributed random variables $\{X_i\}$. Specifically, when $X = X_1 + \cdots + X_{n,}$, then

$$f_X^e(s) = \prod_{i=1}^{n} f_{X_i}^e(s),$$

where $f_{X_i}^e(s)$ the LT of the PDF of X_i. In addition if $\{X_i\}_{i=1}^{n}$ are also *identically* distributed, then

$$f_X^e(s) = (f_{X_1}(s))^n.$$

We observe that the convolution relation becomes a multiplicative relation in the transform domain.

Recall that the CF of a random variable X is denoted by $\varphi_X(t)$. For any distribution, the CF always exists, is continuous and determines the distribution function uniquely. Consequently, the CF is applicable to random variables that are neither continuous nor discrete. Moreover, it exists even when the moment generating function $M_X(t)$ fails to exist. For α, $\beta \in \mathbb{R}$, we define $Y = \alpha + \beta X$. Then

$$\varphi_Y(x) = \varphi_{\alpha+\beta X}(s) = E[\exp(iu(\alpha + \beta X))] = e^{iua} \cdot E[\exp(i(\beta u)X)]$$

$$= e^{iua} \cdot \varphi_X(\beta u). \tag{6.1}$$

To obtain the various moments of X, similar to the derivations used for LT's we will find

$$E(X^n) = \frac{1}{i^k} \frac{d^n}{ds^n} \varphi_X(s) \Big|_{s=0}.$$

When $X = X_1 + \cdots + X_n$, where $\{X_i\}_{i=1}^{n}$ are independently distributed random variables, we have

$$\phi_X(s) = \prod_{i=1}^{n} \varphi_{X_i}(s).$$

In addition if $\{X_i\}_{i=1}^{n}$ are also identically distributed, then it simplifies to

$$\varphi_X(s) = (\varphi_{X_1}(s))^n.$$

We shall show the usage of the above in the sequel.

6.3. Uniform and Exponential Random Variable

There are two elementary building blocks for constructing other variants of continuous random variables. One is the standard uniform random variable and other is the exponential random variable.

Let X follows the following density

$$f_X(x) = 1, \quad 0 < x < 1.$$

Then X is said to assume a standard uniform distribution. We write it as $X \sim \text{unif}(0, 1)$. It follows that its distribution function takes a simple form

$$F(x) = \begin{cases} 0 & \text{if } x \leq 0, \\ x & \text{if } 0 < x < 1, \\ 1 & \text{if } x \geq 1. \end{cases}$$

When the term random number generator is mentioned, typically it refers to sampling from $\text{unif}(0, 1)$.

A generalization of $\text{unif}(0, 1)$ is by a location change along x-axes yielding

$$f_X(x) = \begin{cases} \dfrac{1}{b - a} & a < x < b \\ 0 & \text{otherwise} \end{cases}$$

We use $\text{unif}(a, b)$ to denote the density of the above uniform random variable.

Example 3. Assume X is a continuous random variable and $Y = F_X(X)$. Assume also that the inverse function $F_X^{-1}(\cdot)$ is unique. Then

$$\begin{aligned} F_Y(y) &= P(Y \leq y) \\ &= P(F_X(X) \leq y) \quad \text{substitution} \\ &= P(X \leq F_X^{-1}(y)) \quad \text{monotone transformation} \\ &= F_X(F_X^{-1}(y)) \quad \text{definition of } F_X \\ &= y. \end{aligned}$$

We recognize $Y \sim \text{unif}(0, 1)$.

We can apply the result of this example in testing whether X follows a distribution F_X provided that we know its form. We simply plug values of X in $F_X^{-1}(\cdot)$ and find the corresponding values of Y. We then check whether these Y values are samples from $\text{unif}(0, 1)$. □

Example 4. Let $U \sim \text{unif}(0,1)$. For any continuous random variable X with distribution function F_X, we define

$$X = F_X^{-1}(U).$$

For simplicity, we assume that the inverse function produces a unique value. Then we see

$$F_X(x) = P(X \leq x)$$
$$= P(F_X^{-1}(U) \leq x)$$
$$= P(U \leq F_X(x))$$
$$= F_X(x).$$

Since computers have random number generators for $\text{unif}(0,1)$, this transformation enables us to generate random samples from F_X. $\qquad\square$

We now move to the other building block for constructing variants of many other continuous random variable, namely, the exponential distribution. Let random variable X assumes the following density:

$$f(x) = \begin{cases} \lambda e^{-\lambda x} & \text{if } x > 0, \\ 0 & \text{otherwise.} \end{cases}$$

Then X is said to follow an exponential distribution with parameter $\lambda > 0$, we write $X \sim \text{expo}(\lambda)$. Its distribution function is given by

$$F_X(x) = \int_0^x \lambda e^{-\lambda \tau} d\tau = 1 - e^{-\lambda x}, \quad x > 0.$$

The complementary cumulative distribution is

$$P(X > x) = 1 - F_X(x) = e^{-\lambda x}, \quad x > 0.$$

We see that the exponential random variable X is memoryless in that

$$P(X > x + y | X > x) = \frac{e^{-\lambda(x+y)}}{e^{-\lambda x}} = e^{-\lambda y} = P(X > y).$$

It can be shown it is the only continuous random variable that is memoryless.

The mean of the exponential random variable X is

$$E(X) = \int_0^\infty P(X > x)dx = \int_0^\infty e^{-\lambda x}dx = \frac{1}{\lambda}.$$

To find the variance of X, we first establish an useful identity for the second moment of any continuous random variable $X > 0$

$$E(X^2) = \int_0^\infty y^2 f_X(y)dy = \int_0^\infty \left(2\int_0^y \tau d\tau \right) f_X(y)dy$$

$$= \int_0^\infty 2\tau \left(\int_\tau^\infty f_X(y)dy \right) d\tau = \int_0^\infty 2\tau P(X > \tau)d\tau.$$

Using the above result, we get

$$E(X^2) = \int_0^\infty 2\tau P(X > \tau)d\tau$$

$$= 2\int_0^\infty \tau e^{-\lambda \tau}d\tau$$

$$= \frac{2}{\lambda^2}.$$

Hence we have

$$Var(X) = \frac{2}{\lambda^2} - \frac{1}{\lambda^2} = \frac{1}{\lambda^2}.$$

$\square$

Competing Exponentials. Consider two exponential random variables X_1 and X_2 with respective parameters μ_1 and μ_2. Assume that $X_1 \perp X_2$. Let $X = \min(X_1, X_2)$. We see that

$$P(X > x) = P(\min(X_1, X_2) > x) = P(X_1 > x, \ X_2 > x)$$

$$= e^{-\mu_1 x - \mu_2 x} = e^{-(\mu_1 + \mu_2)x}, \quad x > 0.$$

Thus $X \sim \text{expo}(\mu_1 + \mu_2)$. Assume a rabbit and a squirrel are engaged in a race. Assume, respectively, X_1 and X_2 measure the lengths of time each reaches the destination. We call this the competing exponentials. Define

$$I = \begin{cases} 1 & \text{if } X_1 < X_2 \text{ (i.e., the rabbit wins the race)}, \\ 0 & \text{otherwise.} \end{cases}$$

We see that the two events $\{I = 1 \text{ and } X > x\}$ and $\{x < X_1 < X_2\}$ are equivalent. Thence we have

$$P(I = 1 \text{ and } X > x) = P(x < X_1 < X_2)$$

$$= \iint\limits_{x < x_1 < x_2} \mu_1 e^{-\mu_1 x_1} \mu_2 e^{-\mu_2 x_2} dx_1 dx_2$$

$$= \int_x^\infty \mu_1 e^{-\mu_1 x_1} \int_{x_1}^\infty \mu_2 e^{-\mu_2 x_2} dx_2 dx_1$$

$$= \int_x^\infty \mu_1 e^{-\mu_1 x_1} e^{-\mu_2 x_1} dx_1$$

$$= \frac{\mu_1}{\mu_1 + \mu_2} e^{-(\mu_1 + \mu_2)x}, \quad x > 0.$$

Since $P(X > x) = e^{-(\mu_1 + \mu_2)x}$, we see that the joint probability factors. So we conclude

$$P(I = 1) = \frac{\mu_1}{\mu_1 + \mu_2} \quad \text{and} \quad P(X = 0) = \frac{\mu_2}{\mu_1 + \mu_2}.$$

In other words, in the case of two competing exponentials, the probability for one to win the race and the density for the time for winning to occur are two *independent* events. They are both functions of the exponential parameters μ_1 and μ_2. Moreover, both have nice intuitive interpretations. For the rabbit to win, it is the ratio of the exponential rate (to finish) for the rabbit divided by the sum of the total of rates (to finish). The wait time for the race to finish follows again an exponential distribution with the rate be the sum of the two rates to finish.

Hazard Rate Functions. The exponential distribution is used extensively in the reliability theory and life sciences. A closely connected notion is that of the hazard rate function defined by

$$\lambda_X(t) = \frac{f_X(t)}{F_X^{>}(t)} = \frac{f_X(t)}{P(X > t)}.$$

An insightful way to look at the above term is by writing it as

$$\lambda_X(t)dt = \frac{f_X(t)dt}{P(X > t)} \approx \frac{P(t < X < t + dt)}{P(X > t)}.$$

Thus, the hazard rate at time t gives the likelihood that an item will fail in the forthcoming small interval given that it has lasted till time t. If $X \sim \exp(\lambda)$, then we see

$$\lambda(t) = \frac{\lambda e^{-\lambda t}}{e^{-\lambda t}} = \lambda, \quad t > 0,$$

namely, it has a constant hazard rate of λ.

For a given positive random variable X, there is a one-to-one relation between its hazard rate function $\lambda_X(t)$ and its distribution function $F_X(t)$. The relation is given by

$$F_X(t) = 1 - \exp\left(-\int_0^t \lambda(s)ds\right).$$

This means that if we know the hazard rate function of random variable X, we can recover the corresponding distribution function F_X. To establish this, we integrate $\lambda_X(\cdot)$ from 0 to t. This gives

$$\int_0^t \lambda(s)ds = \int_0^t \frac{f_X(s)}{1 - F_X(s)}ds.$$

We do a change of variable using $u = 1 - F_X(s)$. Thus $du = -f_X(s)ds$ and

$$\int_0^t \lambda(s)ds = \int -\frac{1}{u}du = [\ln(u)]_{1 - F_X(t)}^1 = -\ln(1 - F_X(t)).$$

We exponentiate the above and obtain

$$F_X(t) = 1 - e^{-\int_0^t \lambda(s)ds}.$$

Example 5. Consider an electronic component whose hazard rate function is given by $\lambda_X(t) = \frac{1}{2}t$, i.e., the hazard rate increases linearly in t at the rate of $\frac{1}{2}$ per unit time. What is the lifetime distribution of X?

We see that

$$F_X(x) = 1 - e^{-\int_0^x \frac{1}{2}s\,ds} = 1 - e^{-\frac{1}{4}x^2}, \quad x > 0.$$

Thus, the density of the lifetime follows

$$f_X(x) = \left(\frac{t}{2}\right) e^{-\left(\frac{t}{2}\right)^2}, \quad x > 0.$$

Later, we will see X follows a Weibull distribution with parameters $\alpha = 2$ and $\beta = 2$. □

6.4. Erlang and Gamma Random Variables

If we consider the exponential distribution is the continuous counterpart of the geometric distribution, then the Erlang distribution is the continuous counterpart of the negative binomial distribution in the sense

$$Z_n = X_1 + \cdots + X_n,$$

where $\{X_i\}$ are mutually independent and $X_i \sim \exp(\lambda)$. In other words, $Z_n \sim \text{Erlang}(n, \lambda)$.

The probability density function for Z_n can be obtained through a probabilistic argument. For $Z_n \in (t, t + dt)$, there must be $n - 1$ Poisson arrivals in an interval of length t. This is given by

$$P\{(n-1) \text{ Poisson arrivals in } (0, t)\} = e^{-\lambda t}\frac{(\lambda t)^{n-1}}{(n-1)!}$$

Then the nth arrival must occurs at a rate of λ at time t. Thus

$$P\{Z_n \in (t, t + dt)\} \approx f_{S_n}(t)dt = e^{-\lambda t}\frac{(\lambda t)^{n-1}}{(n-1)!}(\lambda dt) = \frac{\lambda e^{-\lambda t}(\lambda t)^{n-1}}{(n-1)!}dt$$

and consequently

$$f_{Z_n}(t) = \frac{\lambda e^{-\lambda t}(\lambda t)^{n-1}}{(n-1)!}, \quad t > 0.$$

The distribution function of S_n can also be obtained by a probabilistic argument. The two events $\{Z_n \leq t\}$ and $\{N(t) \geq n\}$ are equivalent, i.e., $\{Z_n \le t\} \iff \{N(t) \ge n\}$ where $N(t) \sim \text{Pois}(\lambda t)$. Hence

$$F_{Z_n}(t) = P(Z_n \le t) = P(N(t) \ge n) = \sum_{k=n}^{\infty} e^{-\lambda t} \frac{(\lambda t)^k}{k!}.$$

The PDF of the above can be found by differentiating F_{Z_n} with respect to t. Another approach is via LT with

$$f_{X_i}^e(s) = \int_0^{\infty} e^{-st} \lambda e^{-\lambda t} dt = \frac{\lambda}{s + \lambda}.$$

Since $\{X_i\}_{i=1}^n$ are independently identically distributed random variables, we have

$$f_X^e(s) = \left(\frac{\lambda}{s + \lambda} \right)^n.$$

Inverting the above LT, we find the PDF of $\text{Erlang}(n, \lambda)$ in Illustration 1 along with its mean and variance.

A generalization from the Erlang distribution to the gamma distribution is done with the integer parameter n of the former replaced by a constant $\alpha > 0$. In other words, $X \sim \text{gamma}(\alpha, \lambda)$, where $\alpha > 0$ and $\lambda > 0$, when

$$f_X(t) = \begin{cases} \dfrac{\lambda e^{-\lambda t} (\lambda t)^{\alpha-1}}{\Gamma(\alpha)} & t > 0, \\ 0 & \text{otherwise,} \end{cases}$$

where $\Gamma(\alpha)$ is called the gamma function defined by

$$\Gamma(\alpha) = \int_0^{\infty} e^{-y} y^{\alpha-1} dy.$$

Using integration by parts, it can be shown that

$$\Gamma(\alpha) = (\alpha - 1)\Gamma(\alpha - 1)$$

and also $\Gamma(1) = 1$. When α is a positive integer, we then have $\Gamma(\alpha) = (\alpha - 1)!$ and the gamma and Erlang assume the same distributional form. The parameter α is called the *shape* parameter and λ the *scale* parameter In Illustration 2, we use Mathematica to obtain the characteristic function of gamma(α, λ). There, the CF is shown to assume the form

$$\varphi_X(s) = \left(1 - \frac{is}{\lambda}\right)^{-\alpha} \tag{6.2}$$

and also

$$E(X) = \frac{\alpha}{\lambda}, \quad \text{Var}(X) = \frac{\alpha}{\lambda^2}.$$

We see when $\alpha = 1$, gamma simplifies to exponential; whereas when α is a positive integer, gamma becomes Erlang. Also in the same illustration displays the shapes of gamma as the values of α vary when λ is fixed at $1/3$.

6.5. Weibull Random Variables

The Weibull distribution is a direct descendant of the exponential random variable through a transformation. The Weibull random variable is used widely in reliability, insurance, and life sciences to model failure rates of equipment and mortality of human beings.

Assume random variable X follows an exponential distribution with parameter 1. Then

$$f_X(x) = e^{-x}, \quad x > 0.$$

Define

$$Y = v + \alpha X^{\frac{1}{\beta}},$$

where constants v, $\alpha > 0$, and $\beta > 0$ are called the location, scale, and shape parameters. Y is called a Weibull random variable, and we denote its density by Weibull(v, α, β). Note that

$$x = \left(\frac{y - v}{\alpha}\right)^{\beta} \equiv h(y),$$

it follows that

$$\frac{d}{dy}h(y) = \beta\left(\frac{y - v}{\alpha}\right)^{\beta-1}\frac{1}{\alpha} = \frac{\beta}{\alpha}\left(\frac{y - v}{\alpha}\right)^{\beta-1}.$$

We now apply the formula for the transformation and obtain its density

$$f_Y(y) = f_X(h(y))\frac{dh(y)}{dy}$$

$$= \exp\left(-\left(\frac{y-v}{\alpha}\right)^\beta\right)\frac{\beta}{\alpha}\left(\frac{y-v}{\alpha}\right)^{\beta-1}, \quad y > v.$$

We note that $x > 0$ implies that $y > v$ in the above. Also, $f_Y(y) = 0$, $y \leq v$. The above seemingly complicated density has its modest origin, namely, expo(1).

Example 6. We consider the case of a Weibull random variable Y with location parameter $v = 0$. Then we have

$$Y = \alpha X^{\frac{1}{\beta}}$$

and its density reduces to

$$f_Y(y) = \begin{cases} \left(\frac{\beta}{\alpha}\right)\left(\frac{y}{\alpha}\right)^{\beta-1} e^{-\left(\frac{y}{\alpha}\right)^\beta} & y > 0, \\ 0 & \text{otherwise.} \end{cases}$$

We find its distribution function by integration and obtain

$$F_Y(y) = 1 - e^{-\left(\frac{y}{\alpha}\right)^\beta}, \quad y > 0.$$

Using the gamma function defined earlier, we can show that $E(Y) = \alpha\Gamma(\frac{1}{\beta}+1)$ and $E(Y^2) = \alpha^2\Gamma(\frac{2}{\beta}+1)$.

The above identity is useful for parameter estimation of the Weibull random variables by the *method of moment*. If we know the first two moments, we can first estimate β, and then α separately. This is demonstrated in Illustration 3.   □

The Hazard Rate Function. For a Weibull random variable with location parameter $v = 0$ and parameters $\alpha > 0$ and $\beta > 0$, the hazard rate function is given by

$$\lambda(t) = \frac{f(t)}{1 - F_Y(y)} = \left(\frac{\beta}{\alpha}\right)\left(\frac{t}{\alpha}\right)^{\beta-1}.$$

We see that when $\beta = 1$, the random variable Y reduces to an exponential and it has a constant hazard rate. It is easy to verify that when $\beta > 1$, then the hazard rate is increasing in t. On the other hand, when $0 < \beta < 1$, the hazard rate is decreasing in t.

In Illustration 3, we give a parameter estimation example involving weib$(0, \alpha, \beta)$.

6.6. Normal and Lognormal Random Variables

The normal distribution is perhaps one of the most well-known distributions — due partly to its connection with the central limit theorem, an important subject we will study later. Random variable Z follows a standard normal distribution with parameters $\mu = 0$ and $\sigma = 1$, i.e., $Z \sim N(0, 1)$, if it has the density

$$f_Z(z) = \frac{1}{\sqrt{2\pi}} \exp\left(-\frac{z^2}{2}\right), \quad -\infty < z < \infty.$$

We derive its characteristic function as follows:

$$\varphi_Z(s) = E(e^{isZ}) = \int_{-\infty}^{\infty} e^{isz} f_Z(z) dz$$

$$= \int_{-\infty}^{\infty} e^{isz} \frac{1}{\sqrt{2\pi}} \exp\left(-\frac{z^2}{2}\right) dz$$

$$= \int_{-\infty}^{\infty} \frac{1}{\sqrt{2\pi}} e^{-\frac{1}{2}(z^2 - 2isz + (i^2 s^2))} e^{\frac{1}{2}i^2 s^2} dz \quad \text{complete the squares}$$

$$= e^{-\frac{1}{2}s^2} \int_{-\infty}^{\infty} \frac{1}{\sqrt{2\pi}} e^{-\frac{1}{2}(z - iu)^2} dz \quad \text{integral integrates to 1}$$

$$= e^{-\frac{1}{2}s^2}.$$

In Illustration 4, we use Mathematica to do the same and produces $E(Z) = 0$ and $\text{Var}(Z) = 1$ as expected. The commonly adopted notations for PDF and CDF of $Z \sim (0, 1)$ are $\phi(z)$ and $\Phi(z)$, respectively. We now define $X = \mu + \sigma Z$, where μ and σ are two constants. We use (6.1) to find the characteristic function for random variable X

$$\varphi_X(s) = e^{-i\mu s} \varphi_Z(\sigma s) = e^{i\mu s - \frac{1}{2}\sigma^2 s^2}.$$

Since $E(X) = \mu$ and $\text{Var}(X) = \sigma^2$. We immediately infer that $X \sim N(\mu, \sigma^2)$. A normal random variable is sometimes called the Gaussian random variable. For many, the density is known as the "bell-shaped" curve. It is symmetric around its mean μ.

When $X \sim N(\mu, \sigma^2)$, the density reads

$$f_X(x) = \frac{1}{\sqrt{2\pi\sigma^2}} e^{-\frac{1}{2\sigma^2}(x-\mu)^2}, \quad -\infty < x < \infty.$$

The transformation $Z = (X - \mu)/\sigma$ is called *standardization*.

The Error Function and $\Phi(z)$. When the upper limit of integration of the gamma function defined earlier is a given constant x (instead of infinity), then the function is called the *incomplete* gamma function. The *error function* $erf(z)$ is a special case of the incomplete gamma function. It is defined by

$$\mathrm{erf}(z) = \frac{2}{\sqrt{\pi}} \int_0^z e^{-x^2} dz.$$

The error function is related to the CDF of $N(0, 1)$ in the following way

$$\Phi(z) = \frac{1}{2} + \frac{1}{2}\mathrm{erf}\left(\frac{z}{\sqrt{2}}\right).$$

Sum of Independent Normals. For each i, we assume that $X_i \sim N(\mu_i, \sigma_i^2)$. Define

$$X = \sum_{i=1}^n X_i.$$

If $\{X_i\}_{i=1}^n$ are independently distributed random variables and $W = X_1 + \cdots + X_n$, then

$$\varphi_X(s) = \prod_{i=1}^n \exp\left(i\mu_i s - \frac{\sigma_i^2 s^2}{2}\right).$$

Then, it follows that

$$E(X) = \sum_{i=1}^n \mu_i \quad \text{and} \quad \mathrm{Var}(X) = \sum_{i=1}^n \sigma_i^2$$

provided that $\{X_i\}$ are mutually independent. In Illustration 5, for the case when $n = 2$, we show the use of Mathematica in verifying the above assertions.

Lognormal Distribution. If $X \sim N(\mu, \sigma^2)$ and $Y = g(X) = e^X$, then Y is called the lognormal distribution. We use $Y \sim \text{lognormal}(\mu, \sigma^2)$ to denote it. In the following, we derive the PDF for Y. Since

$$x = g^{-1}(y) = \ln(y) \equiv h(y),$$

we have

$$\frac{d}{dy} h(y) = \frac{1}{y}.$$

Therefore

$$f_Y(y) = f_X(h(y)) \left| \frac{d}{dy} h(y) \right|$$

$$= \frac{1}{\sqrt{2\pi\sigma^2}} \exp\left(-\frac{1}{2\sigma^2} (h(y) - \mu)^2 \right) \frac{1}{y}$$

$$= \frac{1}{\sqrt{2\pi}\sigma y} \exp\left(-\frac{1}{2\sigma^2} (\ln(y) - \mu)^2 \right), \quad y > 0.$$

The lognormal random variable Y is characterized by the parameters μ and σ^2 (yes, the parameters of its "parent"). In other words, if $Y \sim \text{lognormal}$ (μ, σ^2), then $X = \ln(Y)$ and $X \sim N(\mu, \sigma^2)$. We use Mathematica shown in Illustration 6 to evaluate $E(Y)$ and $E(Y^2)$. In each case, we employ the formula $E[g(X)]$. Then $\text{Var}(Y)$ is found using $E(Y^2) - E^2(Y)$. So we have

$$E(Y) = \exp\left(\mu + \frac{1}{2}\sigma^2 \right) \tag{6.3}$$

and

$$\text{Var}(Y) = (\exp(2\mu + \sigma^2))(e^{\sigma^2} - 1). \tag{6.4}$$

In Illustration 6, we show that $E(Y)$ and $\text{Var}(Y)$ are indeed so.

Example 7 (Use of Lognormal to Model Stock Price in Finance).
Using the lognormal distribution to model stock price has a long history in finance. Let S_i denote the price of a stock on day i. Let

$$X_i \equiv \ln\left(\frac{S_i}{S_{i-1}} \right) \approx \left(\frac{S_i}{S_{i-1}} - 1 \right) = \frac{S_i - S_{i-1}}{S_{i-1}}.$$

Then X_i represents the approximate return of the of the stock on day i.

Let

$$X = X_1 + \cdots + X_n = \text{ the sum of the returns over a period of } n \text{ days.}$$

Empirical finance has shown that in many cases the $\{X_i\}$ can be approximated by normal distributions. Moreover, $\{X_i\}$ are mutually independent. Hence, the return of a stock over a given period follows a normal distribution. Now

$$X = \sum_{i=1}^{n} \ln\left(\frac{S_i}{S_{i-1}}\right)$$

$$= \ln\left(\frac{S_1}{S_0} \times \frac{S_2}{S_1} \times \cdots \times \frac{S_n}{S_{n-1}}\right)$$

$$= \ln\left(\frac{S_n}{S_0}\right).$$

Upon exponentiating the above, we find

$$e^X = \frac{S_n}{S_0} \quad \text{or} \quad S_n = S_0 e^X. \tag{6.5}$$

Thus, the price of the stock in an interval of length n follows the lognormal distribution with parameter μ and σ^2, where last two parameters are the mean and variance of the normal return over the corresponding period and S_0 is the initial stock price. $\square$

Example 8 (Price Movement of Stock in Continuous-Time Finance). In continuous-time mathematical finance, one paradigm for modeling the movement of stock price S_t over an interval of length T is by assuming that S_t follows a geometric Brownian motion with parameters μ and σ, called drift and diffusion parameters, respectively. Under this paradigm, we write (6.5) as

$$S_T = S_0 e^{X_T},$$

where X_T follows a normal distribution with $E(S_T) = (\mu - \frac{\sigma^2}{2})T$ and $\text{Var}(S_T) = \sigma^2 T$. Here X_T plays the role of *return* over the interval $(0, T)$.

Since

$$\log S_T = \log S_0 + X_T,$$

we have

$$\mu \equiv E[\log S_T] = \log S_0 + \left(\mu - \frac{\sigma^2}{2}\right)T, \tag{6.6}$$

$$\sigma^2 \equiv Var[\log S_T] = \sigma^2 T. \tag{6.7}$$

Since $e^{\log S_T} = S_T$, the marginal distribution at time T follows a lognormal distribution with parameters (μ, σ^2). By substituting (6.6) and (6.7) in (6.3) and (6.4), we find

$$E(S_T) = \exp\left[\log S_0 + \left(\mu - \frac{\sigma^2}{2}\right)T + \frac{1}{2}\sigma^2 T\right] = S_0 e^{\mu T}$$

and

$$\begin{aligned}
\mathrm{Var}(S_T) &= \left[\exp\left(2\left(\log S_0 + (\mu - \frac{\sigma^2}{2})T\right) + \sigma^2 T\right]\left[e^{\sigma^2 T} - 1\right]\right. \\
&= [\exp(2\log S_0 + 2\mu T]\left[e^{\sigma^2 T} - 1\right] \\
&= S_0^2 e^{2\mu T}\left[e^{\sigma^2 T} - 1\right].
\end{aligned}$$

$\square$

Example 9. Consider a given stock whose current price is 20. Assume that the return over the next year follows a normal distribution with mean $\mu = 0.20$ variance $\sigma^2 = 0.4^2$. Then the price a year from now, denoted by S_1, follows a lognormal distribution with

$$E(S_1) = S_0 e^{\mu T} = 20 e^{0.20(1)} = 24.43$$

and

$$\mathrm{Var}(S_1) = S_0^2 e^{2\mu T}\left[e^{\sigma^2 T} - 1\right] = (20)^2 e^{2(0.2)(1)}\left(e^{0.16(1)} - 1\right) = 103.54.$$

$\square$

6.7. More Continuous Random Variables and Variance Gamma

There are a few other continuous random variables that are of common interest. We will introduce them as follows.

Beta Random Variables. Let X be a beta random variable with parameters a and b if it assumes the following density:

$$f_X(x) = \begin{cases} \dfrac{1}{B(a,b)} x^{a-1}(1-x)^{b-1} & 0 < x < 1, \\ 0 & \text{otherwise,} \end{cases}$$

where

$$B(a,b) = \int_0^1 x^{a-1}(1-x)^{b-1} dx$$

is called the complete beta function. We use beta(a, b) to denote the PDF for random variable X.

The beta function is related to the gamma function shown earlier through

$$B(a,b) = \frac{\Gamma(a)\Gamma(b)}{\Gamma(a+b)}.$$

We see that

$$\begin{aligned} \frac{B(a+1,b)}{B(a,b)} &= \frac{\Gamma(a+1)\Gamma(b)}{\Gamma(a+b+1)} \frac{\Gamma(a+b)}{\Gamma(a)\Gamma(b)} \quad \text{by definition of } B(a,b) \\ &= \frac{a\Gamma(a)\Gamma(b)}{(a+b)\Gamma(a+b)} \frac{\Gamma(a+b)}{\Gamma(a)\Gamma(b)} \quad \text{as } \Gamma(a+1) = a\Gamma(a) \\ &= \frac{a}{a+b}. \end{aligned}$$

Using the above, we compute the first moment of beta as follows:

$$\begin{aligned} E(X) &= \frac{1}{B(a,b)} \int_0^1 x^a(1-x)^{b-1} dx \\ &= \frac{B(a+1,b)}{B(a+b)} \\ &= \frac{a}{a+b}. \end{aligned}$$

We see that when $a = b$, then $E(X) = \frac{1}{2}$. Similarly, we can obtain the second moment

$$E(X^2) = \frac{(a+1)a}{(a+b+1)(a+b)}$$

and the variance can be found from $E(X^2) - E^2(X)$.

The beta random variable assumes values in the interval between 0 and 1. It is frequently used in modelling fractions in Bayesian statistical analysis. Also it finds its uses in survey sampling.

Pareto Random Variables. The Pareto distribution was named after Vilfredo Pareto, an Italian economist. It is used to describe the wealth of a society — particularly the so-called "80-20 rule", namely, 80% of wealth of a society is held by 20% of its population.

If X is an exponential random variable with parameter $\lambda > 0$ and $a > 0$ is a given constant. Then $Y = ae^X$ is said to follow a Pareto density with parameters λ and a. The Pareto density is denoted by Pareto(λ, a). We see

$$P(Y > y) = P\left(e^X > \frac{y}{a}\right)$$
$$= P\left(X > \ln\left(\frac{y}{a}\right)\right)$$
$$= \exp\left(-\lambda \ln\left(\frac{y}{a}\right)\right)$$
$$= \left(\frac{a}{y}\right)^\lambda.$$

As a consequence, we obtain

$$F_Y(y) = 1 - \left(\frac{a}{y}\right)^\lambda, \quad y > a$$

and

$$f_Y(y) = \begin{cases} \lambda a^\lambda \dfrac{1}{y^{\lambda+1}} & y > a, \\ 0 & \text{otherwise.} \end{cases}$$

When $\lambda \le 1$, we see that $F_Y(\infty) < 1$ and hence $E(X) = \infty$. For $\lambda > 1$, we have

$$E(Y) = \int_0^\infty P(Y > y)dy = \int_0^a 1 \cdot dy + \int_a^\infty \left(\frac{a}{y}\right)^\lambda dy = a + \frac{a}{\lambda - 1} = \frac{a\lambda}{\lambda - 1}.$$

Consider the case when $\lambda > 2$. We apply the identity introduced earlier for finding the second moment of a positive random variable to the Pareto random variable:

$$E(Y^2) = \int_0^\infty 2\tau P(Y > \tau) d\tau = \int_0^a 2x dx + \int_a^\infty 2\tau \left(\frac{a}{\tau}\right)^\lambda d\tau$$

$$= a^2 + \frac{2a^2}{\lambda - 2} = \frac{\lambda a^2}{\lambda - 2}.$$

Hence, for $\lambda > 2$, its variance is given by

$$\text{Var}(Y) = \frac{\lambda a^2}{(\lambda - 2)(\lambda - 1)^2}.$$

Chi-Square Random Variables. The chi-square random variable Y with one degree of freedom, commonly stated as χ_1^2 random variable, is related to the standard normal random variable Z by

$$Y = Z^2.$$

From Example 2, we know

$$f_Y(y) = \frac{1}{2\sqrt{y}} \left(f_Z(\sqrt{y}) + f_Z(-\sqrt{y})\right)$$

$$= \frac{1}{2\sqrt{y}} \left(\frac{1}{\sqrt{2\pi}} e^{-\frac{1}{2}y} + \frac{1}{\sqrt{2\pi}} e^{-\frac{1}{2}y}\right)$$

$$= \frac{1}{2\sqrt{y}} \frac{2}{\sqrt{2\pi}} e^{-\frac{1}{2}y}$$

$$= \frac{\frac{1}{2} \exp\left(-\frac{1}{2}y\right) \left(\frac{1}{2}y\right)^{\frac{1}{2}-1}}{\Gamma\left(\frac{1}{2}\right)} \qquad \Gamma\left(\frac{1}{2}\right) = \sqrt{\pi}.$$

W see that $\chi_1^2 \sim \text{gamma}(\alpha, \lambda)$, where $\alpha = \frac{1}{2}$ and $\lambda = \frac{1}{2}$.

Let $Y = X_1^2 + \cdots + X_n^2$, where $\{X_i\}_i^n$ are independently identically distributed random variables with $X_1 \sim N(0, 1)$. Then Y is called a chi-square random variable with n-degrees of freedom. It is denoted by χ_n^2. We will give details about this generalization in the next chapter.

A Variant. Assume now random variable $X \sim N(0, \sigma^2)$. Define $W = X_1^2$. This will give a generalization of the above χ_1^2 random variable. It follows that

$$
\begin{aligned}
f_W(w) &= \frac{1}{2\sqrt{w}} \frac{2}{\sqrt{2\pi}\sigma} e^{-\frac{1}{2\sigma^2}w} \\[2mm]
&= \frac{1}{2^{\frac{1}{2}}\sigma \Gamma\left(\frac{1}{2}\right)} w^{\frac{1}{2}-1} e^{-\frac{1}{2\sigma^2}w} \\[2mm]
&= \frac{1}{\sqrt{\frac{w}{2\sigma^2}}(2\sigma^2)^{\frac{1}{2}} \cdot 2^{\frac{1}{2}}\sigma} e^{-\frac{1}{2\sigma^2}w} \quad \text{for matching with gamma} \\[2mm]
&= \frac{1}{\Gamma\left(\frac{1}{2}\right)}(2\sigma^2)^{-1} \left(\frac{1}{2\sigma^2}w\right)^{\frac{1}{2}-1} e^{-\frac{1}{2\sigma^2}w}.
\end{aligned}
$$

Thus we conclude that W follows a gamma distribution with parameters $\alpha = \frac{1}{2}$ and $\lambda = \frac{1}{2\sigma^2}$.

The F Distribution. A random variable X follows an F distribution, with m and n degrees of freedom, if

$$
X = \frac{Y_1/m}{Y_2/n},
$$

where $Y_1 \sim \chi_m^2$, $Y_2 \sim \chi_n^2$, and $Y_1 \perp Y_2$. We use $F(m, n)$ to denote the F distribution. It plays a prominent role in statistics — specially in analysis of variance. In Section 7.6 of Chapter 7, we will display its somewhat messy PDF and show its simple derivation using the Mellin transform.

Cauchy Random Variables. A standard Cauchy random variable, denoted by $C(0, 1)$, assumes the following density

$$
f_X(x) = \frac{1}{\pi} \frac{1}{1 + x^2}, \quad -\infty < x < \infty.
$$

A Cauchy PDF is displayed in Illustration 7. This density is unusual in that its moments $\{E(X^k)\}$ do not exist but has its place in financial applications. A transformation with $Y = 1/X$ will also show that $Y \sim C(0, 1)$.

We remark that all the even moments of the Cauchy random variable are infinite and that its moment generating function does not exist. However, its characteristic function does exist and is given by

$$\varphi_X(t) = e^{-|t|}, \quad t \in \mathbb{R}.$$

Its derivation is given in Illustration 7. A generalized version of the Cauchy has the density

$$f_C(x) = \frac{1}{\pi} \frac{a}{a^2 + (x - m)^2}, \quad -\infty < x < \infty.$$

It is denoted by $C(m, a)$.

Example 10. Assume that Y is uniformly distributed between $-\pi$ and π. Let $X = \cot(Y)$. We show that $X \sim C(0, 1)$, namely, X follows the Cauchy distribution.

Solution. We have $x = g(y) = \cot(y)$ and $y = g^{-1}(x)$. For $y > 0$, this is one-to-one transformation; for $y < 0$, it is another one-to-one transformation. Moreover, we have

$$f_Y(y) = \frac{1}{2\pi} \quad -\pi < y < \pi$$

and

$$\left| \frac{d}{dx} \cot(x) \right| = \frac{1}{1 + x^2}.$$

Thus, we conclude that

$$f_X(x) = f_Y(g^{-1}(x)) \mathbf{1}_{\{y>0\}} \left| \frac{d}{dx} \cot(x) \right| + f_Y(g^{-1}(x)) \mathbf{1}_{\{y<0\}} \left| \frac{d}{dx} \cot(x) \right|$$

$$= \frac{1}{2\pi} \frac{1}{1 + x^2} + \frac{1}{2\pi} \frac{1}{1 + x^2} = \frac{1}{\pi} \frac{1}{1 + x^2}, \quad -\infty < x < \infty. \qquad \square$$

Variance Gamma Distribution. Let $X \sim \text{gamma}(\alpha, \lambda)$ and $Y \sim \text{gamma}(\alpha, \gamma)$. Assume that $X \perp Y$. We now define $W = X - Y$. Now (6.2) implies that the CFs for the two gamma random variables are, respectively,

$$\varphi_X(s) = \left(1 - \frac{is}{\lambda}\right)^{-\alpha}, \quad \varphi_Y(s) = \left(1 - \frac{is}{\gamma}\right)^{-\alpha}.$$

Applying (6.1) yields

$$\varphi_{-Y}(s) = \left(1 + \frac{is}{\gamma}\right)^{-\alpha}.$$

Since W is the convolution of X and $-Y$, the characteristic function of W is given by

$$\varphi_W(s) = \left(1 - \frac{is}{\lambda}\right)^{-\alpha} \left(1 - \frac{is}{\gamma}\right)^{-\alpha}$$

$$= \left(\frac{\gamma\lambda}{\gamma\lambda + (\lambda - \gamma)i\,s + s^2}\right)^{\alpha}.$$

W is known as the variance gamma (VG) random variable with parameters α, λ, and γ. Using the above we can easily produce $E(X)$ and $\text{Var}(X)$. This is shown in Illustration 8. Moreover, we can show that the kurtosis of W is greater than 3. Thus it exhibits a heavy tail. When $\alpha = 1$ and $\gamma = \lambda$, it reduces to a doubled-sided exponential distribution.

According to Schouten [2], a VG distribution would fit historical daily log-returns better than a lognormal distribution. Hence it has become one of the favorite choices in mathematical finance (e.g., see [2]). □

6.8. Summary

The exponential random variable and its relatives are summarized in the following figure. We can see that the random variables related to the exponential constitute a large family.

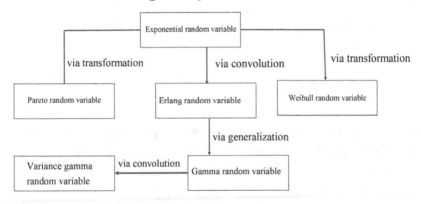

The normal random variable and its descendants are shown below:

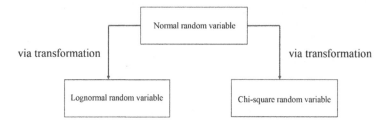

We remark that the chi-square distribution is a distant relative of the gamma family.

6.9. Exploring with Mathematica

Illustration 1. We use the functions for Laplace transform and Inverse Laplace Transform of **Mathematica** to find the PDF, mean, and variance of X, where $X \sim \text{expo}(\lambda)$.

```
fe[s_] = LaplaceTransform[(λ Exp[-λt]), t, s]
```
$$\frac{\lambda}{s+\lambda}$$

```
InverseLaplaceTransform[(fe[s])ⁿ, s, t]
```
$$\frac{e^{-t\lambda} t^{-1+n} \lambda^n}{\text{Gamma}[n]}$$

```
fX[t]    (*   The above is the PDF  of  gamma(n,λ)  *)
```
$$\lambda^{\alpha} (-i z + \lambda)^{-\alpha}$$

```
fc[s_] := (──────)ⁿ;    (*   The n-fold covolution of  f  with itslef *)
           s + λ                                        xᵢ
```
$$fc[s\_] := \left(\frac{\lambda}{s+\lambda}\right)^n;$$

```
EX = -fc'[0]
```
$$\frac{n}{\lambda}$$

```
EX2 = Simplify[fc''[0]];
Var[X] = Simplify[EX2 - EX^2]
```
$$\frac{n}{\lambda^2}$$

□

Illustration 2. We use integration to find CF of $\text{gamma}(\alpha, \lambda)$. Then we solve for $E(X)$ and $Var(X)$. We then display the shapes of $\text{gamma}(\alpha, \lambda)$ while varying the values of the shape parameter α.

```
fn[t_] := λ Exp[-λ t] (λ t)^(α-1) / Gamma[α];    ( * PDF of gamma (α, λ)

g[z_] := Integrate[Exp[i z t] fn[t], {t, 0, ∞}, Assumptions →
    Re[α] > 0 && Im[z] + Re[λ] > 0]

g[z]    (*    This is the CF  of  X  *)
```

$$\lambda^\alpha \, (-i \, z + \lambda)^{-\alpha}$$

```
       1
EX = ── g'[0]
       i
```

$$\frac{\alpha}{\lambda}$$

```
EX2 = Simplify[ 1/i² g''[0] ]
```

$$\frac{\alpha \, (1 + \alpha)}{\lambda^2}$$

```
VarX = Simplify[EX2 - EX²]
```

$$\frac{\alpha}{\lambda^2}$$

□

Note in Mathematica, a gamma distribution is stated as gamma(α, β), whereas in this text we use the convention gamma(α, λ). They are related by $\beta = 1/\lambda$.

```
Plot[Table[PDF[GammaDistribution[α, 3], x], {α, {1, 4, 6}}]
    // Evaluate, {x, 0, 40}, Filling → Axis]
```

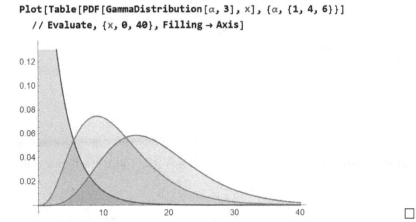

□

The three curves are for gamma(1, 1/3), gamma(4, 36), and gamma (6, 54). Let the doublets denote $(E(X), \mathrm{Var}(X))$. Then they respectively have values (3, 9), (12, 36), and (18, 54).

Illustration 3 (Parameter Estimation of Weibull Parameters). Assume that the certain type of insurance claims occur in a month has $E(Y) = 46,000$ and $\mathrm{Var}(Y) = (12,500)^2$. Assume that the insurance company uses the Weibull distribution to model the monthly claim size. Note

$$\frac{\Gamma(\frac{2}{\beta}+1)}{\Gamma^2(\frac{1}{\beta}+1)} = \frac{\text{Var}(Y)+E^2(Y)}{E^2(Y)}.$$

We see that the term α^2 gets cancelled on the left side of the above. With the right side value of the above given, we can find an approximate numerical value of β using any search scheme. Once we know β, we can recover the value of α using the formula for $E(Y)$. For this example, we find $\alpha = 50648$ and $\beta = 4.14346$. The details are given below.

```
EY = 46000;  VarY = 12500²;  Ratio = (VarY + EY²) / EY² ;

FindRoot[Gamma[2.0/β + 1] / (Gamma[1.0/β + 1.0])² - Ratio == 0, {β, 1}]

{β → 4.14346}

FindRoot[α Gamma[1/4.14346 + 1] - EY == 0, {α, 1}]

{α → 50648.}

Clear[α, β]

α = 50648;  β = 4.14346;

fn[y_] := (β/α) (y/α)^(β-1) Exp[-(y/α)^β];

Plot[fn[y], {y, 0, 90000}, AxesLabel → {"y", "f(y)"}, AxesStyle → {Black, Medium},
  LabelStyle → {Black, Medium}, PlotStyle → {Black, Medium}]
```

Illustration 4. Finding the CF of $Z \sim N(0,1)$ and its mean and variance.

```
f[z_] := 1/√(2π) Exp[-z²/2];

φ[s_] = FourierTransform[f[z], z, s, FourierParameters → {1, 1}]

e^(-s²/2)

EZ = 1/i φ'[0]

0

VarZ = 1/i² φ''[0] - EZ²

1
```

Illustration 5. Let $X_1 \sim N(\mu_1, \sigma_1^2)$ and $X_2 \sim N(\mu_2, \sigma_2^2)$. Assume $X = X_1 + X_2$ and X_1 and X_2 are independent. The following condition shows that $X \sim N(\mu_1 + \mu_2, \sigma_1^1 + \sigma_2^2)$.

```
f1[x_] := 1/Sqrt[2 π σ1²] Exp[- (x - μ1)²/(2 σ1²)];

g1[s_] := FourierTransform[f1[x], x, s, FourierParameters → {1, 1}];

f2[x_] := 1/Sqrt[2 π σ2²] Exp[- (x - μ2)²/(2 σ2²)];

g2[s_] := FourierTransform[f2[x], x, s, FourierParameters → {1, 1}]

φ[s_] := g1[s] * g2[s];

FullSimplify[InverseFourierTransform[φ[s], s, x,
    FourierParameters → {1, 1}], σ1 > 0 && σ2 > 0]
```

$$\frac{e^{-\frac{(-x+\mu_1+\mu_2)^2}{2\left(\sigma_1^2+\sigma_2^2\right)}}}{\sqrt{2\pi}\,\sqrt{\sigma_1^2+\sigma_2^2}}$$

☐

Illustration 6. Derivations of $E(Y)$ and $\text{Var}(Y)$ where $Y \sim$ lognormal (μ, σ^2).

```
fn[x_] := 1/Sqrt[2 π σ] Exp[- 1/(2σ²) (x - μ)²];

EY = Integrate[Exp[x] fn[x], {x, -∞, ∞}, Assumptions → σ > 0]
```

$$e^{\mu + \frac{\sigma^2}{2}}$$

```
EY2 = Integrate[(Exp[x])² fn[x], {x, -∞, ∞}, Assumptions → σ > 0];

VarY = EY2 - EY²
```

$$e^{2\left(\mu+\sigma^2\right)} - e^{2\mu+\sigma^2}$$

☐

Illustration 7. A PDF for a two-sided Cauchy random variable:

```
f2[x_] := 1/(π (1 + x²))        (* Two sided Cuachy distribution *)

Plot[f2[x], {x, -5, 5}]
```

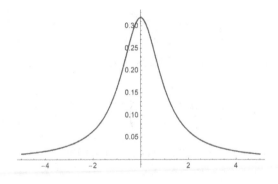

☐

The derivation of the characteristic functions for Cauchy random variable is given below:

$$\int_{-\infty}^{\infty} \texttt{Exp[i t x]} \; \frac{1}{\pi \, (1 + x^2)} \; \texttt{d} x$$

$$\texttt{ConditionalExpression} \left[e^{-\texttt{Abs}[t]}, \; t \in \mathbb{R} \right]$$

□

Illustration 8. The CF for variance gamma distribution with parameters α, λ, and γ and its mean and variance:

$$\varphi[\texttt{s\_}] = \left(\frac{\gamma \lambda}{\gamma \lambda + (\lambda - \gamma) \; \texttt{i} \, s + s^2} \right)^{\alpha};$$

$$\texttt{EX} = \frac{1}{\texttt{i}} \, \varphi'[0]$$

$$-\frac{\alpha \, (-\gamma + \lambda)}{\gamma \lambda}$$

$$\texttt{EX2} = \texttt{Simplify} \left[\frac{1}{\texttt{i}^2} \, \varphi''[0] \right];$$

$$\texttt{VarX} = \texttt{Simplify} \left[\texttt{EX2} - \texttt{EX}^2 \right]$$

$$\alpha \left(\frac{1}{\gamma^2} + \frac{1}{\lambda^2} \right)$$

□

Problems

1. Let X be a random variable with the following density

$$f_X(x) = \begin{cases} \frac{1}{18}(3 + 2x) & \text{if } 2 < x < 4, \\ 0 & \text{otherwise.} \end{cases}$$

(a) Find $P(2 < X < 3)$. (b) Find the CDF of X. (c) Finf $E(X)$.

2. (a) Let $X \sim U(0,1)$ and $a > 0$. Let

$$Y = \frac{aX}{1 - X}.$$

Find $F_Y(y)$. (b) Let $\{Y_i\}_{i=1}^n$ be iid rvs and $W = \min(Y_1, \ldots, Y_n)$. Show that

$$f_W(w) = \frac{n}{a} \left(\frac{a}{w + a} \right)^{n+1}, \quad y > 0,$$

and find $E(Y)$. (c) Let $Z = \max(Y_1, \ldots, Y_n)$. Find $F_Z(z)$.

3. Let $X \sim N(0,1)$. Find the PDF of $Y = |X|$.

4. Let $X \sim U(0,1)$. Define $Y = X^2$. Find $f_Y(y)$.

5. Let $X \sim U(0,1)$. Define $Y = -\log(X)$. Find $f_Y(y)$.

6. Consider the random variable X with PDF

$$f_X(x) = \frac{2}{9}(x+1), \quad -1 < X < 2.$$

Let $Y = X^2$. Find $F_Y(y)$.

7. Assume $X \sim \text{expo}(\alpha)$. Let Y be the integer part of X, i.e., $Y = [X]$. Let W be the fractional part of X, i.e., $W = X - [X]$. (a) Find the PMF of X, i.e., find $p_Y(y)$. (b) Find the PDF of $f_W(w)$. (c) Are Y and W independent random variables?

8. The amount of time needed to wash a car at a car-washing station is exponentially distributed with an expected value of 15 minutes. You arrive at the car-washing station while it is occupied and one other car is waiting for a wash. The owner of the car-washing station informs you that the car in the washing station has already been there for 10 minutes. (a) What is the probability that the car in the washing station will need no more than 5 minutes extra? (b) What is the probability that you have to wait more than 20 minutes before your car can be washed.

9. Suppose that the number of miles that a car run before its battery wears out is exponentially distributed with an average value of 10,000 miles. (a) If a person wants to take a 5000-miles trip with a new battery, what is the probability that the person will be able to complete the trip without having to replace the battery? (b) Given that the battery has already gone through 5000 miles, it the person wants to take a 7000-mile trip, what is the probability that the person is able to complete the trip without having to replace the battery?

10. Best Byte sells a warranty contract to its buyers of laptop computers. It will pay $500 if a laptop fails in the first years of usage, and pay $200 if the laptop fails in the second year of usage. Assume that the lifetime X of a laptop has a hazard rate function $\lambda_X(t) = \frac{1}{4}t$ for $t > 0$. (a) What is the distribution function $F_X(t)$ of X. (b) What is the probability density function $f_X(x)$ of X. (c) Find $E(X)$. (d) What is this random variable X? (e) What is expect cost of the warranty per each laptop?

11. Let X be a random variable with the density $f_X(x) = \frac{\lambda}{2}e^{-\lambda|x|}$.(a) Find $E(X)$. (b) Find $Var(X)$.(c) What kind of random variable is X?

12. Let X denote the lifetime in hours of a computer chip. Assume X has the density $f_X(x)$, where

$$f_X(x) = \begin{cases} \frac{100}{x^2} & \text{if } x > 100, \\ 0 & \text{if } x \leq 0. \end{cases}$$

Assume that an equipment contains three such chips and the lifetimes of the chips are independent. If one chip fails, then the equipment will be down. What is the probability that the equipment will be down after 1000 hours?

13. Let X be the lifetime of a piece of equipment. Assume the hazard rate function of X is

$$\lambda(t) = \frac{t}{1+t}, \quad t > 0.$$

(a) What is the CDF of X? (b) What is the conditional probability $P(X > s + t | X > t)$.

14. The *mean residual life* of a positive and continuous random variable X is defined by

$$m(t) = E(X - t | X > t).$$

(a) Express $m(t)$ as a function of the PDF $f_X(t)$ and CDF $F_X(t)$ of X.
(b) Express $m(t)$ as a function of the CDF $F_X(t)$ of X.

15. Consider the lifetime X (in years) of a drilling bit has a hazard rate function $\lambda_X(t) = \frac{1}{5}\sqrt{t}$ for $t > 0$. Assume that the drilling bit has already lasted for three years. What is the mean residual life of this drilling bit?

16. Let X be the lifetime of a camera for a home-security system measured in hours. Assume that X follows a Weibull distribution with shape parameter $\beta = 2.5$ and scale parameter $\alpha = 1250$. What is the mean residual life of the camera given that it has already survived 500 hours.

17. A medical experts in a paternity suit testifies that the length (in days) of human gestation is approximately normally distributed with parameters $\mu = 272$ and $\sigma^2 = 95$. The defendant in the suit is able to prove that he

was out of the country during the period that began 295 days before the birth of the child and ended 250 days before the birth. If the defendant was, in fact, the father of the child, what is the probability that the mother could have had the very long or very short gestation indicated by the testimony?

18. Consider light bulbs, produced by a machine, whose lifetime X in hours is a random variable obeying an exponential distribution with mean lifetime of 1000 hours. (a) What is the probability that a randomly selected light bulb will have a lifetime greater than 1020 hours. (b) Find the probability that a sample of 100 bulbs selected at random from the output of the machine will contain between 30 and 40 bulbs with a lifetime greater than 1020 hours.

19. Let X be the sedimentation rate at a given stage of pregnancy. Assume that X follows a gamma distribution with $\alpha = 5$ and $\lambda = 0.1$. (a) Find $E(X)$ and $\text{Var}(X)$ (b) $P(10 < X < 90)$. (c) Use Mathematica to verify the result obtained in Part (b).

20. We use a Weibull distribution to model the effect of advertising. Let X be the length of time (in days) after the end of an advertising campaign that a person is able to remember the name of the product being advertised, say a cereal called Los Angles Treat. Assume the parameters have been estimated from empirical data. They are $\beta = 0.98, \alpha = 73.60$, and $v = 1.0$. (a) Find $E(X)$ and $\text{Var}(X)$. (b) Estimate the fraction of people who are expected to remember the name of the brand 20 days after the cessation of the advertisement.

21. Let X be the incubation time in years between infection with HIV virus and the onset of full-blown AIDS. Assume that X follows a Weibull distribution with parameters $v = 0$, $\beta = 2.396$, and $\alpha = 9.2851$. (a) Find $P(X > 10)$. (b) Find $E(X)$.(c) Find $P(10 < X < 20)$. Do the problem in Mathematica.

22. Let X denote the fire losses per incidence in thousands of dollars. Assume that X follows a lognormal distribution with mean 25.2 and variance 100. (a) Find the parameters μ and σ of the lognormal distribution. (b) Find $P(X > 30)$.

23. For $i = 1$ and 2, we let X_i be a lognormal random variable with parameters μ_i and σ_i. Assume $X_1 \perp X_2$. Let $Y = X_1 X_2$. What is the distribution for the random variable Y.

24. Let Z be a normal random variable with parameters $\mu = 0$ and $\sigma = 1$. (a) Find the characteristic function of $Y = Z^2$. (b) Find $E(Y)$ and Var(Y).

25. Let $\{Z_i\}_{i=1}^{n}$ be n mutually independent random variable with a common density $N(0,1)$. Define $Y = Z_1 + \cdots + Z_n$. (a) Find the characteristic function of Y. (a) Find $E(Y)$ and $E(Y)$.

26. Let $X \sim \text{expo}(1)$. Define $Y = \log X$. (a) Find the PDF $f_Y(y)$ of Y. (b) Find the characteristic functions $\varphi_Y(t)$ of Y. (c) Find $E(Y)$ and Var(Y). (d) Find the moment generating function of Y.

27. The Logistic Distribution. A random variable X follows the following PDF:

$$f_X(x) = \frac{e^{-x}}{(1 + e^{-x})^2}, \quad -\infty < x < \infty.$$

(a) Verify the $f_X(x)$ is indeed a PDF. (b) Plot the PDF. (c) Find $E(X)$ and Var(X).

28. Let A be the fraction of people who have been vaccinated against CV-19 by Country A, and B be that for Country B. According to some estimates, A follows beta$(4, 2)$ and B follows beta$(1, 3)$. (a) In terms of vaccination, which country is in a better shape? (b) Plot the CDFs of A and B.

29. Let X be a random variable with the following PDF:

$$f_X(x) = \frac{bx^{b-1}}{a^b}, \quad 0 < x < a,$$

where $a > 0$ and $b > 0$. (a) For $a = 10$, and $b = 0.7$, plot the PDF $f_X(x)$ and verify that it is a legitimate density. (b) Let $Y = \frac{1}{X}$. Find the PDF $f_Y(y)$. (c) Let $Z = \frac{X}{a}$. Find the PDF of $f_Z(z)$. How does it related to the PDF of beta(a, b). (d) Find $E(Z)$ and $E(Z^2)$.

30. Let X be a gamma random variable with parameters α and λ. Thus its PDF is given by

$$f_X(t) = \frac{\lambda e^{-\lambda t}(\lambda t)^{\alpha-1}}{\Gamma(\alpha)}, \quad x > 0.$$

(a) Find the hazard rate function $\lambda_X(t)$. (b) If $\lambda = 3$, $\alpha = 2$, plot $1/\lambda_X(t)$ for $t \in [0, 1]$. How would you speculate the general case when $\alpha > 1$? (c) If $\lambda = 3$, $\alpha = 0.5$, plot $1/\lambda_X(t)$ for $t \in [0, 1]$. How would you speculate the general case when $\alpha < 1$? (d) If $\alpha = 1$, what you can say about $\lambda_X(t)$?

31. Let continuous random variable $X \sim Pareto(\lambda, a)$. Find the conditional probability

$$P(X \geq c | X \geq b)$$

where $a \geq b \geq c$.

32. A student is taking a take-home exam. We assume that the time (in hours) needed to complete the take-home follows a Pareto distribution with parameters $\lambda = 2$ and $a = 1$. This implies that the minimum time to complete the exam is one hour. (a) Given that the student has already spent 3 hours on the exam, what is the expected length for the student to complete the exam. (b) Use Mathematica to verify your result. (c) Use Mathematica to compute $E(Y | Y \geq y)$ for $y \in \{3, 4, 5, 6, 7, 8\}$. (d) What observation can be drawn from Part (c).

33. Let the random variable $X = 10^Y$ where $Y \sim unif(0, 1)$. Find the PDF of X.

34. Let Y denote the livetime of a memory stick. Assume $Y \sim expo(\lambda)$, where $\lambda > 0$. Let $f_\Lambda(\lambda)$ denote the prior PDF of Λ denote the random variable denoting the unknown parameter λ. (a) How to obtain the posterior probability $f_{\Lambda|Y}(\lambda|y)$? (b) Assume that $\Lambda \sim unif(1, 1.5)$. What is the posterior PDF of Λ if we observe that $y = 0.72$? (c) What is the posterior mean of Λ given $y = 0.72$. (d) Give an interpretation of the result obtained in (b).

35. Let X be the lifetime of a microprocessor. Assume that its failure rate function is given by

$$\lambda_X(t) = \frac{\alpha \beta t^{\alpha-1}}{(1 + t^\alpha)} \quad \alpha > 0, \ \beta > 0$$

(a) Plot $\lambda_X(t)$ for the case when $\alpha = 1.5$ and $\beta = 3$. (b) What observations can be made from the shape of the plot. (c) Find the PDF of the lifetime X. (d) Plot the PDF of X.

36. Let X be the lifetime of a car battery. Assume that is failure rate function is given by

$$\lambda_X(t) = \frac{2\alpha(1 - t)(2t - t^2)^{\alpha-1}}{(1 - (2t - t^2)^\alpha)} \quad 0 < t < 1$$

where $0 < \alpha < 1$. (a) Plot $\lambda_X(t)$ for the case for $t \in [0,1]$ when $\alpha = 1.5$. (b) What observations can be made from the shape of the plot. (c) Find the CDF of the lifetime of X. (d) Plot the CDF of X.

Remarks and References

The use of transforms for working with probabilistic models have been elaborated in [3]. The variance gamma distribution is omnipresent in mathematical finance (e.g., see [1, 2, 4, 5]). The ease with which we introduce the variance gamma distribution demonstrates one of the advantages of introducing the notion of CFs.

[1] Barndorff-Nielsen, O. E. and Shiryaev, A., *Change of Time and Change of Measure*, 2nd edn., World Scientific, 2015.
[2] Cont, R. and Tankov, P., *Financial Modelling with Jump Processes*. Chapman & Hall/CRC, 2004.
[3] Kao, E. P., *An Introduction to Stochastic Processes*, Duxbury 1998, and Dover 2019.
[4] Madan, D. and Schoutens, W., *Applied Conic Finance*, Cambridge University Press, 2016.
[5] Schoutens, W., *Lévy Process in Finance: Pricing Financial Derivatives*, Wiley, 2003.

Chapter 7

Jointly Distributed Random Variables

7.1. Introduction and Distribution Functions

Jointly distributed random variables are ubiquitous in applications. As an example, we can have X denoting the height of a person and Y denoting the weight of the person. The (X, Y) represents a bivariate random variable. As another example, we can use X to denote the price of a given stock on a given day and Y to denote the corresponding volume. In general, a random variable of interest can be a vector of dimension n.

Assume X and Y are random variables. Many univariate random variables are resulting from transformations of the types $X + Y$, $X - Y$, $X \cdot Y$, X/Y, and their variants, e.g., a linear transformation. In the final analysis, characteristic functions could help in their derivations. But their specializations, e.g., to the case of positive continuous random variables, make other forms of transformations less involved. As an example, when $X \perp Y$ and both are positive continuous random variables, for finding the density of $X + Y$, using Laplace transform becomes handy. The same can be said about $X \cdot Y$ where a straightforward application of a Mellin transform will produce its density (the Mellin transform will be introduced later in this chapter).

Multivariate Distribution Functions. The first question is how to prescribe the probability distribution for such a multidimensional random variable. For simplicity, we will use a bivariate random variable (X, Y) to start our exposition. We call the term (X, Y) a random variable but in actuality it is a random vector.

The joint distribution function of (X, Y) is defined by

$$F_{X,Y}(x, y) = P(X \leq x, Y \leq y), \quad -\infty < x, y < \infty.$$

When we set one of the arguments to infinity, it yields

$$F_{X,Y}(x, \infty) = P(X \le x, Y \le \infty) = P(X \le x) = F_X(x)$$

and

$$F_{X,Y}(\infty, y) = P(X \le \infty, Y \le y) = P(Y \le y) = F_Y(y).$$

In the context of jointly distributed random variables, the single-variate distribution functions are called the *marginal* distribution functions. If we know the joint distribution function of (X, Y), then we know everything about X and Y — jointly and individually. But in general, if we know the marginal distributions of constituent random variables (e.g., X or Y), we will not able to construct their joint distribution. We will address this when we elaborate on the issue of independence.

Given a joint distribution function, we now consider the computing of the joint probability

$$P(a_1 < X \le a_2, \ b_1 < Y \le b_2),$$

where $a_1 < a_2$ and $b_1 < b_2$. We see that

$$P(a_1 < X \le a_2, b_1 < Y \le b_2)$$
$$= F_{(X,Y)}(a_2, b_2) - F_{(X,Y)}(a_1, b_2) - F_{(X,Y)}(a_2, b_1) + F_{(X,Y)}(a_1, b_1)$$
$$= P(A) - P(B) - P(C) + P(D).$$

The following graph clearly depicts the above computation. The probability we are looking for is the probability being in the region enclosed by the square defined by the four points of $(a_2, b_2), (a_1, b_2), (a_1, b_1)$, and (a_2, b_1). In the following graph, we label the four indicated regions A, B, C, and D. The respective distribution functions give the corresponding probabilities. The reason that we have the term last "$+P(D)$" is that the term $P(D)$ is being subtracted twice. So we need to remedy the double subtractions.

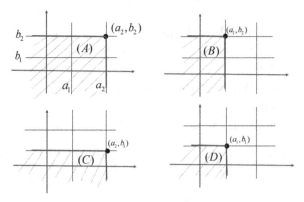

When (X,Y) is discrete, we define the joint probability mass function PMF by

$$p_{X,Y}(x,y) = P(X = x, Y = y).$$

The joint PMF can always be obtained from the joint distribution function. The following two examples illustrates the construction of PMFs based on the experiments involved.

Example 1. Consider an experiment involving the toss of a fair coin once and the roll a fair die once. Assume they are done independently. Let X denote the outcome of the coin toss and Y denote the outcome of rolling the die. Thus we have $X \in \{1, 0\}$ and $Y \in \{1, \ldots, 6\}$, where we let $X = 1$ to denote that we observe heads, and 0, otherwise. By independence, we find the joint PMF:

$$p_{X,Y}(x,y) = p_X(x) \times p_Y(y) = \frac{1}{2} \times \frac{1}{6} = \frac{1}{12}, \quad x = 0, 1, \quad \text{and} \quad y = 1, \ldots, 6.$$
$$\square$$

Example 2. Consider an experiment involving the rolling of a fair die once. Let Y be the outcome of such a roll. If $Y = y$, then a fair coin is tossed y times. Let X denote the number of times heads are observed. Hence $X = x | Y = y \sim$ binomial$(y, \frac{1}{2})$, $x = 0, 1, \ldots, y$ and $P(Y = y) = \frac{1}{6}$, $y = 1, \ldots, 6$. We want to find the joint PMF of (X, Y).

Conditioning on Y, we find the joint PMF as follows:

$$p_{(X,Y)}(x,y) = \begin{cases} P(X = x | Y = y)P(Y = y) & y = 1, \ldots, 6 \quad \text{and} \\ & x = 0, 1, \ldots, y, \\ 0 & \text{otherwise.} \end{cases}$$

The results of the computation is summarized in the following table.

X	Y						$p_X(x)$
	1	2	3	4	5	6	
0	$\frac{1}{12}$	$\frac{1}{24}$	$\frac{1}{48}$	$\frac{1}{96}$	$\frac{1}{192}$	$\frac{1}{384}$	$\frac{63}{384}$
1	$\frac{1}{12}$	$\frac{2}{24}$	$\frac{1}{16}$	$\frac{1}{24}$	$\frac{5}{192}$	$\frac{1}{64}$	$\frac{120}{384}$
2		$\frac{1}{24}$	$\frac{1}{16}$	$\frac{1}{16}$	$\frac{5}{96}$	$\frac{5}{128}$	$\frac{99}{384}$
3			$\frac{1}{48}$	$\frac{1}{24}$	$\frac{5}{96}$	$\frac{5}{96}$	$\frac{64}{384}$
4				$\frac{1}{96}$	$\frac{5}{192}$	$\frac{5}{128}$	$\frac{29}{384}$
5					$\frac{1}{192}$	$\frac{1}{64}$	$\frac{8}{384}$
6						$\frac{1}{384}$	$\frac{1}{384}$
$p_Y(y)$	$\frac{1}{6}$	$\frac{1}{6}$	$\frac{1}{6}$	$\frac{1}{6}$	$\frac{1}{6}$	$\frac{1}{6}$	1

The formula for computing the joint PMF is

$$P_{(X,Y)}(x,y) = p_y(y)p_{X|Y}(x|y) = \frac{1}{6} \times \binom{y}{x}\left(\frac{1}{2}\right)^x\left(\frac{1}{2}\right)^{y-x}$$

$$= \frac{1}{6} \times \binom{y}{x}\left(\frac{1}{2}\right)^y, \quad y = 1,\ldots,6; \quad x = 0,\ldots,y.$$

Using the above, we construct the above table in Illustration 1. □

We now switch to bivariate continuous random variables. The joint probability density function PDF for (X,Y) is denoted by

$$f_{X,Y}(x,y).$$

The interpretation of the joint density is similar to that for a univariate continuous random variable. Specifically, we can consider a following approximation:

$$P(x < X < x + dx, \quad y < Y < y + dy) \approx f_{X,Y}(x,y)dxdy.$$

Thus $f_{X,Y}(x,y)$ can be viewed as the "rate" of likelihood of having (X,Y) located in the neighborhood of (x,y). If we wish to find the probability that (X,Y) is a given region (A,B), we use

$$P\{(X,Y) \in (A,B)\} = \int_{x \in A}\int_{y \in B} f_{(X,Y)}(x,y)dydx \tag{7.1}$$

or more generally, in a region C, then

$$P\{(X,Y) \in C\} = \int_{(X,Y) \in C} f_{(X,Y)}(x,y)dxdy. \tag{7.2}$$

Example 3. Assume (X,Y) follows the joint density

$$f_{X,Y}(x,y) = \begin{cases} 2e^{-x}e^{-2y} & x > 0, \ y > 0, \\ 0 & \text{otherwise.} \end{cases}$$

Find (a) $P(X < Y)$, (b) $P(X > 1, Y < 1)$, and (c) $P(X < a)$.

Solution. To gain some understanding about the shape of the joint density, we produce the 3D plot of $f_{X,Y}(\cdot,\cdot)$ in Illustration 2.

(a) Define the region of integration $C = \{(x,y)|x < y, x > 0, y > 0\}$. We use (7.2) to write

$$
\begin{aligned}
P(X < Y) &= \int_{(X,Y) \in C} 2e^{-x} e^{2y} \, dx \, dy \\
&= \int_0^\infty 2e^{-2y} \left(\int_0^y e^{-x} \, dx \right) dy \\
&= \int_0^\infty 2e^{-2y} \left(1 - e^{-y} \right) dy \\
&= \int_0^\infty 2e^{-2y} \, dy - 2 \int_0^\infty e^{-3y} \, dy \\
&= 1 - \frac{2}{3} = \frac{1}{3}.
\end{aligned}
$$

(b) To find $P(X > 1, Y < 1)$, the region of interest is $(X,Y) \in C$, where $C = \{(x,y)|x > 0, 0 < y < 1\}$. We use (7.1) to write

$$
\begin{aligned}
P\{(X,Y) \in C\} &= \int_1^\infty \int_0^1 2e^{-x} e^{-2y} \, dx \, dy \\
&= \int_1^\infty e^{-x} \, dx \int_0^1 2e^{2y} \, dy \\
&= e^{-1}(1 - e^{-2}).
\end{aligned}
$$

(c) To find the marginal distribution function for X, we integrate out the random variable Y

$$
\begin{aligned}
f_X(x) &= \int_0^\infty f_{X,Y}(x,y) \, dy \\
&= e^{-x} \int_0^\infty 2e^{-2y} \, dy = e^{-x}.
\end{aligned}
$$

Thus, we have

$$
F_X(a) = \int_0^a f_X(\tau) \, d\tau = \int_0^a e^{-\tau} \, d\tau = 1 - e^{-a}. \qquad \square
$$

Example 4 (An Application in Auto Insurance Claims). Let X be the loss due to damage of an automobile and Y be the allocated adjustment expenses associated with the accident. Assume that (X, Y) has the joint PDF

$$f_{X,Y}(x, y) = \frac{5}{10^6} e^{-\frac{x}{1000}}, \quad 0 < 5y < x < \infty.$$

In Illustration 3, we plot the joint density and use Mathematica to find the marginal densities and their respective moments.   □

7.2.   Independent Random Variables

We recall that when two events A and B are independent, then $P(A \cap B) = P(A) \times P(B)$. The same property prevails for jointly distributed random variables. If X and Y are two *independent* random variables, then

$$P(X \in A, Y \in B) = P(X \in A)P(Y \in B),$$

where A and B are two regions. The above implies that for joint CDF, we have

$$F_{X,Y}(x, y) = F_X(x)F_Y(y) \quad \text{for all } x \text{ and } y.$$

It follows that for a discrete bivariate random variable (X, Y), the joint PMS for (X, Y) is multiplicative, i.e.,

$$p_{X,Y}(x, y) = p_X(x)p_Y(y) \quad \text{for all } x \text{ and } y.$$

For a continuous bivariate random variable (X, Y), the joint PDF for (X, Y) is multiplicative, i.e.,

$$f_{X,Y}(x, y) = f_X(x)f_Y(y) \quad \text{for all } x \text{ and } y.$$

The generalization of the above to n-dimensional random vector $\{X_1, \ldots, X_n\}$ is immediate.

Consider a bivariate continuous random variable (X, Y). Assume that X and Y are independent. We compute the expectation of XY as follows:

$$E(XY) = \iint xy f_{(X,Y)}(x, y)dxdy = \iint xy f_X(x)f_Y(y)dxdy$$
$$= \int x f_X(x)dx \int y f_Y(y)dy = E(X)E(Y).$$

An extension of the above is the expectation of the product of the functions of X and Y, namely, $E[g(X)h(Y)]$ is equal to the product of their respective expectations $E[g(X)]$ and $E[h(X)]$.

Another important result is the expectation of the sum of X and Y without the *assumption* of independence. We see that

$$E(X+Y) = \iint (x+y) f_{(X,Y)}(x,y) dxdy$$

$$= \iint x f_{(X,Y)}(x,y) dxdy + \iint y f_{(X,Y)}(x,y) dxdy$$

$$= \int x \left(\int f_{(X,Y)}(x,y) dy \right) dx + \int y \left(\int f_{(X,Y)}(x,y) dx \right) dy$$

$$= \int x f_X(x) dx + \int y f_Y(y) dy$$

$$= E(X) + E(Y).$$

Theorem 1. *Let $\{X_i\}_{i=1}^n$ be a sequence of random variables. Let $Y = X_1 + \cdots + X_n$. Then*

$$E(Y) = E(X_1) + \cdots + E(X_n).$$

Proof. The generalization of the aforementioned result extending to the case where $n \geq 2$. □

The above theorem is quite useful in applications because nothing is said about whether $\{X_i\}_{i=1}^n$ are independent and identically random variables. The following example clearly illustrated its use.

Example 5. Consider Example 3 of Chapter 5 again. There, urn 1 has 5 white and 6 black balls; urn 2 has 8 white and 10 black. We first randomly take two balls from urn 1 and place them in urn 2. We then randomly select three balls from urn 2. Let W denote the number of white balls drawn from urn 2. We want to find $E(W)$.

Solution: We call the first-stage of sampling the act of taking two balls from urn 1; the second-stage the act of removing three balls from urn 2. For each white balls in urn 1, to have it included in the second-stage of sampling requiring it being passing through the two stages. For each white ball originally in urn 2, to have it included in the second-stage does not involve the first-stage transfer.

With the above prelude, the stage is set for its solution. For each white ball in urn 1, we define

$$X_i = \begin{cases} 1 & \text{if the } i\text{th white ball is selected in stage two} \\ 0 & \text{otherwise.} \end{cases}$$

It is clear that

$$P(X_i = 1) = \frac{\binom{1}{1}\binom{10}{1}}{\binom{11}{2}} \times \frac{\binom{1}{1}\binom{19}{2}}{\binom{20}{3}} = 0.027273$$

and $E(X_i) = P(X_i = 1) = 0.027273$.

For each white ball originally in urn 2, we define

$$Y_i = \begin{cases} 1 & \text{if the } i\text{th white ball is selected in stage two,} \\ 0 & \text{otherwise.} \end{cases}$$

We note that in stage two, the urn 2 has a total of 20 balls now. If follows

$$P(Y_i = 1) = \frac{\binom{1}{1}\binom{19}{2}}{\binom{20}{3}} = 0.15$$

and $E(Y_i) = 0.15$.

Since $W = X_1 + \cdots + X_5 + Y_1 + \cdots + Y_8$, taking the expectation yields

$$E(W) = 5E(X_1) + 8E(Y_1) = 5(0.027273) + 8(0.15) = 1.33636.$$

This answer is what we found earlier in Chapter 5 where a direct approach was used. It is important to recognize the fact the absence of independence assumption enables us to do expectation. □

Example 6. Assume $X \sim U(0,1)$ and $Y \sim U(0,1)$. Moreover, $X \perp Y$. Find (a) $P(|X - Y| < 0.5)$ and (b) $P\left(\left|\frac{X}{Y} - 1\right| < 0.5\right)$.

Solution. (a) By independence, we see

$$f_{X,Y}(x,y) = \begin{cases} 1 & 0 < x < 1, 0 < y < 1, \\ 0 & \text{otherwise.} \end{cases}$$

Now

$$P(|X - Y| < 0.5) = P(-0.5 < X - Y < 0.5)$$

$$= \int_0^{0.5} \int_0^{x+0.5} dy\,dx + \int_{0.5}^1 \int_{x-0.5}^1 dy\,dx$$

$$= 0.375 + 0.375 = 0.75.$$

(b)

$$P\left(\left|\frac{X}{Y} - 1\right| < 0.5\right) = P\left(-0.5 < \frac{X}{Y} - 1 < 0.5\right)$$

$$= P\left(0.5 < \frac{X}{Y} < 1.5\right) = P\left(0.5Y < X < 1.5Y\right)$$

$$= \int_0^{0.5} \int_{\frac{2}{3}x}^{2x} dy\,dx + \int_{0.5}^1 \int_{\frac{2}{3}x}^1 dy\,dx = 0.41667.$$

To specify the limits of integration, it is helpful to draw a picture of the region of integration. This is shown below:

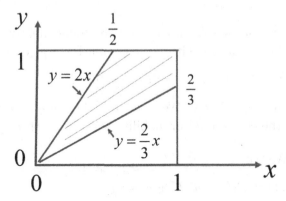

For this problem, we can "eyeball" for the solution. The probability is given by the shaded region of integration. Hence

$$P(A) = 1 - \frac{1}{2}\left(\frac{1}{2} + \frac{2}{3}\right) = \frac{5}{12} = 0.41667. \qquad \square$$

Example 7. Let X be the number of male drivers incurring an accident on a given day and Y is the number of female drivers incurring an accident on a given day. Assume $X \sim \text{pois}(\lambda_1)$ and $Y \sim \text{pois}(\lambda_2)$. We want to find the joint PMF of the random variable $Z = X + Y$, the total number of traffic accidents occurring on a given day. We note that Z is the *convolution* of X and Y. Thus we have

$$P(Z = n) = \sum_{k=0}^{n} P(X = k, Y = n - k)$$

$$= \sum_{k=0}^{n} P(X = k)P(Y = n - k) \quad \text{(by independence)}$$

$$= \sum_{k=0}^{n} e^{-\lambda_1} \frac{\lambda_1^k}{k!} e^{-\lambda_2} \frac{\lambda_2^{n-k}}{(n - k)!}$$

$$= \frac{e^{-(\lambda_1+\lambda_2)}}{n!} \sum_{k=0}^{n} \binom{n}{k} \lambda_1^k \lambda_2^{n-k}$$

$$= \frac{e^{-(\lambda_1+\lambda_2)}}{n!} (\lambda_1 + \lambda_2)^n.$$

Therefore, we conclude that $Z \sim \text{pois}(\lambda_1 + \lambda_2)$. We recall in Example 9 of Chapter 5, we used a probability generating function approach to deal with this convolution problem. $\square$

Example 8. Suppose that U and V are two independently and identically distributed random variables with the common PMF given by

$$g(k) = p(1 - p)^k, \quad k = 0, 1, 2, \ldots,$$

where $p > 0$. Define $Z = U + V$. We want to find (a) $p_{(U,Z)}(u, z) = P(U = u, Z = z)$ and (b) the conditional probability mass function $p_{(U|Z)}(u|z)$.

Solution. (a) We see

$$p_{U,Z}(u, z) = P(U = u, \ U + V = z)$$

$$= P(U = u, V = z - u)$$

$$= P(U = u) \times P(V = z - u)$$

$$= p(1 - p)^u \times p(1 - p)^{z-u}$$

$$= p^2(1 - p)^z, \quad u = 0, 1, 2, \ldots, z; \ z = 0, 1, 2, \ldots.$$

(b) First, we find

$$p_Z(z) = \sum_{u=0}^{z} p_{(U,Z)}(u, z) = \sum_{u=0}^{z} p^2(1 - p)^z = (z + 1)p^2(1 - p)^z.$$

Since Z is a convolution of U and V, in Illustration 4 we demonstrate the use of Z-transform (i.e., the PGF) to find $p_Z(z)$.

Using the definition of conditional probability, we obtain

$$p_{U|Z}(u|z) = \frac{p_{(U,Z)}(u,z)}{p_Z(z)} = \frac{p^2(1-p)^z}{(z+1)p^2(1-p)^z} = \frac{1}{1+z}, \quad u = 0, 1, 2, \dots, z$$

i.e., the conditional random variable $U|Z = z$ follows a discrete uniform distribution over the set of integers $0, 1, \dots, z$. $\square$

7.3. Order Statistics

Let $\{X_i\}_{i=1}^n$ be independent identically distributed random variables where $X \sim f_X(\cdot)$. Let $\{X_{(i)}\}_{i=1}^n$ denote the order statistics associated with $\{X_i\}_{i=1}^n$, where $X_{(i)}$ represents the nth smallest of the $\{X_i\}_{i=1}^n$. We first find the joint distribution of $\{X_{(i)}\}_{i=1}^n$. By independence, the joint density is the product of the n individual densities. However, for each realization of $\{X_{(i)}\}_{i=1}^n$ it contains $n!$ duplicated copies of the same random variables $\{X_i\}$. For example, consider $n = 2$, the realization that $X_{(1)} = 12$ and $X_{(2)} = 15$ can be resulted from either $\{X_1 = 12, X_2 = 15\}$ or $\{X_1 = 15, X_2 = 12\}$. Hence the joint density of the order statistics is given by

$$f_{X_{(1)}, \dots, X_{(n)}}(x_1, \dots, x_n) = n!\, f_X(x_1) \cdots f_X(x_n), \quad x_1 \le \cdots \le x_n.$$

From the joint density, we now find the marginal density of $X_{(j)}$ by a straightforward probabilistic argument using the definition of distribution function. This means we have

$$f_{X_{(j)}}(x)dx = \binom{n}{j-1, 1, n-j} F_X(x)^{j-1}(1 - F(x))^{n-j} f_X(x)dx.$$

To justify the above assertion, we consider three cases: (i) there must be $j - 1$ observations with $X_i \le x$; the probability for this to occur is $F_X(x)^{j-1}$; (ii) there must be $n - j$ observations with $X_i > x$; the probability of this to occur is $(1 - F_X(x))^{n-j}$; and (iii) the probability that one observation occurs in $(x; x + dx)$ is approximately $f(x)dx$. The multinomial coefficient accounts for the number of ways the n samples can fall into the three cases. We remark that the preceding derivations rely on the i.i.d. assumption of $\{X_i\}_{i=1}^n$. When all these probabilities are multiplied together, we establish the needed result.

A similarly line of reasoning leads to the joint density for $(X_{(i)}, X_{(j)})$, with $i < j$,

$$f_{X_{(i)}, X_{(j)}}(x_i, x_j) = \binom{n}{i-1, 1, j-i-1, 1, n-j}$$
$$\times (F_X(x_i))^{i-1}(F_X(x_j) - F_X(x_i))^{j-i-1}$$
$$\times (1 - F_X(x_j))^{n-j} f_X(x_i) f_X(x_j),$$

where $x_i < x_j$.

Example 9 (Range of a Random Sample). Let $\{X_i\}_{i=1}^n$ be a random sample of size n from a distribution $F_X(\cdot)$. The range of the sample is defined by $R = X_{(n)} - X_{(1)}$. We want to find the CDF for R. The joint density of $X_{(1)}$ and $X_{(n)}$ is

$$f_{X_{(1)}, X_{(n)}}(x_1, x_n) = \binom{n}{n-2, 1, 1} (F_X(x_n) - F_X(x_1))^{n-2} f_X(x_1) f_X(x_n).$$

We find the density for R as follows:

$$P(R \le r) = P\left(X_{(n)} - X_{(1)} \le r\right) = \iint\limits_{x_n - x_1 \le r} f_{X_{(1)}, X_{(n)}}(x_1, x_n) dx_n dx_1$$

$$= \int_{-\infty}^{\infty} \int_{x_1}^{x_1+r} \frac{n!}{(n-2)!} (F_X(x_n) - F_X(x_1))^{n-2}$$
$$\times f_X(x_1) f_X(x_n) dx_n dx_1.$$

We look at the inner integral

$$\int_{x_1}^{x_1+r} \frac{n!}{(n-2)!} (F_X(x_n) - F_X(x_1))^{n-2} f_X(x_n) dx_n.$$

With x_1 fixed at a specific value, we do a change of variable using

$$y = F_X(x_n) - F_X(x_1) \quad \Longrightarrow \quad dy = f_X(x_n) dx_n.$$

The limits of integration are given by

x_n	y
x_1	0
$x_1 + r$	$F_X(x_1 + r) - F_X(x_1)$

Hence the inner integral reads

$$n(n-1)\int_0^{F_X(x_1+r)-F_X(x_1)} y^{n-2}dy = n\left(F_X(x_1+r) - F_X(x_1)\right)^{n-1}$$

and

$$P(R \le r) = n\int_{-\infty}^{\infty} \left(F_X(x_1+r) - F_X(x_1)\right)^{n-1} f(x_1)dx_1. \qquad \square$$

Example 10 (Range of a Random Sample from Standard Uniform Density). Assume $\{X_i\}_{i=1}^n$ are n independent samples from $U(0,1)$. We want to find the density for its range $R = X_{(n)} - X_{(1)}$. Specializing the result from Example 9 to this case, we have

$$P(R \le r) = n\int_0^1 \left(F_X(x_1+r) - F_X(x_1)\right)^{n-1} dx_1.$$

We observe that

$$F_X(x_1+r) - F_X(x_1) = \begin{cases} x_1 + r - x_1 & \text{if } 0 < x_1 < 1-r, \\ 1 - x_1 & \text{if } 1-r < x_1 < 1. \end{cases}$$

Therefore

$$P(R \le r) = n\int_0^{1-r} r^{n-1}dx_1 + n\int_{1-r}^1 (1-x_1)^{n-1}dx_1$$

$$= nr^{n-1}(1-r) + r^n$$

and

$$f_R(r) = n(n-1)r^{n-2}(1-r), \quad 0 < r < 1.$$

We recall the beta density

$$f(x) = \frac{1}{B(a,b)}x^{a-1}(1-x)^{b-1}, \quad \text{where } B(a,b) = \frac{\Gamma(a)\Gamma(b)}{\Gamma(a+b)}.$$

Thus

$$f_R(r) \sim \frac{1}{B(n-1,2)}r^{(n-1)-1}(1-r)^{2-1}.$$

This means the range follows a beta distribution with parameters $(n-1, 2)$. $\qquad \square$

7.4. Conditional Random Variables

For a bivariate random variable (X, Y), the distribution function for the conditional random variable $X|Y = y$ is defined by

$$F_{X|Y}(x|y) = P(X \le x|Y = y).$$

For discrete random variables X and Y, the PMF for the conditional random variable $X|Y = y$ is given by

$$p_{X|Y}(x|y) = P(X = x|Y = y) = \frac{p_{X,Y}(x, y)}{p_Y(y)}$$

provided that $p_Y(y) > 0$ for all y.

For continuous random variables X and Y, the PDF for the conditional random variable $X|Y = y$ is given by

$$f_{X|Y}(x|y) = \frac{f_{(X,Y)}(x, y)}{f_Y(y)}$$

provided that $f_Y(y) > 0$ for all y.

Consider the case of continuous random variables X and Y. Since $X|Y = y$ is a random variable, then what is the expectation of this conditional random variable? We know that the random variable $X|Y = y$ has a density $f_{X|Y}(\cdot|y)$, we use the definition of expectation and write

$$E(X|Y = y) = \int x f_{X|Y}(x|y) dx.$$

If we "uncondition" on Y and write

$$\int E(X|Y = y) f_Y(y) dy = \iint x f_{X|Y}(x|y) f_Y(y) dy dx$$

$$= \int x \left(\int f_{X|Y}(x|y) f_Y(y) dy \right) dx$$

$$= \int x \left(\int f_{X,Y}(x, y) dy \right) dy$$

$$= \int x f_X(x) dx$$

$$= E(X),$$

then we obtain an useful identity

$$E[X] = E[E[X|Y]],$$

where the "exterior" expectation is with respect to Y. Sometimes, the identity is called the "law of iterated expectations". The relation holds for discrete random variables as well.

Example 11. Let $X \sim \text{pois}(\lambda_1)$ and $Y \sim \text{pois}(\lambda_2)$. Assume $X \perp Y$. Define $Z = X + Y$. We are interested in finding the PMF of the conditional random variable $X \mid Z = n$.

We see

$$P(X = k|Z = n) = \frac{P(X = k, \ Z = n)}{P(Z = n)}$$

$$= \frac{P(X = k, Y = n - k)}{P(Z = n)} = \frac{P(X = k)P(Y = n - k)}{P(Z = n)}.$$

Since we know that $Z \sim \text{pois}(\lambda_1 + \lambda_2)$, we have

$$P(X = k|Z = n) = \frac{e^{-\lambda_1}(\lambda_1)^k e^{-\lambda_2}(\lambda_2)^{n-k}}{k!(n-k)!} \left(\frac{e^{-(\lambda_1+\lambda_2)}(\lambda_1 + \lambda_2)^n}{n!} \right)^{-1}$$

$$= \binom{n}{k} \left(\frac{\lambda_1}{\lambda_1 + \lambda_2} \right)^k \left(\frac{\lambda_2}{\lambda_1 + \lambda_2} \right)^{n-k}.$$

We see that the condition random variable $X|Z = n \sim \text{bino}(n, \frac{\lambda_1}{\lambda_1+\lambda_2})$. □

Example 12. Let X and Y be two continuous random variables with a joint PDF:

$$f_{X,Y}(x, y) = \begin{cases} \dfrac{e^{-\frac{x}{y}}e^{-y}}{y} & x > 0 \text{ and } y > 0, \\ 0 & \text{otherwise.} \end{cases}$$

We want to find the conditional density for random variable $X|Y = y$. We see that

$$f_Y(y) = \int_0^\infty f_{(X,Y)}(x, y)dx$$

$$= \int_0^\infty \frac{e^{-\frac{x}{y}}e^{-y}}{y} dx$$

$$= e^{-y} \int_0^\infty \frac{1}{y} e^{-\frac{x}{y}} dx$$

$$= e^{-y},$$

where we recognize that the last equality is due to the fact that the last integral is the integral over an exponential density with parameter $\frac{1}{y}$. Since the integral has a value one, then result follows. Now

$$f_{X|Y}(x|y) = \frac{f_{(X,Y)}(x,y)}{f_Y(y)} = \frac{e^{-\frac{x}{y}}e^{-y}}{y}e^y = \frac{1}{y}e^{-\frac{x}{y}}, \quad x > 0 \text{ and } y > 0.$$

Thus the conditional random variable $X|Y = y$ is again exponential with parameter $\frac{1}{y}$.

It is instructive to see the joint PDF of (X, Y) from the perspective of

$$f_{X,Y}(x,y) = f_{X|Y}(x|y)f_Y(y),$$

namely, the parameter of the conditional density $f_{(X|Y)}(x|y)$ is being generated by an exponential source $f_Y(y)$. ☐

7.5. Functions of Jointly Distributed Random Variables

As stated in the introduction of this chapter, there are many transformations of random variables leading to new random variables. We first consider the sum of random variables X and Y. Let $Z \equiv g(X, Y) = X + Y$.

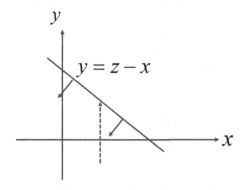

From the above graph depicting the region of integration, we write

$$F_Z(z) = P(X + Y \le z)$$

$$= \iint_{\{(x,y):(x+y)\le z\}} f_{X,Y}(x,y)dxdy$$

$$= \int_{-\infty}^{\infty} dx \int_{-\infty}^{z-x} dy\, f_{X,Y}(x,y)$$

$$= \int_{-\infty}^{\infty} dx \int_{-\infty}^{z} d\tau f_{X,Y}(x, \tau - x), \quad \tau = x + y.$$

We differentiate the above with respect to z and obtain the density

$$f_Z(z) = \int_{-\infty}^{\infty} dx \, f_{X,Y}(x, z - x) = \int_{-\infty}^{\infty} dy \, f_{X,Y}(z - y, y). \qquad (7.3)$$

If X and Y are *independent*, then for any z, we have

$$f_Z(z) = \int_{-\infty}^{\infty} dx \, f_X(x) f_Y(z - x) = \int_{-\infty}^{\infty} dy \, f_X(z - y) f_Y(y). \qquad (7.4)$$

Similar results can be obtained when X and Y are discrete random variables.

Example 13. Consider $X \sim \text{unif}(0, 1)$ and $Y \sim \text{expo}(1)$. Also, we assume that $X \perp Y$. Let $Z = X + Y$. We want to find the PDF of Z.

As $Z > 0$, we use (7.4) to write

$$f_Z(z) = \int_0^{\infty} f_X(z - y) f_Y(y) dy$$

$$= \int_0^{\infty} \mathbf{1}\{0 < z - y < 1\} e^{-y} dy$$

$$= \int_0^{\infty} \mathbf{1}\{-1 < y - z < 0\} e^{-y} dy$$

$$= \int_0^{\infty} \mathbf{1}\{z - 1 < y < z\} e^{-y} dy$$

$$= \int_{(z-1)+}^{z} e^{-y} dy$$

$$= \left[-e^{-y}\right]_{y=(z-1)+}^{z}.$$

We see that

$$(z - 1)^+ = \begin{cases} 0 & \text{if } 0 < z < 1, \\ z - 1 & \text{if } z \geq 1. \end{cases}$$

Thence we have

$$f_Z(z) = \begin{cases} 1 - e^z & \text{if } 0 < z < 1, \\ e^{1-z} - e^{-z} & \text{if } z \geq 1. \end{cases}$$

We see that the $f_Z(z)$ is continuous at $z = 1$. But it is easy to verify that it does not have the same derivative at $z = 1$. In Illustration 5, we display the PDF of Z. □

When $Z = X - Y$. We can make use of the above results by noting $Z = X + (-Y)$ and make the corresponding substitutions. For the general case, we find

$$f_Z(z) = \int_{-\infty}^{\infty} dy \, f_{(X,Y)}(z+y, y) = \int_{-\infty}^{\infty} dx \, f_{(X,Y)}(x, x-z).$$

If $X \perp Y$, then the joint density splits in two parts and the convolution formula becomes

$$f_Z(z) = \int_{-\infty}^{\infty} f_X(z+y) f_Y(y) dy = \int_{-\infty}^{\infty} f_X(x) f_Y(x-z) dx.$$

Consider now $Z = XY$ where X and Y are two continuous random variables. Mimicking the previous derivations, we first draw the region of integration below:

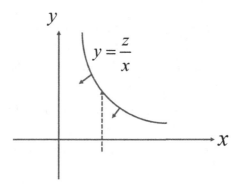

It follows that

$$F_Z(XY \le z) = \iint_{((X,Y):\, xy \le z)} f_{X,Y}(x, y) dx dy$$

$$= \int_{-\infty}^{\infty} dx \int_{-\infty}^{\frac{z}{x}} dy \, f_{X,Y}(x, y)$$

$$= \int_{-\infty}^{\infty} dx \int_{0}^{z} d\tau \frac{1}{|x|} f_{X,Y}\left(x, \frac{z}{x}\right), \quad \tau = xy.$$

We now differentiate the above with respect to z and obtain the density

$$f_Z(z) = \int_{-\infty}^{\infty} dx \frac{1}{|x|} f_{X,Y}\left(x, \frac{z}{x}\right) = \int_{-\infty}^{\infty} dx \frac{1}{|x|} f_{X,Y}\left(\frac{y}{x}, x\right). \qquad (7.5)$$

A similar approach will yield the above result when we have $Z = X/Y$:

$$f_Z(z) = \int_{-\infty}^{\infty} dx \, |x| \, f_{X,Y}(zx, x). \tag{7.6}$$

Mellin Transforms:

The Mellin transform is handy for transformations of the types $Z = X \cdot Y$ and $X = X/Y$, when X and Y are continuous, positive, and independent random variables.

Let X be a continuous positive random variable with PDF $f_X(x)$. The Mellin transform is defined by

$$\mathcal{M}_X(s) = E\left[x^{s-1}\right] = \int_0^{\infty} x^{s-1} f_X(x) dx.$$

The transform is a variant of a Fourier transform. If $Z = aX$, where a is a constant, then it follows that

$$\mathcal{M}_Z(s) = a^{s-1} \mathcal{M}_X(s). \tag{7.7}$$

When Y is also a continuous random variable and $X \perp Y$, the joint PDF $f_{X,Y}$ factors and (7.5) simplifies. This would lead to a nice result that

$$\mathcal{M}_Z(s) = \mathcal{M}_X(s)\mathcal{M}_Y(s), \tag{7.8}$$

where $Z = X \times Y$. Moreover, if $W = X/Y$, then

$$\mathcal{M}_W(s) = \mathcal{M}_X(s)\mathcal{M}_Y(2 - s). \tag{7.9}$$

There are other related relations shown in the cited references given at the end of the chapter.

Example 14. Let $X \sim \text{expo}(1)$ and $Y \sim \text{expo}(2)$. Moreover, $X \perp Y$. Define $Z = X \times Y$. In Illustration 6, we show that we can easily obtain the PDF of Z. We show numerically in Illustration 6 that $E(Z) = E(X)E(Y) = 0.5$. $\square$

Example 15. Let X and Y be two gamma random variables with parameters (s, λ) and (t, λ), respectively. Let $Z = X + Y$. We would like to find the density $f_Z(z)$.

Recall when $X \sim \text{gamma}(\alpha, \lambda)$, we have

$$f_X(x) = \begin{cases} \dfrac{\lambda e^{-\lambda t}(\lambda t)^{\alpha-1}}{\Gamma(\alpha)} & x > 0, \\ 0 & \text{otherwise.} \end{cases}$$

Since

$$f_Z(z) = \int dx\, f_X(x) f_Y(z-x),$$

this implies that

$$f_Z(z) = \frac{1}{\Gamma(s)\Gamma(t)} \int_0^z dx\, \lambda e^{-\lambda x}(\lambda x)^{s-1} \lambda e^{-\lambda(z-x)} \left(\lambda(z-x)\right)^{t-1}$$

$$= \frac{\lambda^{s+t}}{\Gamma(s)\Gamma(t)} e^{-\lambda z} \int_0^z x^{s-1}(z-x)^{t-1} dx$$

$$= \frac{\lambda^{s+t}}{\Gamma(s)\Gamma(t)} e^{-\lambda z} \int_0^1 (yz)^{s-1}(z - yz)^{t-1} z\, dy$$

$$\times \text{(change of variable with } y = x/z)$$

$$= \frac{\lambda^{s+t}}{\Gamma(s)\Gamma(t)} e^{-\lambda z} z^{s+t-1} \int_0^1 y^{s-1}(1-y)^{t-1} dy$$

$$= \frac{\lambda^{s+t}}{\Gamma(s)\Gamma(t)} e^{-\lambda z} z^{s+t-1} B(s,t)$$

$$= \frac{\lambda^{s+t}}{\Gamma(s)\Gamma(t)} e^{-\lambda z} z^{s+t-1} \frac{\Gamma(s)\Gamma(t)}{\Gamma(s+t)}$$

$$= \frac{\lambda e^{-\lambda z}(\lambda z)^{s+t-1}}{\Gamma(s+t)}.$$

Thus, we have just shown that $X + Y$ is a gamma random variable with parameters $(s + t, \lambda)$. In Illustration 7, the above derivation is done via Laplace transforms.

It is straightforward to generalize the above to the case when $\{X\}_{i=1}^n$ are independent gamma random variables with respective parameters (t_i, λ) and $Z = X_1 + \cdots + X_n$. Then Z follows a gamma distribution with parameters $(t_1 + \cdots + t_n, \lambda)$. □

Example 16. Let X and Y be two random variables with a joint density

$$f_{X,Y}(x,y) = \begin{cases} e^{-(x+y)} & x > 0 \text{ and } y > 0, \\ 0 & \text{otherwise.} \end{cases}$$

Let $Z = \frac{X}{Y}$. We are interested in finding $f_Z(z)$.

Using (7.6), we can write

$$f_Z(z) = \int_0^\infty x f_{(X,Y)}(zx, x) dx$$

$$= \int_0^\infty x e^{-(z+1)x} dx$$

$$= \int_0^\infty \frac{1}{(z+1)^2} w^{2-1} e^{-w} dw \quad (\text{let } w = (z+1)x)$$

$$= \frac{1}{(z+1)^2} \int_0^\infty w^{2-1} e^{-w} dw$$

$$= \frac{1}{(z+1)^2} \qquad (z > 0),$$

where the last equality is due to the fact that the integral is the complete gamma function of order 2, namely, $\Gamma(2)=1\times\Gamma(1) = 1$. In Illustration 9, the above result is verified via the use of Mellin transform. $\qquad\square$

We now present a general method for working with functions of jointly distributed continuous random variables. Let $X_1, \ldots, X_n$ be the n jointly distributed random variables with a density $f_{(X_1,\ldots,X_n)}(x_1, \ldots, x_n)$. Define

$$Y_1 = g_1(X_1, \ldots, X_n),$$

$$\cdots$$

$$Y_n = g_n(X_1, \ldots, X_n).$$

We assume the functions $\{g_i\}$ have continuous partial derivative and that the Jacobian determinant

$$J(x_1, \ldots, x_n) = \begin{vmatrix} \dfrac{\partial g_1}{\partial x_1} & \cdots & \dfrac{\partial g_1}{\partial x_n} \\ \vdots & & \vdots \\ \dfrac{\partial g_n}{\partial x_1} & \cdots & \dfrac{\partial g_n}{\partial x_n} \end{vmatrix} \neq 0$$

at all points $(x_1, \ldots, x_n)$. Assume also that the system of equations

$$y_1 = g_1(x_1, \ldots, x_n)$$

$$\vdots$$

$$y_n = g_n(x_1, \ldots, x_n)$$

has a unique solution, $x_i = h_i(y_1, \ldots, y_n)$, $i = 1, \ldots, n$. Then, the joint density of $(Y_1, \ldots, Y_n)$ is given by

$$f_{(Y_1, \ldots, Y_n)}(y_1, \ldots, y_n) = f_{(X_1, \ldots, X_n)}(x_1, \ldots, x_n) |J(x_1, \ldots, x_n)|^{-1}$$

where $x_i = h_i(y_1, \ldots, y_n)$, $i = 1, \ldots, n$.

Variant. Sometimes, finding the partial derivatives of $\{h_i\}$ with respect to $\{y_i\}$ is easier. Then the following variant can be used. Here

$$J(y_1, \ldots, y_n) = \begin{vmatrix} \dfrac{\partial h_1}{\partial y_1} & \cdots & \dfrac{\partial h_1}{\partial y_n} \\ \vdots & & \vdots \\ \dfrac{\partial h_n}{\partial y_1} & \cdots & \dfrac{\partial h_n}{\partial y_n} \end{vmatrix}$$

provided that $J \neq 0$. Then, the joint density of $(Y_1, \ldots, Y_n)$ is given by

$$f_{(Y_1, \ldots, Y_n)}(y_1, \ldots, y_n) = f_{(X_1, \ldots, X_n)}(y_1, \ldots, y_n) |J(y_1, \ldots, y_n)|$$

Example 17. Let X_1 and X_2 be two independently identically distributed exponential random variables with parameter λ. Define

$$Y_1 = X_1 + X_2 \quad \text{and} \quad Y_2 = \frac{X_1}{X_2}.$$

We know that Y_1 follows Erlang distribution with parameters $n = 2$ and $\lambda > 0$ and density of Y_2 has just been derived in Example 11. We want to demonstrate that the aforementioned approach yields the same results.

Let T be the transformation from (x_1, x_2) onto (y_1, y_2) with

$$y_1 = g_1(x_1, x_2) = x_1 + x_2, \quad y_2 = g_2(x_1, x_2) = x_1/x_2.$$

The inverse transformation that maps (y_1, y_2) onto (x_1, x_2) is given by

$$x_1 = h_1(y_1, y_2) = \frac{y_1 y_2}{1 + y_2}, \quad x_2 = h_2(y_1, y_2) = \frac{y_1}{1 + y_2}.$$

For the Jacobian matrix, we have

$$J(x_1, x_2) = \begin{vmatrix} \dfrac{\partial g_1}{\partial x_1} & \dfrac{\partial g_1}{\partial x_2} \\ \dfrac{\partial g_2}{\partial x_1} & \dfrac{\partial g_2}{\partial x_2} \end{vmatrix} = \begin{vmatrix} 1 & 1 \\ \dfrac{1}{x_2} & -\dfrac{x_1}{x_2^2} \end{vmatrix} = -\left(\frac{x_1 + x_2}{x_2^2} \right)$$

or

$$|J(y_1, y_2)| = \frac{y_1}{(1 + y_2)^2}.$$

Since X_1 and X_2 are independent, we have

$$f_{X_1,X_2}(x_1, x_2) = f_{X_1}(x)f_{X_2}(x_2) = \lambda^2 e^{-\lambda(x_1+x_2)}, \quad x_1 > 0 \text{ and } x_2 > 0.$$

Thus, we have

$$f_{Y_1,Y_2}(y_1, y_2) = \lambda^2 e^{-\lambda y_1} \left(\frac{y_1}{(1 + y_2)^2} \right)$$

$$= \lambda^2 y_1 e^{-\lambda y_1} \left(\frac{1}{(1 + y_2)^2} \right)$$

$$= \left(\frac{\lambda e^{-\lambda_1 y}(\lambda y_1)^{2-1}}{(2 - 1)!} \right) \left(\frac{1}{(1 + y_2)^2} \right).$$

We see the above is the product of an Erlang$(2, \lambda)$ density and the density given in Example 15. We conclude that Y_1 and Y_2 are two independent random variables. □

Example 18. Let $\{X_i\}_{i=1}^{3}$ be three independently identically distributed standard normal random variables. Let T be the linear transformation from (x_1, x_2, x_3) onto (y_1, y_2, y_3) with

$$y_1 = g_1(x_1, x_2, x_3) = x_1 + x_2 + x_3,$$
$$y_2 = g_2(x_1, x_2, x_3) = x_1 - x_2,$$
$$y_3 = g_3(x_1, x_2, x_3) = y_1 - y_3.$$

The inverse transformation T^{-1} from (y_1, y_2, y_3) onto (x_1, x_2, x_3) is given by

$$x_1 = h_1(y_1, y_2, y_3) = \frac{y_1 + y_2 + y_3}{3},$$
$$x_2 = h_2(y_1, y_2, y_3) = \frac{y_1 - 2y_2 + y_3}{3},$$
$$x_3 = h_3(y_1, y_2, y_3) = \frac{y_1 + y_2 - 2y_3}{3}.$$

For the Jacobian matrix, we have

$$
J(x_1, x_2, x_3) = \begin{vmatrix} \dfrac{\partial g_1}{\partial x_1} & \dfrac{\partial g_1}{\partial x_2} & \dfrac{\partial g_1}{\partial x_3} \\[2mm] \dfrac{\partial g_2}{\partial x_1} & \dfrac{\partial g_2}{\partial x_2} & \dfrac{\partial g_2}{\partial x_3} \\[2mm] \dfrac{\partial g_3}{\partial x_1} & \dfrac{\partial g_3}{\partial x_2} & \dfrac{\partial g_3}{\partial x_3} \end{vmatrix} = \begin{vmatrix} 1 & 1 & 1 \\ 1 & -1 & 0 \\ 1 & 0 & -1 \end{vmatrix} = 3
$$

and

$$
|J(y_1, y_2, y_3)| = \frac{1}{3}.
$$

Since $\{X_i\}_{i=1}^3$ are independent and $N(0, 1)$, we have

$$
f_{X_1, X_2, X_3}(x_1, x_2, x_3) = \frac{1}{(2\pi)^{\frac{3}{2}}} \exp\left(\sum_{i=1}^{3} \frac{x_i^2}{2} \right),
$$

it follows that

$$
f_{Y_1, Y_2, Y_3}(y_1, y_2, y_3) = \frac{1}{(2\pi)^{\frac{3}{2}}} \exp\left(\sum_{i=1}^{3} \frac{h_i^2}{2} \right) \left(\frac{1}{3} \right),
$$

where $\{h_i\}_{i=1}^3$ were defined as a function of y_1, y_2, and y_3 earlier. $\square$

7.6. More Well-Known Distributions

We now present some well-known distributions which are functions of multivariate random variables. Many have extensive use in statistics.

The Chi-Squared Distribution. Let $\{Z_i\}_{i=1}^n$ be independently identically distributed standard normal random variables. Define

$$
Y = Z_1^2 + \cdots + Z_n^2.
$$

Define $Y_i = Z_i^2$, $i = 1, \ldots, n$. Then $Y = Y_1 + \cdots + Y_n$. Recall Section 6.6 of Chapter 6, we have shown that Y_i follows a gamma distribution with parameter $(\alpha, \lambda) = (\frac{1}{2}, \frac{1}{2})$. Using the generalization of the result from Example 9, we conclude that Y follows a gamma distribution with parameter

$(n/2,\ 1/2)$ with its density given by

$$f_Y(y) = \frac{\frac{1}{2}e^{-\frac{y}{2}}\left(\frac{y}{2}\right)^{\frac{n}{2}-1}}{\Gamma\left(\frac{n}{2}\right)} \quad (y > 0)$$

$$= \frac{e^{-\frac{y}{2}}y^{\frac{n}{2}-1}}{2^{\frac{n}{2}}\Gamma\left(\frac{n}{2}\right)} \quad (y > 0).$$

The above density is known as the chi-square distribution with n degrees of freedom. We use χ_n^2 to denote it.

A Variant. For $i = 1, \ldots, n$, assume now random variable $X_i \sim N(0, \sigma^2)$. Moreover, assume that $\{X_i\}$ are mutually independent. Define

$$W = X_1^2 + \cdots + X_n^2.$$

Recall that $X_i \sim \text{gamma}\left(\frac{1}{2}, \frac{1}{2\sigma^2}\right)$ for each i. By the generalization given in Example 9, we conclude that $W \sim \text{gamma}\left(\frac{n}{2}, \frac{1}{2\sigma^2}\right)$. The density of W is given by

$$f_W(w) = \begin{cases} \dfrac{1}{2^{\frac{n}{2}}\sigma^n\Gamma\left(\frac{n}{2}\right)}w^{\left(\frac{n}{2}\right)-1}e^{-\frac{w}{2\sigma^2}} & w > 0, \\ 0 & \text{otherwise.} \end{cases}$$

The above is known in statistics as the chi-square distribution with parameters n and σ.

Another Variant. Let W the chi-square distribution defined in the above variant with parameters n and σ. Define

$$Y = \sqrt{\frac{W}{n}}.$$

We would like to obtain the density of W. Set $y = g(w) = \sqrt{w/n}$ and $w = g^{-1}(y) = ny^2$. We do a change of variable using

$$f_Y(y) = f_W(g^{-1}(x))\left|\frac{d}{dy}g^{-1}(y)\right|.$$

Thus

$$f_Y(y) = \frac{1}{2^{\frac{n}{2}}\sigma^n\Gamma\left(\frac{n}{2}\right)}(ny^2)^{\left(\frac{n}{2}\right)-1}e^{-\frac{ny^2}{2\sigma^2}}\,2ny.$$

Hence

$$f_Y(y) = \begin{cases} \dfrac{2\left(\frac{n}{2}\right)^{\frac{n}{2}}}{\sigma^n \Gamma\left(\frac{n}{2}\right)} y^{n-1} e^{-\left(\frac{n}{2\sigma^2}\right)y^2} & y > 0, \\ 0 & \text{otherwise.} \end{cases}$$

The above is called the χ distribution with parameters n and σ.

The F-distribution

A positive continuous random variable X follows an F-distribution, with m and n degrees of freedom, i.e., $X \sim F(m, n)$ if

$$X = \frac{Y_1/m}{Y_2/n},$$

where $Y_1 \sim \chi_m^2$, $Y_2 \sim \chi_n^2$, and $Y_1 \perp Y_2$. The PDF of X is given by

$$f_X(x) = cx^{\frac{m}{2}-1}\left(1 + \frac{m}{n}x\right)^{-\left(\frac{n+m}{2}\right)}, \quad x > 0,$$

where the normalization constant c is defined as

$$c = \left(\frac{m}{n}\right)^{\frac{m}{2}} \frac{1}{Beta\left(\frac{n}{2}, \frac{m}{2}\right)} = \left(\frac{m}{n}\right)^{\frac{m}{2}} \frac{\Gamma(\frac{1}{2}(n+m))}{\Gamma\left(\frac{n}{2}\right)\Gamma\left(\frac{m}{2}\right)}.$$

We see that

$$X = \left(\frac{n}{m}\right)\left(\frac{Y_1}{Y_2}\right).$$

Since Y_1 and Y_2 are both positive, continuous, and independent random variables, the PDF for X can readily be found by Mellin transforms (7.7) and (7.9). This is done in Illustration 9.

The Student's t-distribution. Students in rudimentary statistics will invariably encounter this distribution. So what is it? Let X be a normal random variable with mean 0 and variance σ^2. Assume random variable Y follows a χ distribution with parameters n and σ. Moreover, assume that $X \perp Y$. Define

$$Z = \frac{X}{Y}.$$

As it terms out, Z follows a Student's t-distribution with n degrees of freedom, i.e., $Z \sim t(n)$.

To derive its density, we use (7.6) to write

$$f_Z(z) = \int_{-\infty}^{\infty} dx \, |x| \, f_{(X,Y)}(zx, x) = \int_{-\infty}^{\infty} dx \, |x| \, f_X(zx) f_Y(x).$$

Thus we have

$$f_Z(z) = \int_0^{\infty} dx \, x \, \frac{1}{\sqrt{2\pi}\sigma} e^{-\frac{1}{2}\left(\frac{zx}{\sigma}\right)^2} \frac{2\left(\frac{n}{2}\right)^{\frac{n}{2}}}{\sigma^n \Gamma\left(\frac{n}{2}\right)} x^{n-1} e^{-\left(\frac{n}{2\sigma^2}\right)x^2}$$

$$= \frac{K}{\sigma^{n+1}} \int_0^{\infty} dx \, x \, e^{-\frac{1}{2}\left(\frac{zx}{\sigma}\right)^2} x^{n-1} e^{-\frac{n}{2}\left(\frac{x}{\sigma}\right)^x},$$

where

$$K = \frac{2\left(\frac{n}{2}\right)^{\frac{n}{2}}}{\Gamma\left(\frac{n}{2}\right)\sqrt{2\pi}}.$$

We do a change of variable with

$$u = \frac{x\sqrt{z^2+n}}{\sigma}.$$

Then we have

$$f_Z(z) = K(y^2+n)^{-\frac{(n+1)}{2}} \int_0^{\infty} du \, u^n \, e^{-\frac{1}{2}u^2}$$

$$= K(y^2+n)^{-\frac{n+1}{2}} 2^{\frac{(n-1)}{2}} \Gamma\left(\frac{n+1}{2}\right).$$

The above is the t-distribution with n degrees of freedom. □

7.7. Mixtures of Random Variables

In real-world applications of probability, observed distributions many times are multi-modaled and/or heavy-tailed. These are manifested themselves in terms of nonzero skewness and/or large kurtosis. One plausible explanation of these phenomena is that this might be caused by the fact sources that generate variability are varied. As an example, we may have a set of data measuring heights of a population. Different ethnic groups may have different height attributes. This would contribute to the heterogeneity of the distribution.

Deterministic Mixture of Distributions. Assume we have random variables $\{X_i\}_{i=1}^n$ where $E(X_i) = \mu_i$ and $Var(X_i) = \sigma_i^2$, for $i = 1, \ldots, n$. Let p_i denote the probability that random variable X_i is in effect. We assume

$p_1 + \cdots + p_n = 1$. Given a set of data, estimation of parameters $\{\mu_i, \sigma_i^2\}_{i=1}^n$ and $\{p_i\}_{i=1}^n$ is a challenging task in statistics.

Let X denote the random variable which is a probabilistic mixture of $\{X_i\}_{i=}^n$, where $X_i \sim f_i(\cdot)$. The PDF of X is given by

$$f_X(x) = \sum_{i=1}^n p_i f_i(x)$$

with mean

$$E(X) = \int x f_X(x) dx = \sum_{i=1}^n p_i \int x f_i(x) dx = \sum_{i=1}^n p_i E(X_i) = \sum_{i=1}^n p_i \mu_i \equiv \mu.$$

For the variance of X, we see that

$$\mathrm{Var}(X) = \int (x - \mu)^2 f_X(x) dx$$

$$= \int (x - \mu)^2 \sum_{i=1}^n p_i f_i(x) dx$$

$$= \sum_{i=1}^n p_i \int (x - \mu)^2 f_i(x) dx$$

$$= \sum_{i=1}^n p_i \int (x - \mu_i + \mu_i - \mu)^2 f_i(x) dx$$

$$= \sum_{i=1}^n p_i \int (x - \mu_i)^2 f_i(x) dx + \sum_{i=1}^n p_i \int (\mu_i - \mu)^2 f_i(x) dx$$

$$= \sum_{i=1}^n p_i \sigma_i^2 + \sum_{i=1}^n p_i (\mu_i - \mu)^2. \tag{7.10}$$

Example 19. In a study of Swedish non-industry fire insurance claims by Cramér (e.g., see [1]), it has been found that the claim size X follows a probabilistic mixture of exponential distribution with

$$f_X(x) = \sum_{i=1}^3 p_i \lambda_i e^{-\lambda_i x},$$

where

$$\lambda_1 = 0.014631, \quad \lambda_2 = 0.190206, \quad \lambda_3 = 5.514588$$

$$p_1 = 0.0039793, \quad p_2 = 0.1078392, \quad p_3 = 0.8881815.$$

In Illustration 10, we find $E(X)$ and $\text{Var}(X)$ by Laplace transform. However, (7.10) could have been used to do the same instead. □

Example 20. Consider a probabilistic mixture of three normals $N(2, 4^2)$, $N(1, 3^2)$, and $N(4, 6^2)$ with respective weights 0.3, 0.4, and 0.3. In Illustration 11, we will use Mathematica to illustrate the computation of the four moments by a cumulant generating function. There we see that the skewness is 1.47614 and the excess kurtosis is 4.19556. □

Random Mixture of Distributions. A random mixture of distributions is more complicated that a deterministic mixture of distribution. In Section 6.7, we have seen an example involving the variance gamma. An alternative construction of the variance gamma involves a random mixture of normal distributions in which the mixing variable in a gamma random variable. This is illustrated in the following example.

Example 21. Consider a normal random variable $X \sim N(\mu G, \sigma^2 G)$ where G is a random variable following gamma(λ, σ). To be specific, we assume that $\mu = 0.3$, $\sigma = 0.2$, and $\lambda = \sigma = 1$. Thus we have

$$f_{X|G}(x|g) = \frac{1}{\sqrt{2\pi\sigma^2 g}} \exp\left[-\frac{1}{2\sigma^2 g}(x - \mu g)^2\right]$$

$$= \frac{1}{\sqrt{2\pi(0.2)^2 g}} \exp\left[-\frac{1}{2(0.2)^2 g}(x - 0.3g)^2\right]$$

and

$$f_G(g) = \frac{\lambda e^{-\lambda t}(\lambda t)^{\alpha-1}}{\Gamma(\alpha)} = e^{-t}.$$

We see that

$$E(G) = \frac{\alpha}{\lambda} = 1 \quad \text{Var}(G) = \frac{\alpha}{\lambda^2} = 1.$$

Note that the parameters of the normal random variable X are driven by the exponential random variable G.

To find the distribution of X, we use the law of total probability and write

$$f_X(x) = \int_0^\infty f_{X|G}(x|g)f_G(g)dg.$$

In Illustration 12 , we produce numerical results and show that the skewness and excess kurtosis of X are 1.9201 and 8.716, respectively. □

Example 22. In this example, we present two handy Mathematica functions: *"esc" dist "esc"* ("esc" means pressing the escape key, typing the name of the probability distribution, and then pressing the escape key again), *"esc" cond "esc"* (means the same, except typing the condition(s) needed). To illustrate the use of the two functions, let $X \sim \phi(0,1)$. We want to compute $E(X|X > 1.96)$. Illustration 13 clearly demonstrates the usual way to do the computation and the use of Mathematica to do the same. $\square$

7.8.   Exploring with Mathematica

Illustration 1. The following computes the Table shown in Example 2.

$$fn[x\_, y\_] := \frac{1}{6} * Binomial[y, x] * \left(\frac{1}{2}\right)^y ;$$

```
Table[fn[x, y], {y, 1, 6}, {x, 0, 6}]
```

$$\left\{\left\{\frac{1}{12}, \frac{1}{12}, 0, 0, 0, 0, 0\right\}, \left\{\frac{1}{24}, \frac{1}{12}, \frac{1}{24}, 0, 0, 0, 0\right\}, \left\{\frac{1}{48}, \frac{1}{16}, \frac{1}{16}, \frac{1}{48}, 0, 0, 0\right\},\right.$$
$$\left\{\frac{1}{96}, \frac{1}{24}, \frac{1}{16}, \frac{1}{24}, \frac{1}{96}, 0, 0\right\}, \left\{\frac{1}{192}, \frac{5}{192}, \frac{5}{96}, \frac{5}{96}, \frac{5}{192}, \frac{1}{192}, 0\right\},$$
$$\left.\left\{\frac{1}{384}, \frac{1}{64}, \frac{5}{128}, \frac{5}{96}, \frac{5}{128}, \frac{1}{64}, \frac{1}{384}\right\}\right\}$$

$\square$

Illustration 2. A plot of the joint PDF for Example 3.

```
f[x_, y_] := 2 * Exp[-x] * Exp[-2 y];

Plot3D[f[x, y], {x, 0, 2}, {y, 0, 0.5}, AxesLabel → {"x", "y", "f(x,y)"}]
```

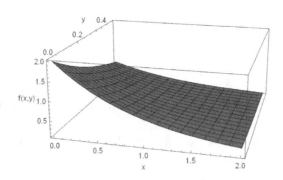

$\square$

Illustration 3 (An Application in Auto Insurance Claim). Let X be the loss due to damage of an automobile and Y be the allocated adjustment expenses associated with the accident. In Example 3, we assume that (X, Y)

has the joint PDF

$$f_{X,Y}(x,y) = \frac{5}{10^6} e^{-\frac{x}{1000}}, \quad 0 < 5y < x < \infty.$$

We plot the density below and verify that it indeed integrates to one:

$$fn[x\_, y\_] = \frac{5}{10^6} \, Exp\left[-\frac{x}{1000}\right];$$

$$\int_0^\infty \int_0^{0.2x} fn[x, y] \, dy \, dx$$

1.

```
Plot3D[fn[x, y], {x, 0, 1500}, {y, 0, 300},
  AxesLabel → {"x", "y", "f(x,y)"    "}]
```

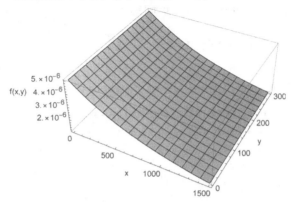

The following finds the marginal densities f_X and f_Y and $E(X)$ and $Var(X)$. We see that $Y \sim expo(\lambda)$, where $\lambda = 1/200$.

$$fx[x\_] = \int_0^{\frac{x}{5}} fn[x, y] \, dy$$

$$\frac{e^{-x/1000} \, x}{1\,000\,000}$$

$$EX = \int_0^\infty x \, fx[x] \, dx$$

2000

$$VarX = \left(\int_0^\infty x^2 \, fx[x] \, dx\right) - EX^2$$

2 000 000

$$fy[y\_] = \int_{5y}^\infty fn[x, y] \, dx$$

$$\frac{e^{-y/200}}{200}$$

We recognize that $X \sim \mathrm{Erlang}(n, \lambda)$, where $n = 2$ and $\lambda = 1/1000$. Thus, $E(X)$ and $\mathrm{Var}(X)$ could have been obtained from the results given in Section 6.3. □

Illustration 4. Example 4 illustrates the use of the Z-transform to do convolution:

```
g[s_] := ZTransform[p (1 - p)^k, k, s];

h[s_] = g[s]^2
```

$$\frac{p^2 s^2}{(-1 + p + s)^2}$$

```
InverseZTransform[%, s, z]
```

$$(1 - p)^z p^2 (1 + z)$$ □

Illustration 5. A plot of the PDF of Z, where Z is defined in Example 13:

```
fn[z_] := Exp[1 - z] - Exp[-z] /; z ≥ 1
fn[z_] := 1 - Exp[-z] /; 0 ≤ z < 1
Plot[fn[z], {z, -2, 5}]
```

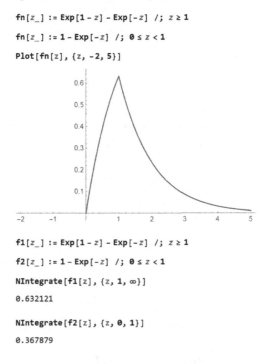

```
f1[z_] := Exp[1 - z] - Exp[-z] /; z ≥ 1
f2[z_] := 1 - Exp[-z] /; 0 ≤ z < 1
NIntegrate[f1[z], {z, 1, ∞}]
0.632121

NIntegrate[f2[z], {z, 0, 1}]
0.367879
```
          □

Illustration 6. This is related to Example 14, the use of Mellin transform to find density of $Z = X/Y$, where $X \sim \mathrm{expo}(1)$, $Y \sim \mathrm{expo}(2)$, and $X \perp Y$.

```
m1[s_] := MellinTransform[Exp[-x], x, s];

m2[s_] := MellinTransform[2 Exp[-2 x], x, s];

mz[s] = m1[s] × m2[s]
2^(1-s) Gamma[s]^2

InverseMellinTransform[%, s, x]
4 BesselK[0, 2 √2 √x]

f[x_] = 4 BesselK[0, 2 √2 √x];

NIntegrate[f[x], {x, 0, 1000}]
1.

NIntegrate[x f[x], {x, 0, 1000}]
0.5                                                    □
```

Illustration 7. This is related to Example 15, where we use Laplace transforms to do the convolution of two gamma random variables.

$$fx[x\_] := \frac{\lambda \, Exp[-\lambda \, x] \, (\lambda \, x)^{s-1}}{Gamma[s]}; \quad fy[x\_] := \frac{\lambda \, Exp[-\lambda \, x] \, (\lambda \, x)^{t-1}}{Gamma[t]};$$

```
m1[v_] = LaplaceTransform[fx[x], x, v];

m2[v_] := LaplaceTransform[fy[x], x, v];

mz[v_] = m1[v] × m2[v];

InverseLaplaceTransform[%, v, x]
```

$$\frac{e^{-x\lambda} x^{-1+s+t} \lambda^{s+t}}{Gamma[s+t]}$$
                                                       □

Illustration 8. This example is about Example 15, where $Z = X/Y$. We use Mellin transform to obtained the PDF for Z.

```
f[s_] = MellinTransform[Exp[-x], x, s];

g[s_] = MellinTransform[Exp[-x], x, s];

g1[s_] = g[2 - s];

h[s_] = f[s] × g1[s];

InverseMellinTransform[h[s], s, x]
```

$$\frac{1}{(1+x)^2}$$
                                                       □

Illustration 9. The following illustrates the derivation of PDF for the *F* distribution using the Mellin transform.

$$f[s\_] = \text{MellinTransform}\left[\frac{\frac{1}{2} \text{Exp}\left[-\frac{x}{2}\right]\left(\frac{x}{2}\right)^{\frac{m}{2} - 1}}{\text{Gamma}\left[\frac{m}{2}\right]}, x, s\right];$$

$$g[s\_] = \text{MellinTransform}\left[\frac{\frac{1}{2} \text{Exp}\left[-\frac{x}{2}\right]\left(\frac{x}{2}\right)^{\frac{n}{2} - 1}}{\text{Gamma}\left[\frac{n}{2}\right]}, x, s\right];$$

$$h[s\_] = \left(\frac{n}{m}\right)^{s-1} f[s] \times g[2 - s];$$

$$fx[x\_] = \text{InverseMellinTransform}[h[s], s, x]$$

$$\frac{m\left(\frac{m}{n}\right)^{m/2} x^{m/2}\left(1 + \frac{m\,x}{n}\right)^{\frac{1}{2}(-m-n)} \text{Gamma}\left[\frac{1}{2}(2 + m + n)\right]}{(n + m\,x)\,\text{Gamma}\left[\frac{m}{2}\right] \text{Gamma}\left[\frac{n}{2}\right]}$$

$$gx[x\_] = \left(\frac{m}{n}\right)^{\frac{m}{2}} \frac{1}{\text{Beta}\left[\frac{n}{2}, \frac{m}{2}\right]} x^{\frac{m}{2} - 1}\left(1 + \frac{m}{n} x\right)^{-\left(\frac{n+m}{2}\right)};$$

$$\text{TrueQ}[fx[x] == gx[x]]$$

True   □

Illustration 10. The computation relating to Example 19 — the Swedish insurance claims.

p1 = 0.0039793; p2 = 0.1078392; p3 = 0.8881815;

λ1 = 0.014631; λ2 = 0.190206; λ3 = 0.514588;

The following is the Laplace transform of the probabilistic mixture of three exponential densities

$$fn[s\_] := p1\left(\frac{\lambda 1}{s + \lambda 1}\right) + p2\left(\frac{\lambda 2}{s + \lambda 2}\right) + p3\left(\frac{\lambda 3}{s + \lambda 3}\right); \ |$$

EX1 = (-1) fn'[s] /. s → 0

2.56494

EX2 = (-1)² fn''[s] /. s → 0

49.8481

VarX = EX2 - EX1²

43.2691

In the following, we show the use of Mathematica function "Mixture-Distribution" to construct the displayed PDF of the exponential mixtures.

We also use the function "RandomVariate" to simulate a sample of 1000 from the exponential mixture.

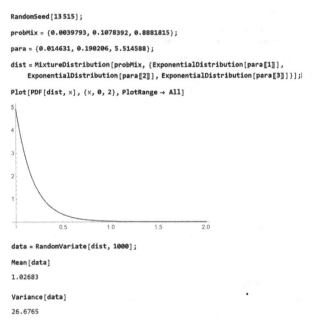

```
RandomSeed[13515];

probMix = {0.0039793, 0.1078392, 0.8881815};

para = {0.014631, 0.190206, 5.514588};

dist = MixtureDistribution[probMix, {ExponentialDistribution[para[[1]]],
    ExponentialDistribution[para[[2]]], ExponentialDistribution[para[[3]]]}];

Plot[PDF[dist, x], {x, 0, 2}, PlotRange → All]
```

```
data = RandomVariate[dist, 1000];

Mean[data]
```
1.02683

```
Variance[data]
```
26.6765

□

Illustration 11. Example 20 demonstrates the deterministic mixture of three normal random variables. It shows the use of the cumulant generating function to find higher moments of random variables.

```
cf[s_, μ_, σ_] := CharacteristicFunction[NormalDistribution[μ, σ], s];

cfmix[s_] = 0.3 cf[s, 2, 4] + 0.4 cf[s, 1, 3] + 0.3 cf[s, 4, 6];

cumgen[s_] := Log[cfmix[s]];
```

$$EX = \mathrm{Re}\left[\frac{1}{i}\, D[\text{cfmix}[s], \{s, 1\}]\ /.\ s \to 0\right]$$
2.2

$$VarX = \frac{1}{i^2}\, D[\text{cfmix}[s], \{s, 2\}]\ /.\ s \to 0$$
25.6

$$u3 = \mathrm{Re}\left[\frac{1}{i^3}\, D[\text{cfmix}[s], \{s, 3\}]\ /.\ s \to 0\right];$$

$$u4 = \frac{1}{i^4}\, D[\text{cfmix}[s], \{s, 4\}]\ /.\ s \to 0;$$

$$shewX = \frac{u3}{VarX^{3/2}}$$
1.47614

$$kurtoisisX = \frac{u4}{VarX^2}$$
4.19556

Also, we use the Mathematica's function "MixtureDistribution" to construct the PDF of the normal mixture.

```
RandomSeed[1234];

dist = MixtureDistribution[{0.3, 0.4, 0.3}, {NormalDistribution[2, 4],
    NormalDistribution[1, 3], NormalDistribution[4, 6]}];

Plot[PDF[dist, x], {x, -20, 25}, PlotRange → All]
```

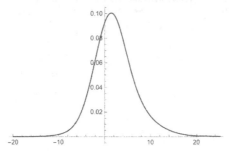

```
data = RandomVariate[dist, 1000];

Mean[data]
2.20259

Variance[data]
21.5948
```
□

Illustration 12. The following produces numerical results of Example 20.

$$fx[x\_, \mu\_, \sigma\_, g\_] := \frac{1}{\sqrt{2\pi\sigma^2 g}} Exp\left[-\frac{1}{2\sigma^2 g}(x-\mu g)^2\right];$$

$$gx[g\_, \alpha\_, \lambda\_] := \frac{\lambda Exp[-\lambda g] (\lambda g)^{\alpha-1}}{Gamma[\alpha]};$$

$$hx[x\_] := \int_0^\infty fx[x, 0.3, 0.2, g] \times gx[g, 1, 1] dg;$$

```
Plot[hx[x], {x, -1.5, 1.5}, PlotRange → All]
```

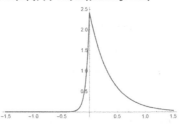

```
ex = Table[NIntegrate[(x - 0.3)^k hx[x], {x, -∞, ∞}], {k, 1, 4}]
{-6.93889×10^-18, 0.13, 0.09, 0.1473}

VarX = ex[[2]]
0.13

skewX = ex[[3]] / VarX^(3/2)
1.92012

KurX = ex[[4]] / VarX^2
8.71598
```
□

Illustration 13. Compute the conditional expectation in two ways:

(i) The usual way to compute the conditional expectation E(X|X>1.96)

$$\frac{\int_{1.96}^{\infty} \frac{x}{2\pi} \mathsf{Exp}\left[-\frac{1}{2}x^2\right]\mathsf{d}x}{\int_{1.96}^{\infty} \frac{1}{2\pi} \mathsf{Exp}\left[-\frac{1}{2}x^2\right]\mathsf{d}x} \quad // \text{ N}$$

2.33783

(ii) The MATHEMATICA way to do the same

NExpectation[x ⫯ x > 1.96, x ≈ NormalDistribution[]]

2.33783

Note the use of the two unusual symbols, the first is "condition on" and the second is "distributed as". □

Problems

1. Let $\{X_i\}_{i=1}^3$ be iid rvs with $X_1 \sim U(0,1)$. Let $\{X_{(i)}\}_{i=1}^3$ be the three order statistics from $\{X_i\}_{i=1}^3$. Find

$$P\left(X_{(3)} > X_{(2)} + d, \ X_{(2)} > X_{(1)} + d\right)$$
$$= P(X_{(3)} - X_{(2)} > d, \ X_{(2)} - X_{(1)} > d)$$

where $d \leq \frac{1}{2}$.

2. Consider a triangle characterized by $C = \{(x, y) : x \geq 0, y \geq 0, x + y \leq 2\}$.

(a) Assume any point in this triangle will occur with equal probability, what is the joint density $f_{X,Y}(x, y)$ for $(x, y) \in C$. (b) Find $f_X(x)$. (c) Find $f_{Y|X}(y|x)$. (d) Find $P(Y > 1|X = x)$.

3. Let $X \sim U(0,1)$ and $Y|X = p \sim \text{bino}(n, p)$. What is $p_Y(y)$?

4. Use a probabilistic argument to show that

$$\int_0^1 p^i(1-p)^{n-i}dp = \frac{i!(n-i)!}{(n+1)!}.$$

5. Consider an urn that contains n red and m blue balls. A random sample of size X is chosen from the urn, where $P(X = i) = \frac{1}{n}$, $i = 1, 2, \ldots, n$. Define the event $R = \{$all balls in the sample are red$\}$. What is $P(R)$?

6. The joint PDF of X and Y is

$$f_{X,Y}(x, y) = a^2 e^{-a(x+y)}, \quad x > 0, y > 0.$$

Let $U = X + Y$. What is the density of $f_U(u)$?

7. Let X and Y be two random variables with joint PDF

$$f_{X,Y}(x,y) = 1, \quad 0 < x < 1, \quad 0 < y < 1.$$

(a) Find $P(X < \frac{1}{2}, Y < \frac{1}{2})$. (b) Find $P(X+Y < 1)$. (c) Find $P(X+Y > 1)$. (d) Find $P(X > 2Y)$. (e) Find $P(X < Y \mid Y > 1/2)$

8. The joint PDF of X and Y is

$$f_{X,Y}(x,y) = e^{-(x+y)}, \quad x > 0, \quad y > 0.$$

(a) Find $P(X > 1)$. (b) Find $P(a < X + Y < b)$, given $0 < a < b$. (c) Find $P(X < Y \mid X < 2Y)$. (d) Find a such that $(X + Y < a) = 0.5$.

9. The joint PDF of X and Y is

$$f_{X,Y}(x,y) = e^{-(x+y)}, \quad x > 0, \quad y > 0.$$

We take three random samples from the above density. What is the probability at least one sample will be inside of the unit square?

10. Assume that the joint PDF of X and Y is given by

$$f_{X,Y}(x,y) = 4x(1-y), \quad 0 < x < 1, \quad 0 < y < 1.$$

Find $f_{X|Y}(x|y < 0.5)$.

11. The joint PDF of random variables X and Y is

$$f_{X,Y}(x,y) = c(y^2 - x^2)e^{-y}, \quad -y < x < y, \quad 0 < y < \infty.$$

(a) Find c. (b) Find $f_X(x)$ and $f_Y(y)$

12. The joint PDF of random variables X and Y is

$$f_{X,Y}(x,y) = \frac{6}{7}\left(x^2 + \frac{xy}{2}\right), \quad 0 < x < 1, \quad 0 < y < 2.$$

(a) Verify that the above function is indeed joint PDF. (b) Find $f_X(x)$. (c) Find $P(X > Y)$. (d) Find $P(Y > \frac{1}{2} \mid X < \frac{1}{2})$.

13. An ambulance travels back and forth along a road of length L at a constant speed. Let X be the location of the ambulance at any time. Thus, we may assume $X \sim U(0, L)$. Similarly, let Y denote the location of an accident along the road at any time. We also assume that $Y \sim U(0, L)$. Assuming that the distance, denoted by Z, between the ambulance's position and the accident's location at the time of accident

is independent. (a) What is $F_Z(z)$? (b) What is $f_Z(z)$? (c) What is $E(Z)$?

14. The joint PDF of random variables X and Y is given by

$$f_{X,Y}(x, y) = \begin{cases} \dfrac{1}{\pi} & x^2 + y^2 \leq 1, \\ 0 & \text{otherwise.} \end{cases}$$

Let

$$U = \sqrt{X^2 + Y^2}, \quad V = \tan^{-1}\left(\frac{Y}{X}\right).$$

(a) Find $f_{U,V}(u, v)$. (b) Find $F_U(u)$ and $F_V(v)$ (c) Are U and V two independent random variables?

15. Let $X \sim N(0, 1)$, $Y \sim N(0, 1)$, and $X \perp Y$. Let $U = X + Y$ and $V = X - Y$. Find $f_U(u)$ and $f_V(v)$. Are $U \perp V$?

16. X and Y are independent and identically distributed random variables with a common density expo(1). Define

$$U = \frac{X}{X + Y}, \quad V = X + Y.$$

(a) Find $f_U(u)$ and $f_V(v)$ (b) Are $U \perp V$?

17. Let $X \sim N(0, 1)$ and $Y \sim \text{expo}(1)$. Assume that $X \perp Y$. Let $W = X\sqrt{2Y}$. Find $F_Z(z)$.

18. Assume that three balls are drawn without replacement from an urn containing 5 white, and 8 red balls. Let

$$X_i = \begin{cases} 1 & \text{if the } i\text{th chosen ball is white,} \\ 0 & \text{otherwise.} \end{cases}$$

Find the joint PMF of (a) $p_{X_1,X_2}(x_1, x_2)$ and (b) $p_{X_1,X_2,X_3}(x_1, x_2, x_3)$.

19. The number of people entering a store in a given hour is a Poisson random variable with parameter $\lambda = 10$. Compute the conditional probability that at most 4 men entered the drugstore, given that 8 women entered in that hour. Assume that there is an equal chance that a given customer is a man or a woman.

20. The joint PDF of random variables X and Y is given by

$$f_{X,Y}(x,y) = \begin{cases} x+y & 0 < x < 1, \quad 0 < y < 1, \\ 0 & \text{otherwise.} \end{cases}$$

(a) Are X and Y independent? (b) What is the PDF of X? (c) Find $P(X + Y < 1)$.

21. The joint PDF of random variables X and Y is given by

$$f_{X,Y}(x,y) = \begin{cases} 12xy(1-x) & 0 < x < 1, 0 < y < 1, \\ 0 & \text{otherwise.} \end{cases}$$

(a) Are X and Y independent? (b) Find $E(X)$ and $E(Y)$. (c) Find $\text{Var}(X)$ and $\text{Var}(Y)$.

22. The daily revenue of Bistro Italiano follows a normal distribution with mean 3000 and standard deviation 540. What is the probability that (a) the total revenue in six days exceeds 20,000? (b) the daily revenue exceeds 3200 in at least 2 of the next 3 days? Assuming independence in revenues from one day to the next.

23. Assume 30 percent of men never eat lunch and 45 percent of women never eat lunch. A random sample of 300 men and 300 women are selected. Find the approximate probability that (a) at least 240 of the sampled people never eat lunch, (b) the number of men who never eat lunch is at least half of the number of women who never eat lunch.

24. The joint PDF of random variables X and Y is given by

$$f_{X,Y}(x,y) = xe^{-x(y+1)}, \quad x > 0, y > 0.$$

(a) Find $f_{X|Y}(x|y)$. (b) Find $f_{Y|X}(y|x)$. (c) Let $Z = XY$, find $f_Z(z)$. (d) Can you use the Mellin transform to solve for (c).

25. We take a random sample of size 5 for $U(0,1)$. Find $P\left(\frac{1}{4} < \text{mean} < \frac{3}{4}\right)$.

26. The joint PDF of random variables X and Y is given by

$$f_{X,Y}(x,y) = \begin{cases} \dfrac{1}{y} & y > x > 0, 1 > y > 0, \\ 0 & \text{otherwise.} \end{cases} \qquad y > x > 0, \quad 1 > y > 0.$$

Find (a) $E(XY)$, (b) $E(X)$, and $E(Y)$.

27. We roll a fair die 10 times. Let X be the sum of the 10 rolls. Find $E(X)$.

28. The joint PDF of random variables X and Y is given by

$$f_{X,Y}(x,y) = \frac{e^{-\frac{x}{y}}e^{-y}}{y}, \quad x > 0, \quad y > 0.$$

Find $E(X^2|Y)$.

29. Suppose that a point (X, Y) is chosen according to the uniform distribution on the triangle with vertices $(0,0), (0,1), (1,0)$. (a) Find $E(X)$ and $\text{Var}(X)$. (b) the conditional mean and variance of X given that $Y = \frac{1}{3}$.

30. Assume that X and Y are random variables with a joint density

$$f_{X,Y}(x,y) = \begin{cases} \dfrac{2}{5}(2x + 3y) & \text{if } 0 < x < 1 \text{ and } 0 < y < 1, \\ 0 & \text{otherwise.} \end{cases}$$

Let $Z = 2X + 3Y$. (a) What is the distribution function $F_Z(z)$? (b) Find $E(X)$ and $E(Y)$, and then $E(Z)$. (c) Use $F_Z(z)$ to find $E(Z)$.

31. Assume that X and Y are random variables with a joint density

$$f_{X,Y}(x,y) = \begin{cases} \dfrac{x}{5} + \dfrac{y}{20} & \text{if } 0 < x < 1 \text{ and } 0 < y < 5, \\ 0 & \text{otherwise.} \end{cases}$$

(a) Are X and Y independent? (b) Find $P(X + Y) > 3)$. (c) Let $U = X + Y$ and $V = X - Y$. Find the joint density $f_{X,Y}(x,y)$ of X and Y.

32. A complex machine is able to operate effectively as long as at least 3 of its 5 motors are functioning. If each motor independently functions for a random amount of time with density $f_X(x) = xe^{-x}$, $x > 0$, compute the density of the length of time that the machine functions.

33. If X_1, X_2, X_3 are independent random variables that are uniformly distributed over $(0, 1)$, compute the probability that the largest of the three is greater than the sum of the other two.

34. Let X and Y be two random variables with joint density

$$f_{XY}(x,y) = \begin{cases} \lambda xy^2 & \text{if } 0 < x < y < 1, \\ 0 & \text{otherwise.} \end{cases}$$

(a) What should be the value of λ so that the density is a legitimate joint PDF? (b) What are the marginal densities for X and Y?

35. Let X and Y be two random variables with joint PDF

$$f_{X,Y}(x, y) = \begin{cases} \sqrt{\dfrac{2}{\pi}} y e^{-xy} e^{-\frac{y^2}{2}} & \text{if } x > 0, \quad y > 0, \\ 0 & \text{otherwise.} \end{cases}$$

(b) What is $f_Y(y)$? (b) Based on Part (b), what is $f_{X|Y}(x|y)$? (c) Compute $P(X < 1|Y = 1)$.

36. Epsilon Airline is an one-aircraft tiny airline flying between Houston and Las Vegas. The single Boeing 737-800 it has 162 economy seats and 12 business seats. For this problem, we focus on the passengers who are flying economy. Epsilon estimates the number of customers who requests a ticket for a given flight follows a Poisson distribution with mean 175. Epsilon decides that for a given flight, they will book a maximum of 170. Past no-show probability p is estimates at 0.09. (a) Let S denote the number of ticket holders showing up for a given flight. Find $E(S)$. (b) Let B denote the number of passengers who are denied boarding for a given flight due to overbooking?

37. Let X and Y be two random variables with joint PDF

$$f_{X,Y}(x, y) = c x(y - x)e^{-y} \quad 0 < x < y < \infty$$

(a) Find c. (b) Find the marginal PDF of X and Y. (c) Find the condition PDF $F_{X|Y}(x|y)$ and $F_{Y|X}(y|x)$. (d) Find $E(X|Y)$ and $E(Y|X)$.

38. Let $\{X_n\}$ be a sequence of independent random variables with a common PDF $unif(0, 1)$. Define

$$N = \min\{n \geq 1, X_1 + \cdots + X_n > x\}$$

and

$$G_n(x) = P(N > n).$$

and $G_0(x)$ for $x \in (0, 1)$. (a) By conditioning on X_1, express $G_n(x)$ as a function of $G_{n-1}(\cdot)$. (b) Find the PMF of N. (c) Find the *PGF* of N (d) Find $E(N)$ and $Var(N)$.

39. Consider a sequence of independent Bernoulli trials with p denoting the probability that a trial is a success and $q = 1 - p$. Let X_3 denote the number of trails needed to obtain 3 successes for the first time and Y_5 denote the number of trials need to obtain 5 failures for the first time.

(a) Find the joint probability mass function $P(X_3 = i, Y_5 = j)$ for $i < j$.
(b) Do the same for $i > j$. (c) Use Mathematica to compute the PMF for $i = 3, \ldots, 40$ and $j = 5, \ldots, 40$ and see if the PMF sums to 1.

40. Let X and Y are two i.i.d. random variables with the common density $expo(1/\lambda)$, where $\lambda > 0$. Let

$$Z = \frac{X}{1+Y}$$

Find the PDF of Z.

41. Consider a positive random variable X. Let φ_X denote the characteristic function of X and $\mathcal{M}$ denote the Mellin transform of X. Show that

$$M_X(s) = \varphi_{\log X}\left(\frac{s-1}{i}\right).$$

42. Let X and Y be two continuous random variables. Define $Z = X \cdot Y$. Use the approach via the Jacobian matrix to derive the PDF for Z. Compare the result with Equation (7.5).

Remarks and References

Illustration 3 is based on [4]. Applications of probability in the insurance sector are abundant, e.g., see [2]. The Mellin transform is covered in [1, 3]. Example 20 and Illustration 12 are related to those appeared in [5] about variance gamma. As pointed out in [5], the approach shown in Example 20 and that was given in Section 6.7, while different, produce the same variance gamma.

[1] Debnath, L. and Bhatta, D. *Integral Transforms and Their Applications*, 2nd edition, Chapman and Hall/CRC, 2007.
[2] Grandell, J. *Aspects of Risk Theory*, Springer-Verlag, 1991.
[3] Giffin, W. K. *Transform Techniques for Probability Modeling*, Academic Press, 1975.
[4] Hogg, R. V. and Klugman, S. A. *Loss Distributions*, Wiley, 1984.
[5] Madan, D. and Schoutens, W. *Applied Conic Finance*, Cambridge University Press, 2016.

Chapter 8

Dependence and More on Expectations

8.1. Introduction

When we study probability, many times we have one thing in mind — whether things are related and more specifically, in which manner they are related. As an example, in Chapter 3, we look at several cases about drawing a diamond from a standard deck without replacement. We have found that it is always 1/4 at each one of the first four draws — why so? What about fifth draws and so on? We have noted the distinction between binomial and hypergeometric random variables. Again, sampling with and without replacement plays an pivotal role there. The notion of exchangeable random variables would shed some insights on this issue. Covariance and correlation are used in assessing linear relations among random variables. Many times they are embedded within the framework of multivariate normal distributions. We shall study bivariate normal distributions in detail. We also explore approaches to do conditioning with respect to mean and variance.

8.2. Exchangeable Random Variables

Let $\{X_i\}_{i=1}^n$ be random variables with joint PMF

$$p_{X_1,\ldots,X_n}(x_1,\ldots,x_n) = P(X_1 = x_1,\ldots,X_n = x_n).$$

The joint PMF is *symmetric* if $p_{X_1,\ldots,X_n}(x_1,\ldots,x_n)$ is a symmetric function of $\{x_i\}_{i=1}^n$. In other words, all $n!$ possible ordering of $\{X_i\}_{i=1}^n$ have the same joint PMF. In this case, we call $\{X_i\}_{i=1}^n$ *exchangeable*.

Lemma 1. *Every subset of exchangeable random variables $\{X_i\}_{i=1}^n$ also is exchangeable.*

Proof. We prove the case when $k < n$ and specifically for $k = 2$. Since $\{X_i\}_{i=1}^n$ is exchangeable, we have

$$P\left(X_1 = x_1, \ldots, X_{i_1} = x_{i_1}, \ldots, X_{i_2} = x_{i_2}, \ldots, X_n = x_n\right)$$
$$= P\left(X_1 = x_1, \ldots, X_{i_2} = x_{i_1}, \ldots, X_{i_1} = x_{i_2}, \ldots, X_n = x_n\right).$$

We now sum over all probability masses associated with X_k's, where $k \neq i_1$ and i_2. This yields

$$P(X_{i_1} = x_{i_1}, X_{i_2} = x_{i_2}) = P(X_{i_2} = x_{i_1}, X_{i_1} = x_{i_2}).$$

Hence the joint PMF of X_{i_1} and X_{i_2} is symmetric and the two random variables are exchangeable. A specialization (to $k = 1$) and generalization of the proof follow a similar argument. $\qquad\square$

Example 1. Assume $\{X_i\}_{i=1}^n$ are independent identically distributed random variables with a common PMF $p_X(\cdot)$. Are $\{X_i\}_{i=1}^n$ exchangeable random variables?

By independence, we have

$$p_{X_1, \ldots, X_n}(x_1, \ldots, x_n) = \prod_{i=1}^n p_X(x_i).$$

The above holds for all $n!$ permutations of $\{X_i\}_{i=1}^n$, hence $\{X_i\}_{i=1}^n$ are exchangeable random variables. $\qquad\square$

Example 2. Assume that we have an urn that contains n red balls and m blue balls. Assume that we withdraw $n + m$ balls one at a time randomly from the urn *without* replacement. Let $\{X_i\}_{i=1}^{n+m}$ be the outcome from the random samples, where X_i denote the result from the ith draw, where

$$X_i = \begin{cases} 1 & \text{the } i\text{th draw is a red ball,} \\ 0 & \text{otherwise.} \end{cases}$$

Are $\{X_i\}_{i=1}^{n+m}$ exchangeable random variables?

Solution. We observe that any realization of $\{X_i\}$ will be a sequence of n ones and m zeros. Since there are $\binom{n+m}{n}$ random positionings of the n reds out of $n + m$ possible spots in a linear ordering, we conclude that each such realization occurs with probability

$$\frac{1}{\binom{n+m}{n}} = \frac{n!m!}{(n+m)!}.$$

Thus the corresponding PMF is symmetric with respect to its arguments and $\{X_i\}_{i=1}^{n+m}$ are exchangeable. $\qquad\square$

Example 3. Two cards are randomly drawn from a standard card deck without replacement. What is the probability that the fifth and tenth cards are diamond?

Solution. Since sampling without replacement results in exchangeable random variables, the probability is the same as the first and second card drawn are diamonds. Thus the answer is $\frac{13}{52} \times \frac{12}{51}$. □

Example 4. An urn contains 40 red balls and 30 black balls. We take 10 balls from the urn without replacement. Assume that the seventh ball is black and the ninth ball is red. What is the probability that the third is black?

Solution. Let

$$X_i = \begin{cases} 1 & \text{the } i\text{th draw is a black ball,} \\ 0 & \text{otherwise.} \end{cases}$$

Because of exchangeability of $\{X_i\}_{i=1}^{10}$, we have

$$P(X_3 = 1 | X_7 = 1, X_9 = 0) = P(X_3 = 1 | X_1 = 1, X_2 = 0) = \frac{29}{68}. \quad \square$$

8.3. Dependence

If two random variables are independent, they can be treated separately. When this is not case, we want to know how they are related and to what extent they are related. One way to accomplish this is through the use of covariance. For random variables X and Y, we define the covariance between the two by

$$\text{Cov}(X, Y) = E\left[(X - \mu_X)(Y - \mu_Y)\right] = E(XY) - \mu_X\mu_Y,$$

where $\mu_X = E(X)$ and $\mu_Y = E(Y)$. When $Y \longleftarrow X$, then $\text{Cov}(X, X) = \text{Var}(X)$. So we need to evaluate

$$E(XY) = \sum_{x,y} xy \times p_{X,Y}(x, y) \qquad \text{(for the discrete case)}$$

$$= \iint_{x,y} xy f_{X,Y}(x, y) dx dy \quad \text{(for the continuous case)}.$$

Example 5. For Example 2 of Chapter 7, we find that

$$\mathrm{Cov}(X, Y) = E(XY) - E(X) \cdot E(Y) = \frac{35}{24}.$$

The computation yielding the above is given in Illustration 1.   □

The question we have is how to give an interpretation of $\mathrm{Cov}(X, Y)$? In what way X and Y are dependent? What is the extent of this dependence?

Recall, we have shown earlier that if random variables X and Y are independent, then $E(XY) = E(X)E(Y)$. This implies that $\mathrm{Cov}(X, Y) = 0$. On the other hand, if $\mathrm{Cov}(X, Y) = 0$, it is not necessary that X and Y are independent. The following examples provides some illustrations.

Example 6. Assume $P(X = 0) = P(X = 1) = P(X = -1) = \frac{1}{3}$. Also, assume that

$$Y = \begin{cases} 0 & \text{if } X \neq 0, \\ 1 & \text{if } X = 0. \end{cases}$$

Then $E(XY) = 0$ and $E(X) = 0$. Thus $\mathrm{Cov}(X, Y) = 0$. In this example, X and Y are clearly dependent.   □

Example 7. Assume $p_{X,Y}(1, 0) = p_{X,Y}(-1, 0) = p_{x,y}(0, 1) = p_{X,Y}(0, -1) = 1/4$. Then it is easy to verify that $\mathrm{Cov}(X, Y) = 0$. But the dependence between X and Y is obvious.   □

For constants a, b, and c, and random variables X, Y, $\{X_i\}_{i=1}^n$, and $\{Y_j\}_{j=1}^m$, the following are salient properties of covariances:

(1) $\mathrm{Cov}(aX, Y) = a\mathrm{Cov}(Y, X)$;
(2) $\mathrm{Cov}(aX, bY) = ab\, \mathrm{Cov}(X, Y)$;
(3) $\mathrm{Cov}(aX + b, cY + d) = ac\, \mathrm{Cov}(X, Y)$;
(4) $\mathrm{Cov}\left(\sum_{i=1}^n X_i, \sum_{j=1}^m Y_j\right) = \sum_{i=1}^n \sum_{j=1}^m \mathrm{Cov}(X_i, Y_j)$.

The proofs for the above results can be established by setting each component in the pairs of parentheses as a single random variable and applying the definition of expectation and use the linearity property of the expectation operator.

A immediate consequence of (4) is the following useful result:

$$\operatorname{Var}\left(\sum_{i=1}^{n} X_i\right) = \operatorname{Cov}\left(\sum_{i=1}^{n} X_i, \sum_{i=1}^{n} X_i\right) = \sum_{i=1}^{n}\sum_{j=1}^{n}\operatorname{Cov}(X_i X_j)$$

$$= \sum_{i=1}^{n}\operatorname{Var}(X_i) + 2\sum\sum_{i<j}\operatorname{Cov}(X_i X_j). \tag{8.1}$$

It is easy to verify that

$$\operatorname{Var}\left(\sum_{i=1}^{n} X_i\right) = [1, \ldots, 1]\begin{bmatrix} \operatorname{Cov}(X_1, X_1) & \cdots & \operatorname{Cov}(X_1, X_n) \\ \vdots & \ddots & \vdots \\ \operatorname{Cov}(X_n, X_1) & \cdots & \operatorname{Cov}(X_n, X_n) \end{bmatrix}\begin{bmatrix} 1 \\ \vdots \\ 1 \end{bmatrix}.$$

We call the above $n \times n$ square matrix Σ, the *covariance matrix* for $\{X_i\}_{i=1}^{n}$. When $\{X_i\}_{i=1}^{n}$ are mutually independent, the off-diagonal elements of Σ vanish. Let $\mathbf{a} = \{a_1, \ldots, a_n\}$ and $\mathbf{X} = \{X_1, \ldots, X_n\}$ be row vectors. Then $\mathbf{a}\mathbf{X}^t = a_1 X_1 + \cdots + a X_n$. The variance of $\mathbf{a}\mathbf{X}^t$ is given by $\operatorname{Var}(\mathbf{a}\mathbf{X}^t) = \mathbf{a}\sum\mathbf{a}^t$.

Remarks. With Σ denoting the covariance matrix, we require the matrix Σ be positive semi-definite. This means that $\mathbf{a}\Sigma\mathbf{a}^t \geq 0$ for all real-valued vectors $\mathbf{a}$.

Example 8 (The Minimum-Variance Portfolio). Consider a mutual fund holds n different stocks $\{S_1, \ldots, S_n\}$. By a portfolio we mean a specific choice of weights $\{w_i\}_{i=1}^{n}$ on $\{S_i\}_{i=1}^{n}$. Assume that we have estimated the mean returns $\{R_i\}_{i=1}^{n}$ and the covariance matrix $\Sigma = \{\operatorname{Cov}(R_i, R_j)\}$ of these returns. We would like to find the portfolio that minimize the variance of the portfolio return subject to a targeted mean portfolio return $\widehat{R}$. This is known as a quadratic programming problem:

$$\text{minimize } \operatorname{Var}(\mathbf{w}\mathbf{X}^t) = \mathbf{w}\sum\mathbf{w}^t = \sum_{i=1}^{n}\sum_{j=1}^{n} w_i w_j \operatorname{Cov}(R_i R_j)$$

subject to

$$\mathbf{w}\mathbf{R}^t = \widehat{R}, \quad \mathbf{w}\mathbf{e}^t = 1.$$

where $\mathbf{w} = \{w_i\}_{i=1}^{n}$ is a row vector, $\mathbf{R}^t = \{R_i\}_{i=}^{n}$ is a column vector, and $\mathbf{e}^t$ is a column vector of ones. In Illustration 2, we present a numerical example. □

In the sequel, we will see that covariance measures linear relation between two random variables. But $\operatorname{Cov}(X, Y)$ depends on the unit of measurement used for each random variable. For example, X could represent the temperature in Houston on a given day and Y the same in Chicago. Depending on whether Fahrenheit or Celsius is used, the larger the covariance does not necessarily mean the relation is "more linear". This lead to a way to measure linear relation which is independent of the unit of measurement used.

Correlation Coefficient. Let X and Y be two random variables. We have

$$
\begin{aligned}
\mathrm{Var}(X+Y) &= E\left[((X+Y)-(\mu_X+\mu_Y))^2\right] \\
&= E\left[((X-\mu_X)+(Y-\mu_Y))^2\right] \\
&= E\left[(X-\mu_X)^2\right] + E[(Y-\mu_Y))^2] + 2E[(X-\mu_X)(Y-\mu_Y)] \\
&= \mathrm{Var}(X) + \mathrm{Var}(Y) + 2\,\mathrm{Cov}(X,Y).
\end{aligned}
$$

We now define the correlation coefficient between X and Y by

$$
\rho(X,Y) = \frac{\mathrm{Cov}(X,Y)}{\sqrt{\mathrm{Var}(X)\,\mathrm{Var}(Y)}} = \frac{\mathrm{Cov}(X,Y)}{\sigma_X\sigma_Y}.
$$

Lemma 2. *We have*

$$
-1 \le \rho(X,Y) \le 1.
$$

Proof. Let $Z_1 = \frac{X}{\sigma_X}$ and $Z_2 = \frac{Y}{\sigma_Y}$. We see that

$$
\begin{aligned}
\mathrm{Var}(Z_1+Z_2) &= \mathrm{Var}(Z_1) + \mathrm{Var}(Z_2) + 2\,\mathrm{Cov}(Z_1,Z_2) \\
&= \mathrm{Var}\left(\frac{X}{\sigma_X}\right) + \mathrm{Var}\left(\frac{Y}{\sigma_Y}\right) + \frac{2}{\sigma_X\sigma_Y}\mathrm{Cov}(X,Y) \\
&= 2 + 2\rho(X,Y) \\
&= 2(1+\rho(X,Y)) \ge 0.
\end{aligned}
$$

This means we have

$$
\rho(X,Y) \ge -1.
$$

Also, we observe

$$
\mathrm{Var}(Z_1-Z_2) = 2(1-\rho(X,Y)) \ge 0.
$$

Thus

$$
\rho(X,Y) \le 1.
$$

To summarize, we have $-1 \le \rho(X,Y) \le 1$. $\qquad\square$

Lemma 3. *(a) If $Y = a + bX$ where a and b is are two given constants, then*

$$\rho(X, Y) = \begin{cases} +1 & \text{if } b > 0, \\ -1 & \text{if } b < 0. \end{cases}$$

(b) If X and Y are two random variables, then the converse of the above holds.

Proof. (a) We see that

$$\text{Var}(Y) = b^2 \text{Var}(X)$$

and

$$\text{Cov}(X, Y) = \text{Cov}(X, a + bX) = b \text{Var}(X).$$

Then

$$\rho(X, Y) = \frac{\text{Cov}(X, Y)}{\sqrt{\text{Var}(X)\text{Var}(Y)}} = \frac{b \text{Var}(X)}{\sqrt{\text{Var}(X) b^2 (\text{Var}(X)}} = \frac{b}{|b|} = |1|.$$

(b) Assume $\rho(X, Y) = 1$, from Lemma 2, we know

$$\text{Var}\left(\frac{X}{\sigma_X} - \frac{Y}{\sigma_Y}\right) = 0 \implies \frac{X}{\sigma_X} - \frac{Y}{\sigma_Y} \overset{a.s.}{\to} E\left(\frac{X}{\sigma_X} - \frac{Y}{\sigma_Y}\right).$$

We write the last expression as

$$\frac{X}{\sigma_X} - \frac{Y}{\sigma_Y} = c,$$

where c is a constant. Multiplying both sides by σ_Y yields

$$Y = -c\sigma_Y + \frac{\sigma_Y}{\sigma_X} X \equiv a + bX,$$

where $b > 0$. When $\rho(X, Y) = -1$, a similar argument yields $b < 0$. □

The above two lemmas shows that the correlation coefficient measures the degree of linear dependency. If Y and X has a perfect linear relationship, then the correlation coefficient will be ± 1.

Example 9. In Illustration 3, we compute the correlation between the daily stock price of Apple and S&P500 Index. We see that the estimated sample correlation coefficient is approximately 0.8988. □

Covariances about Events: Let A and B be two events. Let indicator random variables I_A and I_B be defined for A and B. Then we have $E(I_A) = P(A)$, $E(I_B) = P(B)$, and $E(I_A I_B) = P(AB)$. Using the definition of covariance we have

$$\text{Cov}(I_A, I_B) = P(AB) - P(A)P(B)$$
$$= P(B)\left[P(A|B) - P(A)\right] = P(A)\left[P(B|A) - P(B)\right].$$

We see when $\text{Cov}(I_A, I_B) = 0$, we have $P(AB) - P(A)P(B) = 0$, or $P(AB) = P(A)P(B)$ and thus A and B are independent. But it is not necessarily so for random variables.

When $\text{Cov}(I_A, I_B) > 0$, it follows that $P(A|B) > P(A)$ and $P(B|A) > P(B)$. We say the two events are *positively dependent*. Knowing the occurrence of one event will increase the conditional probability of the occurrence of the other. Similarly, when $\text{Cov}(I_A, I_B) < 0$, it follows that $P(A|B) < P(A)$ and $P(B|A) < P(B)$. We say the two events are *negatively dependent*. Knowing the occurrence of one event will decrease the conditional probability of the occurrence of the other.

8.4. Bivariate Normal Distributions

Standardized Bivariate Normal. Assume $X \sim N(0,1)$ and $Y|X = x \sim N(\rho x, 1 - \rho^2)$. What is the joint density $f_{X,Y}(x,y)$? We see

$$f_{X,Y}(x,y) = f_{Y|X}(y|x)f_X(x)$$
$$= \frac{1}{\sqrt{2\pi}\sqrt{1-\rho^2}} \exp\left(-\left(\frac{(y - \rho x)^2}{2(1-\rho^2)}\right)\right) \times \frac{1}{\sqrt{2\pi}} \exp\left(-\frac{x^2}{2}\right)$$
$$= \frac{1}{2\pi\sqrt{1-\rho^2}} \exp\left(-\left(\frac{y^2 - 2\rho xy + \rho^2 x^2}{2(1-\rho^2)} + \frac{x^2}{2}\right)\right)$$
$$= \frac{1}{2\pi\sqrt{1-\rho^2}} \exp\left(-\frac{1}{2(1-\rho^2)}\left(x^2 - 2\rho xy + y^2\right)\right).$$

By symmetry, assume that $Y \sim N(0,1)$ and $X|Y = y \sim N(\rho y, 1 - \rho^2)$. The joint density $f_{X,Y}(x,y)$ will again assume the above form. We call the above the standard bivariate normal density and use $\text{BVN}((0,1),(0,1),\rho)$ to denote it.

Generate Standardized Bivariate Normal from Independent Normals: Given $X \perp Z$, $X \sim N(0,1)$, and $Z \sim N(0,1)$. Define

$$Y = \rho X + \sqrt{1 - \rho^2} Z.$$

We see that Y is a linear transformation of two independent standard normals. We now show that $f_{X,Y}(z, y)$ follows the standard bivariate normal density, namely, $(X, Y) \sim \text{BVN}((0,1), (0,1), \rho)$.

We write

$$\begin{bmatrix} X \\ Y \end{bmatrix} = \begin{bmatrix} 1 & 0 \\ \rho & \sqrt{1-\rho^2} \end{bmatrix} \begin{bmatrix} X \\ Z \end{bmatrix}.$$

We see that T is a linear transformation from (x, z) onto (x, y) with

$$x = g_1(x, z) = x,$$
$$y = g_2(x, z) = \rho x + \sqrt{1 - \rho^2} z.$$

The inverse transformation from (x, y) onto (x, z) is given by

$$x = h_1(x, y) = x,$$
$$z = h_2(x, y) = -\frac{\rho}{\sqrt{1-\rho^2}} x + \frac{1}{\sqrt{1-\rho^2}} y.$$

The Jacobian matrix is

$$J(x, z) = \begin{vmatrix} \dfrac{\partial g_1}{\partial x} & \dfrac{\partial g_1}{\partial z} \\ \dfrac{\partial g_2}{\partial x} & \dfrac{\partial g_2}{\partial z} \end{vmatrix} = \begin{vmatrix} 1 & 0 \\ \rho & \sqrt{1-\rho^2} \end{vmatrix} = \sqrt{1 - \rho^2}.$$

This implies

$$|J(x, z)|^{-1} = \frac{1}{\sqrt{1-\rho^2}}.$$

We now make the needed substitutions and find

$$f_{X,Y}(x, y) = \frac{1}{2\pi\sqrt{1-\rho^2}} \exp\left\{ -\frac{1}{2}\left(x^2 + \left(-\frac{\rho}{\sqrt{1-\rho^2}} x + \frac{1}{\sqrt{1-\rho^2}} y \right)^2 \right) \right\}$$

$$= \frac{1}{2\pi\sqrt{1-\rho^2}} \exp\left\{ -\frac{1}{2(1-\rho^2)}\left(x^2 - 2\rho x y + y^2 \right) \right\}.$$

This is the density of BVN$((0,1),(0,1),\rho)$. Among many of its implications, the above relation is useful in Monte-Carlo simulation for generating BVN random variables.

Example 10. In Illustration 4, we illustrate the generation of a sequence of correlated standard normal samples from the simple linear transformation of two sequences of independent standard normals. □

Example 11. Let $X \sim N(0,1)$ and $Y \sim N(0,1)$. Assume $X \perp Y$ and

$$X_\theta = X \cos \theta + Y \sin \theta.$$

We want to know $f_{X_\theta}(\cdot)$. Let T be the transformation maps (x,y) onto (x,x_θ) with

$$x = g_1(x,y) = x, \quad x_\theta = g_2(x,y) = x \cos \theta + y \sin \theta,$$

i.e.,

$$\begin{bmatrix} X \\ X_\theta \end{bmatrix} = \begin{bmatrix} 1 & 0 \\ \cos \theta & \sin \theta \end{bmatrix} \begin{bmatrix} X \\ Y \end{bmatrix}.$$

The inverse transform T^{-1} maps (x, x_θ) onto (x, y) with

$$x = h_1(x, x_\theta) = x,$$

$$y = h_2(x, x_\theta) = -x \frac{\cos \theta}{\sin \theta} + \frac{1}{\sin \theta} x_\theta.$$

For the Jacobian matrix, we have

$$J(x,y) = \begin{vmatrix} \dfrac{\partial g_1}{\partial x} & \dfrac{\partial g_1}{\partial y} \\ \dfrac{\partial g_2}{\partial x} & \dfrac{\partial g_2}{\partial y} \end{vmatrix} = \begin{vmatrix} 1 & 0 \\ \cos \theta & \sin \theta \end{vmatrix} = \sin \theta$$

and

$$|J(x,y)|^{-1} = \frac{1}{\sin \theta}.$$

Recall the density of the independent BVN

$$f_{X,Y}(x,y) = \frac{1}{2\pi} \exp\left(-\frac{1}{2}(x^2 + y^2)\right).$$

We make the needed substitutions and obtain

$$f_{X,X_\theta}(x, x_\theta) = \frac{1}{2\pi} \exp\left(-\frac{1}{2}\left(x^2 + \left(-x\frac{\cos\theta}{\sin\theta} + \frac{1}{\sin\theta}x_\theta\right)^2\right)\right)$$

$$= \frac{1}{2\pi \sin\theta} \exp\left(-\frac{1}{2\sin^2\theta}\left(x^2 - 2\left(\cos\theta\right)xx_\theta + x_\theta^2\right)\right).$$

Define $\rho = \cos\theta$. We see that

$$\sqrt{1 - \rho^2} = \sqrt{1 - \cos^2\theta} = \sin\theta.$$

Thus

$$f_{X,X_\theta}(x, x_\theta) = \frac{1}{2\pi\sqrt{1 - \rho^2}} \exp\left(-\frac{1}{2(1 - \rho^2)}\left(x^2 - 2\rho xx_\theta + x_\theta^2\right)\right).$$

Using the above joint density, we find the marginal density for X_θ as follows:

$$f_{X_\theta}(x_\theta) = \int_{-\infty}^{\infty} \frac{1}{2\pi\sqrt{1 - \rho^2}} \exp\left(-\frac{1}{2(1 - \rho^2)}\left(x^2 - 2\rho xx_\theta + x_\theta^2\right)\right) dx$$

$$= \frac{1}{2\pi\sqrt{1 - \rho^2}} \int_{-\infty}^{\infty} \exp\left(-\frac{1}{2(1 - \rho^2)}(x^2 - 2\rho xx_\theta + \rho^2 x_\theta^2\right.$$

$$\left. + (1 - \rho^2)x_\theta^2\right) dx$$

$$= \int_{-\infty}^{\infty} \frac{1}{\sqrt{2\pi}\sqrt{1 - \rho^2}} \exp\left(-\frac{1}{2(1 - \rho^2)}(x - \rho x_\theta)^2\right) dx$$

$$\times \frac{1}{\sqrt{2\pi}} \exp\left(-\frac{1}{2(1 - \rho^2)}(1 - \rho^2)x_\theta^2\right)$$

$$= \frac{1}{\sqrt{2\pi}} \exp\left(-\frac{1}{2}x_\theta^2\right).$$

Thus we conclude that $X_\theta \sim N(0, 1)$. Moreover, $f_{X|X_\theta}(x|x_\theta) = f_{X,X_\theta}(x, x_\theta)/f_{X_\theta}(x_\theta)$ implies that $X|X_\theta = x_\theta \sim N(\rho x_\theta, 1 - \rho^2)$. $\quad\square$

The aforementioned examples clearly demonstrate that the BVN possesses the property of *rotational symmetry* as displayed in the following two diagrams:

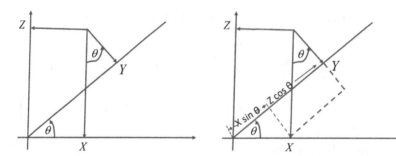

The left-hand diagram shows that X and Z are perpendicular to each other as it ought to be since they are independent. The angle between X and Y is θ. Let Y be the projection of (X, Z) at an angle θ to the X-axis as shown in the left-hand diagram. By geometry show in the left-hand diagram, we have

$$Y = X \cos \theta + Z \sin \theta.$$

We note that

$$\rho(X, Y) = E(XY) = E(X(X \cos \theta + Z \sin \theta))$$
$$= E(X^2) \cos \theta + E(XZ) \sin \theta$$
$$= \cos \theta.$$

We see when $\rho = -1$, we have $\theta = \pi$. This means $Y = -X$, i.e., Y moves in opposite direction of X. Also, when $\rho = 0$, we have $\theta = \frac{\pi}{2}$. This means Y is independent of X. Recall that, in general, $Cov(X, Y) = 0$ does not necessarily imply independence. But for BVN, it does. Also we note that notational symmetry implies that

$$P(x \leq X \leq x + \Delta x) = P(x \leq X_\theta \leq x + \Delta x), \quad 0 \leq \theta \leq 2\pi. \qquad \square$$

Example 12. In Illustration 5, we use the cosine transformation to generate correlated standard normals from two streams of independent standard normals. $\qquad \square$

Example 13. Francis Galton's student Karl Pearson measured the heights of 1078 of fathers and sons. Let X denote father's height and Y son's height. His findings were

$$E(X) = 5'9'', \quad Sdv(X) = 2'',$$
$$E(Y) = 5'10'', \quad Sdv(X) = 2'',$$

and $\rho(X, Y) = 0.5$. Given the height of a father who is $6'2''$ tall, what is the predicted height of his son.

Solution. Since $E(Y|X = x) = \rho x$, with $X = 6'2''$, the standardized value of x is

$$\frac{6'2'' - E(X)}{Sdv(X)} = \frac{6'2'' - 5'9''}{2''} = 2.5.$$

Thus $E(Y|X = 2.5) = 0.5 \times 2.5 = 1.25$ standardized unit. Now

$$\frac{Y - E(Y)}{Sdv(Y)} = \frac{Y - 5'10''}{2''} = 1.25$$

This implies

$$Y = 5'10'' + 1.25 \times 2'' = 6'0.5''.$$

We observe that the father's height is $6'2''$ but the son's height is predicted to be $6'0.5''$ tall. Galton called this phenomenon *regression to the mean*.

In Illustration 6, we display the joint PDF of random variables X and Y. Moreover, we plot the condition density of $f_{Y|X}(y|74)$. □

Example 14. Consider Example 13 once more. What is the probability that your prediction will be off by more that 1 inch?

Solution. We note that 1 inch is 0.5 times the $Sdv(X) = 2''$ and $|Y - \rho X|$ is the difference between the actual and predicted value (all in standardized units). The question becomes that of finding

$$P\left(|Y - \rho X|\right) > 0.5 \,|\, X = 2.5).$$

Noting $Y - \rho X = \sqrt{1 - \rho^2}Z$ and $Z \perp X$, we conclude that $Y - \rho X \perp X$. Now $Y - \rho X \sim N(0, 1 - \rho^2)$ where $\sqrt{1 - 0.5^2} = 0.866$. Thus

$$P\left(|Y - \rho X|\right) > 0.5 \mid X = 2.5) = P\left((|Y - \rho X|) > 0.5\right) \quad \text{(by independence)}$$

$$= P\left(\frac{|Y - \rho X| - 0}{0.866} > \frac{0.5 - 0}{0.866}\right)$$

$$= P((|Z| > 0.577) = 2\left(1 - \Phi(0.577)\right)$$

$$= 0.5639. \qquad □$$

Example 15. Consider Example 13 one more time. What is the probability that for all father–son pairs, both father and son were above the average height? In other words, using standardized values for X and Y, we ask for $P(X \geq 0, Y \geq 0)$.

Solution. *Approach* 1: A direct way to find the answer is by a brute-force integration:

$$P(X \geq 0, Y \geq 0) = P(X \geq 0, X_\theta \geq 0) = \int_0^\infty \int_0^\infty f_{X,X_\theta}(x, x_\theta) dx dx_\theta$$

$$= \int_0^\infty \int_0^\infty \frac{1}{2\pi\sqrt{1 - \rho^2}}$$

$$\times \exp\left(-\frac{1}{2(1 - \rho^2)}\left(x^2 - 2\rho x x_\theta + x_\theta^2\right)\right) dx dx_\theta,$$

where $\rho = \cos\theta = 0.5$. In Illustration 7, we find $P(X \geq 0, Y \geq 0) = \frac{1}{3}$.

Approach 2: We now give a geometric solution to the problem. Since $Y = \rho X + \sqrt{1 - \rho^2}Z$ and $X \perp Z$, we have

$$P(X \geq 0, Y \geq 0) = P(X \geq 0, \ \rho X + \sqrt{1 - \rho^2}Z \geq 0)$$

$$= P\left(X \geq 0, \ Z \geq \frac{-\rho}{\sqrt{1 - \rho^2}}X\right).$$

We sketch the region of integration below.

The diagram shows the (X, Z) plane.

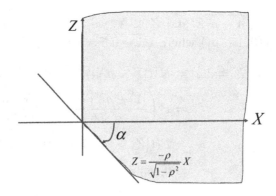

The line

$$Z = \frac{-\rho}{\sqrt{1 - \rho^2}}X \equiv bZ$$

implies that

$$\tan\alpha = \frac{-\rho}{\sqrt{1-\rho^2}} = \frac{-\frac{1}{2}}{\sqrt{1-\frac{1}{4}}} = -\frac{1}{3}\sqrt{3} \implies \alpha = \tan^{-1}\left(-\frac{1}{3}\sqrt{3}\right)$$

$$= -\frac{\pi}{6} = -30°.$$

Thus the angle β of the shaded area is $\frac{30°+90°}{360°} = \frac{1}{3}$. This means $P(X \geq 0, Y \geq 0) = \frac{1}{3}$. □

Bivariate Normal. Let $X \sim N(\mu_x, \sigma_x)$ and $Y \sim N(\mu_y, \sigma_y)$ and $\rho(X,Y) = \rho$ be the correlation coefficient. Then (X,Y) are known as the bivariate normal. We use the notation $(X,Y) \sim \text{BVN}((\mu_x, \sigma_x), (\mu_y, \sigma_y)), \rho)$. We state its density below:

$$f_{X,Y}(x,y) = \frac{1}{2\pi\sigma_x\sigma_y\sqrt{1-\rho^2}}$$

$$\times \exp\left\{-\frac{1}{2(1-\rho^2)}\left(\left(\frac{x-\mu_x}{\sigma_x}\right)^2 + \left(\frac{y-\mu_y}{\sigma_y}\right)^2\right.\right.$$

$$\left.\left.-2\rho\left(\frac{x-\mu_x}{\sigma_x}\right)\left(\frac{y-\mu_y}{\sigma_y}\right)\right)\right\}.$$

We now do a transformation of the above density using

$$Z = \frac{X-\mu_x}{\sigma_x} \quad W = \frac{Y-\mu_y}{\sigma_y}$$

The Jabobian of the above transformation $J(Z,W) = \sigma_x\sigma_y$. Hence

$$f_{Z,W}(z,w) = \frac{1}{2\pi\sqrt{1-\rho^2}}\exp\left(-\frac{1}{2(1-\rho^2)}(z^2+w^2-2\rho zw)\right)$$

$$= \frac{1}{2\pi\sqrt{1-\rho^2}}\exp\left(\frac{1}{2(1-\rho^2)}(w-\rho z)^2\right) \cdot \frac{1}{\sqrt{2\pi}}\exp\left(-\frac{z^2}{2}\right)$$

$$= f_{W|Z}(w|z)f_Z(z)$$

where $W|Z \sim N(\rho z, 1-\rho^2)$ and $Z \sim N(0,1)$.

Since $E(W|Z) = \rho Z$, it follows that

$$E\left(\frac{Y - \mu_y}{\sigma_y}\bigg| X\right) = \rho\left(\frac{X - \mu_x}{\sigma_x}\right).$$

We multiply both sides by σ_y and obtain

$$E(Y|X) = \mu_y + \frac{\rho\sigma_Y}{\sigma_X}(X - \mu_X). \tag{8.2}$$

Moreover, since $Var(W|Z) = 1 - \rho^2$, it follows that

$$Var\left(\frac{Y - \mu_y}{\sigma_y}\bigg| X\right) = 1 - \rho^2.$$

We multiply both sides by σ_y and obtain

$$Var(Y|X) = \sigma_Y^2(1 - \rho^2). \tag{8.3}$$

Based on the above, we conclude that the conditional random variable $Y|X$ is normal with its mean given by $E(Y|X)$ and variance given by $\text{Var}(Y|X)$.

Example 16 A (Simple Linear Regression). Let (X_i, Y_i) be the score earned by student i in the midterm and final exams, respectively. In Illustration 8, we use Mathematica to do a simple linear regression $Y = a + bX$. The computation using (8.2) yields precisely the fitted results shown in the illustration. Also we show the prediction is of the mean and variance of the final score Y given a midterm score X is done via (8.2) and (8.3). $\square$

Generalization to Multivariate Normal. Let $\mathbf{X}$ denote an n-dimensional random vector whose elements are $\{X_i\}_{i=1}^n$. We say $\mathbf{X}$ follows a multivariate normal (MVN) with parameters specified by the mean vector μ and covariance matrix Σ. The elements of μ are $\{\mu_i\}_{i=1}^n$, where $\mu_i = E(X_i)$. The elements of Σ are $\{\text{Cov}(X_i, X_j)\}_{i,j=1}^n$. For the case when $n = 2$, it simplifies to a BVN.

Let $\mathbf{Y} = b + BX$, where Y is a column vector of dimension m, b is a column vector of dimension m, and B is a matrix of dimension $m \times n$. Then $E(\mathbf{Y}) = b + B\mu$ and $\text{Cov}(Y) = B\Sigma B^t$, where B^t is a transpose of B. Moreover, $\mathbf{Y}$ follows MVN with the mean vector $E(\mathbf{Y})$ and covariance matrix $B\Sigma B^t$. We cite the above important results here but leave details to the literature (e.g., see, [2] or [6]).

8.5. More on Expectations and Related Subjects

In Section 7.4, we present a very useful identity:

$$E[X] = E[E[X||Y]].$$

It is sometimes known as the law of iterated expectations. Stated simply, if we want to find $E(X)$ it may be easier to condition on another related random variable Y. This might make its solution simpler. We illustrate this in the following example.

Example 17 (Craps Game Revisited). Recall Example 13 of Chapter 3. Let R denote the number of times of a roll of two dice needed to end a game of crabs. Assume we want to find (a) $E(R)$, (b) $E(R \mid$ the shooter wins), and (c) $E(R \mid$ the shooter loses).

Solution. (a) Let S denote the sum in a roll of two dice. Define $P_i = P(S = i)$. From the table shown in Example 13 of Chapter 3, we note that

$$P_i = P_{14-i} = \frac{i-1}{36}, \quad i = 2, \ldots, 7.$$

We claim that

$$E(R = i) = \begin{cases} 1 & \text{if } i = 2, 3, 7, 11, 12, \\ 1 + \dfrac{1}{P_i + P_7} & \text{otherwise.} \end{cases}$$

To see that the above assertion holds, we note the "1" accounts for the roll in the first stage of the game. In the second stage, the probability that the game terminates in one roll is $p \equiv P_i + P_7$. Thus the number of rolls needed for the game to end in the second stage follows a geometric distribution with mean $\frac{1}{p}$. The assertion follows. Thus we have

$$E(R) = 1 + \sum_{i=4}^{6} E(R|i)P_i + \sum_{i=8}^{10} E(R|i)$$

$$= 1 + \sum_{i=4}^{6} \left(\frac{P_i}{P_i + P_7} \right) + \sum_{i=8}^{10} \left(\frac{P_i}{P_i + P_7} \right)$$

$$= 1 + 2 \times \left(\frac{\frac{3}{36}}{\frac{3}{36} + \frac{6}{36}} + \frac{\frac{4}{36}}{\frac{4}{36} + \frac{6}{36}} + \frac{\frac{5}{36}}{\frac{5}{36} + \frac{6}{36}} \right)$$

$$= \frac{557}{165} = 3.3758.$$

(b) Recall that we define $B_i = \{\{i\} \text{ occurs before } \{7\}\}$ in the second stage of the game and

$$P(B_i) = \frac{P_i}{P_i + P_7}.$$

Thus the computation of winning probability $P(A)$ shown as a short-cut in Example 13 of Chapter 3 reads

$$P(A) = \sum_{i=2}^{12} P(A|S = i)P_i$$

$$= P_7 + P_{11} + \sum_{i=4}^{6}\left(\frac{P_i}{P_i + P_7}\right) + \sum_{i=8}^{10}\left(\frac{P_i}{P_i + P_7}\right) = 0.4929.$$

Now we see that

$$E(R|A) = \sum_i E(R_i|A \text{ and } S = i) \times P(A \text{ and } S = i \mid A)$$

$$= \sum_i E(R_i|A \text{ and } S = i) \times P(S = i \mid A)$$

$$= 1 + \sum_{i=4,5,6,8,9,10} E(R_i|A \text{ and } S = i) \times Q_i,$$

where $Q_i = P(S = i \mid A)$ and

$$Q_i = \frac{P(S = i, A)}{P(A)} = \frac{P_i \frac{P_i}{P_i + P_7}}{P(A)} = \frac{P_i^2}{P(A)(P_i + P_7)}, \quad i = 4, 5, 6, 8, 9, 10.$$

In the above, we see that the term $E(R|A)$ contains an "1" and all extra rolls are caused by additional rolls incurred in the second stage and yielding a win for the shooter. Using the fact that the expectation of the geometric random variable is $\frac{1}{P_i + P_7}$ where i is the *point* and the conditional probabilities $Q_i's$, we find

$$E(R|A) = 1 + \sum_{i=4}^{6}\left(\frac{1}{P_i + P_7}\right)Q_i + \sum_{i=8}^{10}\left(\frac{1}{P_i + P_7}\right)Q_i.$$

Based on the above, we find $Q_4 = Q_{10} = 0.056356$, $Q_5 = Q_9 = 0.09017$, and $Q_6 = Q_8 = 0.1281$. Hence

$$E(R|A) = 1 + 1.93842 = 2.93842.$$

(c) To find $E(R|A^c)$, we see that

$$E(R) = E(R|A)P(A) + E(R|A^c)(1 - P(A^c)).$$

Thus

$$E(R|A^c) = \frac{E(R) - E(R|A)P(A)}{1 - P(A)} = \frac{3.3758 - 2.93842 \times (0.4929)}{1 - 0.4929} = 3.8009.$$

We observe that if the shooter wins, the average number of rolls needed is about one roll shorter. $\square$

Conditional Variance. In the context of bivariate normals, we have touched on the notation of conditional variance. We now consider it a bit more. Recall the definition

$$\mathrm{Var}(X) = E[(X - E(X))^2].$$

Now the conditional random variable X given Y, denoted by $X|Y$ calls for the following modification:

$$\mathrm{Var}(X|Y) = E\left[(X - E(X|Y))^2|Y\right].$$

The above can be viewed as the formal definition of conditional variance. This definition leads to the following identity:

$$\mathrm{Var}(X) = E[\mathrm{Var}(X|Y)] + \mathrm{Var}[E(X|Y)]. \tag{8.4}$$

Due to the symmetric appearances of the expectation operator and variance operator, the above conditional variance formula is easy to remember. To show the formula holds, we recall $\mathrm{Var}(X) = E(X^2) - E^2(X)$. We now mimic the identity for the case of $X|Y$. Then it should read

$$\mathrm{Var}(X|Y) = E\left(X^2|Y\right) - E^2(X|Y).$$

We take expectations on both side of the above and write

$$\begin{aligned} \text{Term } 1 &= E\left[\mathrm{Var}(X|Y)\right] = E\left[E\left(X^2|Y\right)\right] - E\left[E^2(X|Y)\right] \\ &= E(X^2) - E\left[E^2(X|Y)\right]. \end{aligned}$$

Now we recognize $E(X|Y)$ is a random variable. Its mean is $E[E[X|Y]] = E[X]$ and its second moment is $E\left[E^2(X|Y)\right]$. Thus

$$\text{Term } 2 = \mathrm{Var}[E(X|Y)] = E\left[E^2(X|Y)\right] - E^2(X).$$

Adding the two terms, we obtain $\mathrm{Var}(X) = E[\mathrm{Var}(X|Y)] + \mathrm{Var}[E(X|Y)]$.

In (8.4), when Y is a discrete random variable with $P(Y = i) \equiv p_i$, then (8.4) reduces to (7.5). However (8.4) is applicable when Y is a continuous random variable.

Example 18. Let $N(t)$ denote the number of arrival to an airport shuttle bus in an interval $(0, t)$ follows a Poisson distribution with parameter λt. The bus arrival time Y is a random variable over the interval $[0, T]$ following an uniform distribution. Then $N(Y)$ is the number of waiting passengers when the bus arrives. We assume $N \perp Y$. We see $N(Y)$ has two random sources, namely, the uncertainties associated with the customer arrival process and the bus arrival process. We want to find (a) $E[N(Y)]$ and $\text{Var}[N(Y)]$.

Solution. (a) We see that

$$E[N(Y)|Y = t] = E[N(t)|Y = t]$$
$$= E[N(t)] = \lambda t \quad \text{(since } N \perp Y\text{)}.$$

Thus

$$E[N(Y)] = \lambda Y.$$

We take expectation with respect to Y and find

$$E[N(Y)] = E\left[E[N(Y)]\right] = E[\lambda Y] = \lambda E[Y] = \frac{\lambda T}{2}$$

as $E[Y] = \frac{T}{2}$.

(b) We use the identity that $\text{Var}[N(Y)] = E[\text{Var}[N(Y)|Y]] + \text{Var}[E[N(Y)|Y]]$. For the first term, we see

$$\text{Var}[N(Y)|Y = t] = \text{Var}[N(t)|Y = t]$$
$$= \text{Var}[N(t)] = \lambda t \quad \text{(since } N \perp Y\text{)}.$$

Thus

$$\text{Var}[N(Y)] = \lambda Y.$$

We take expectation with respect to Y and find

$$E[\text{Var}[N(Y)|Y]] = E[\lambda Y] = \frac{\lambda T}{2}.$$

For the second term, from (a) we know

$$E[N(Y)|Y] = \lambda Y.$$

Thus

$$\text{Var}\left[E[N(Y)|Y]\right] = \text{Var}\left[\lambda Y\right] = \lambda^2 \text{Var}[Y] = \lambda^2 \frac{T^2}{12}$$

as $\text{Var}[Y] = \frac{T^2}{12}$. Thus we conclude that

$$\text{Var}[N(Y)] = \frac{\lambda T}{2} + \frac{(\lambda T)^2}{12}. \qquad \square$$

Example 19 (Multinomial Distribution). Consider m independent trials each resulting in any one of the r possible outcomes with probabilities $p_1, \ldots, p_r$, where $p_1 + \cdots + p_r = 1$. Let random variable N_i denote the number of trials yielding outcome i. Then $(N_1, \ldots, N_r)$ follows a multinomial distribution with joint probability mass function

$$p_{N_1,\ldots,N_r}(n_1, \ldots, n_r) = \binom{m}{n_1, \ldots, n_r} p_1^{n_1} \cdots p_r^{n_r},$$

$$n_i = 0, 1, ..m, \text{ and } n_1 + \cdots + n_r = m.$$

Define

$$N_i(k) = \sum_{k=1}^{m} I_i(k),$$

where

$$I_i(k) = \begin{cases} 1 & \text{if trial } k \text{ yields outcome } i, \\ 0 & \text{otherwise.} \end{cases}$$

Define $q_i = 1 - p_i$. Then $N_i \sim \text{binomial}(m, p_i)$ with for $i = 1, \ldots, r$

$$E(N_i) = E\left(\sum_{k=1}^{m} I_i(k)\right) = mE\left(I_1(k)\right) = mp_i,$$

$$\text{Var}(N_i) = \text{Var}\left(\sum_{k=1}^{m} I_i(k)\right) = m\,\text{Var}(I_1(k)) = mp_i q_i.$$

We now want to obtain $\text{Cov}(N_i, N_j)$, for $i \neq j$. We see that N_i and N_j are negatively dependent. When N_i is large, N_j is expected to be small because $N_1 + \cdots + N_r = m$.

The direct way to find $\mathrm{Cov}(N_i, N_j)$ is messy as it involves first finding $E(N_i N_j)$. An indirect way is using

$$\mathrm{Var}(N_i + N_j) = \mathrm{Var}(N_i) + \mathrm{Var}(N_j) + 2\,\mathrm{Cov}(N_i, N_j).$$

Let $Z = N_i + N_j$. We see $Z \sim \mathrm{binomial}(m, p_i + p_j)$. Thus $\mathrm{Var}(Z) = \mathrm{Var}(N_i + N_j) = m(p_i + p_j)(1 - p_i - p_j)$. Hence

$$\mathrm{Cov}(N_i, N_j) = \frac{1}{2}\left[m(p_i + p_j)(1 - p_i - p_j) - mp_i(1 - p_i) - mp_j(1 - p_j)\right]$$

$$= -mp_i p_j.$$

Another way is by using the identity

$$\mathrm{Cov}(N_i, N_j) = \mathrm{Cov}\left(\sum_{k=1}^{m} I_i(k), \sum_{l=1}^{m} I_j(l)\right) = \sum_{k=1}^{m}\sum_{l=1}^{m} \mathrm{Cov}(I_i(k), I_j(l)).$$

Since for $k \neq l$, trails k and l are independent, and hence $\mathrm{Cov}(I_i(k), I_j(l)) = 0$. For $k = l$, $E(I_i(l), I_j(l)) = 0$ since it is not possible that trial l have two distinct outcomes. Moreover, $E(I_i(l)) = p_i$ and $E(I_j(l)) = p_j$. Thence $\mathrm{Cov}(I_i(l), I_j(l)) = 0 - p_i p_j$ for $l = 1, \ldots, m$. Therefore

$$\mathrm{Cov}(N_i, N_j) = -mp_i p_j. \qquad \square$$

Example 20 Minimum Variance Hedge Ratio (using Futures Contracts). Assume that a manufacturer of electronic parts wants to purchase W tons of copper a year from now. Let $t = 0$ denote the present and $t = T$ a year from now. Without hedging, the manufacturer's cash flow at time T will be $x = -xW S_T$, where S_T is the spot price of copper at time T. To protect the manufacturer against the price fluctuation of copper over time, the manufacturer decides to use futures contracts to hedge. Let F_t denote the price of a futures contract at time t for the delivery of W tons of copper at time T. We assume that the profits (or losses) of the futures account are settled at T and interest payments on margin accounts are negligible.

If the manufacturer decides to use futures contracts to hedge price fluctuations, the manufacturer's cash flow will be manufacturer will be

$$X = x + (F_T - F_0)h,$$

where h is the number of contracts chosen for hedging and X deenotes the cashflow at time T. We note that if $F_T > F_0$, the manufacturer has a saving of $F_T - F_0$. Otherwise it has to pay the difference $F_0 - F_T$. At time 0, only

F_0 is known and x and F_T are random variables. The manufacturer wants to choose h so as to minimize the variance of X.

Taking expectations yields

$$E(X) = E(x) + E(F_T) - F_0)h$$

Hence

$$
\begin{aligned}
\text{Var}(X) &= E[(X - E(X))^2] \\
&= E\left[(x + (F_T - F_0)h - E(x) - (E(F_T) - F_0)h)^2\right] \\
&= E\left[((x - E(x)) + (F_T - E(F_T)h)^2\right] \\
&= E\left[((x - E(x))^2 - 2((x - E(x))(F_T - E(F_T)h) \right. \\
&\quad \left. + (F_T - E(F_T)h)^2]\text{Var}((x - E(x)^2) + 2\text{Cov}(x, F_T) + h^2\text{Var}(F_T) \right. \\
&= \text{Var}(x) + 2h\,\text{Cov}(x, F_T) + h^2\,\text{Var}(F_T)
\end{aligned}
$$

To minimize $\text{Var}(X)$, we differentiate it with respect to h

$$\frac{d}{dh}\text{Var}(X) = 2\text{Cov}(x, F_T) + 2h\text{Var}(F_T)$$

Setting the above to zero yields

$$h^* = -\frac{\text{Cov}(x, F_T)}{\text{Var}(F_T)}$$

also we see the above gives the minimum as

$$\frac{d^2}{dh^2}\text{Var}(X) = 2\,\text{Var}(F_T) > 0.$$

Using h^* in $\text{Var}(X)$, we obtain

$$\text{Var}(X) = \text{Var}(x) + 2\left(-\frac{\text{Cov}(x, F_T)^2}{\text{Var}(F_T)}\right) + \left(-\frac{\text{Cov}(x, F_T)}{\text{Var}(F_T)}\right)^2\text{Var}(F_T)$$

$$= \text{Var}(x) - \frac{\text{Cov}(x, F_T)^2}{\text{Var}(F_T)}$$

If we do a regression $Y = a + bX$ with $Y \longleftarrow X$ and $X \longleftarrow F_T$ and apply (8.2), we obtain

$$E(S_T | F_T) = E(x) + \frac{\rho_{x,F_T} \sigma_x}{\sigma_{F_T}} (F_T - E(F_T))$$

$$= E(x) + \frac{\mathrm{Cov}(x, F_T) \sigma_{S_T}}{\sigma_x \sigma_{S_T}} (F_T - E(F_T))$$

$$= E(x) + h^*(F_T - E(F_T)).$$

The above show that the minimum variance hedge ratio is actually the regression coefficient. From (8.3), we see that

$$\mathrm{Var}\,(X | F_T) = \mathrm{Var}(x)(1 - \rho_{x,FT}^2).$$

Thus the reduction in variance is achieved by having a large $\rho_{x,FT}^2$.   □

Example 21 (A Binomial Tree for Pricing European Call Options).
In Example 10 of Chapter 4, we have looked at a single period model for pricing a European option whose underlying asset is the stock S. We now extend the model to a multi-period paradigm. For a very brief exposition on such a call option, readers are referred to the description given in the example given there. We are at time 0. Now we let L denote the expiration date of the call option. We divide L into n subintervals of length $\Delta t = \frac{L}{n}$ and let the time indices be $\{0, 1, \ldots, T\}$. Then S_T is the stock price at time T and the time T payoff of the call is

$$g(T) = (S_T - K)^+,$$

where K is the strike price. Using (4.1), the expected value of the above function of random variable S_T is given by

$$E[g(S_T)] = E\left[(S_T - K)^+\right].$$

Since S_T is a random variable, we use the binomial-tree paradigm of Cox, Ross, and Rubinstein (CCR) [1] to obtain the PMF of S_T. The sample path of moving from S_0 to S_T is identical to that of the random walk model shown in Example 14 of Chapter 3. However, in the classical random-walk model, reaching S_T is through an *additive* process as illustrated in the example. On the other hand the CCR model, reaching S_T from S_0 is a *multiplicative* process. Specifically, for $t = 1, \ldots, T$, we have

$$S_t = \begin{cases} S_{t-1}u & \text{with probability } p, \\ S_{t-1}d & \text{with probability } 1 - p. \end{cases}$$

The following graph depicts the price movement of a two-period model:

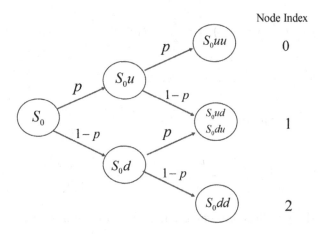

The price movement in a two-period CCR model.

In the above graph, the nodes at the last column are indexed so that they count the number of downward moves leading to the column. For a T-period CCR tree, the nodes are indexed by $\{0, 1, \ldots, T\}$.

According to CCR, we have

$$p = \frac{1 + r - d}{u - d}, \quad u = e^{\sigma\sqrt{\Delta t}}, \quad d = e^{-\sigma\sqrt{\Delta t}},$$

where σ is the parameter in the assumption that $S_T \sim \text{lognormal}(\mu, \sigma^2)$ (cf., Example 7 of Chapter 6) and r is the single-period risk-free interest rate. Moreover, p is called the *risk-neutral (RN)* probability (the idea of RN will become clear at the end of this example). In addition, we note

$$\frac{S_t - S_{t-1}}{S_{t-1}} = \frac{S_{t-1}u - S_{t-1}}{S_{t-1}} = S_{t-1}(u - 1), \quad t = 1, \ldots, T.$$

Thus $u - 1$ is the percentage of change in the stock price over Δt.

The PMF of S_T is given by bino (T, p), where

$$p_{S_T}(n) \equiv P\left(S_T = S_0 u^n d^{T-n}\right) = \binom{T}{n} p^n (1-p)^{T-n}, \quad n = 0, 1, \ldots, T.$$

Let c_t denote the price of the European call at time t. When the option expires at time T, its payoff is given by

$$c_T = E\left[(S_T - K)^+\right] = \sum_{n=0}^{T} (S_0 u^n d^{T-n} - K)^+ p_{S_T}(n).$$

When $S_T > K$, $(S_T - K)^+ = S_T - K$. Now

$$S_0 u^n d^{T-n} - K \iff \left(\frac{u}{d}\right)^n > \frac{K}{S_0 d^T},$$

$$n \log\left(\frac{u}{d}\right) > \log\left(\frac{K}{S_0 d^T}\right) \iff n > \frac{\log\left(\frac{K}{S_0 d^T}\right)}{\log\left(\frac{u}{d}\right)} \equiv x.$$

Define n^* as the smallest integer greater than or equal to x. Then

$$E\left[(S_T - K)^+\right] = \sum_{n=n^*}^{T} \binom{T}{n} p^n (1-p)^{T-n} (S_0 u^n d^{T-n} - K)$$

$$= S_0 \sum_{n=n^*}^{T} \binom{T}{n} (pu)^n ((1-p)d)^{T-n}$$

$$-K \sum_{n=n^*}^{T} \binom{T}{n} (p)^n ((1-p))^{T-n}.$$

Since we are interested in finding the call price at time 0, we discount the time T payoff back to time 0 using $(1+r)^{-T}$. Thus we have

$c_0 \equiv$ price of the call at time 0

$$= \frac{1}{(1+r)^T}\left[S_0 \sum_{n=n^*}^{T} \binom{T}{n} (pu)^n ((1-p)d)^{T-n} \right.$$

$$\left. -K \sum_{n=n^*}^{T} \binom{T}{n} (p)^n ((1-p))^{T-n} \right]$$

$$= S_0 \sum_{n=n^*}^{T} \binom{T}{n} \left(\frac{pu}{1+r}\right)^n \left(\frac{(1-p)d}{1+r}\right)^{T-n}$$

$$- \frac{K}{(1+r)^T} \sum_{n=n^*}^{T} \binom{T}{n} (p)^n ((1-p))^{T-n}$$

$$= S_0 \sum_{n=n^*}^{T} \binom{T}{n} q^n (1-q)^{T-n} - \frac{K}{(1+r)^T} \sum_{n=n^*}^{T} \binom{T}{n} (p)^n ((1-p))^{T-n},$$

where we define

$$q \equiv \frac{pu}{1+r} > 0, \quad q^* \equiv \frac{(1-p)d}{1+r} > 0.$$

Using the fact that $p = (1+r-d)/(u-d)$, or $p(u-d) = 1+r$, we have

$$q + q^* = \frac{p(u-d)+d}{1+r} = \frac{1+r-d+d}{1+r} = 1,$$

i.e., $q^* = 1 - q$. In other words, the PMF is $\text{bino}(T, q)$. To summarize we obtain

$$c_0 = S_0 \sum_{n=n^*}^{T} \binom{T}{n} q^n (1-q)^{T-n} - \frac{K}{(1+r)^T} \sum_{n=n^*}^{T} \binom{T}{n} (p)^n ((1-p))^{T-n}.$$

$$(8.5)$$

In addition to the fact that (8.2) can be used for pricing the call at time 0, the expression has a nice economical interpretation. Before we do, we first introduce the notion of an *Arrow–Debreu (AD)* security. Let A be an event and let

$$I_A(\omega) = \begin{cases} 1 & \text{if } \omega \in A, \\ 0 & \text{otherwise}, \end{cases}$$

where ω is a realization of a sample path leading to the end of the path. Then $E(I_A) = P(\omega \in A) \equiv p_A$. The price of an AD security at time 0 is equal to the value of p_A discounted back to the start of the path. It gives the expected value at time 0 of a random income of \$1 received at the start.

We now apply the notion of an AD security to the CCR binomial tree. For the first term on the right side of (8.2), we let $A_i \equiv \{S_T = \{i\}\} = \{S_T = S_0 u^{T-i} d^i\}$, where i is the node index. Then the price of an AD security for node i is given by

$$\text{AD}_i \equiv \frac{P(S_T = \{i\})}{(1+r)^T}.$$

For the second term on the right side of (8.2), we let $B = \{\mathbf{1}_{\{S_T > k\}}\}$. The price of an AD security for the event B is

$$\text{AD}_B = \frac{P(B)}{(1+r)^T}.$$

Consequently

$$c_0 = S_0 \sum_{i \in ITM}^{T} \text{AD}_i - K \times \text{AD}_B \tag{8.6}$$

where ITM represents the set of node indices where $S_T > K$.

With the above digression, we are ready to give a simple economic interpretation of (8.4). At time T, if $S_T > K$, the holder of the call option takes the ownership of stock but has to pay the strike price for exercising the call. So the right-hand side of (8.5) denotes the (holder's) discounted expected cash flow at time 0. In other words, this is the expected worth of the call to the buyer of the contract. At time 0, the seller use this amount to hedge the risk associated with the call. Hence, there is no risk to both parties and the fair price of the call at time 0 is c_0. The term risk neutral follows accordingly. An numerical example is given in Illustration 9. □

8.6. Sampling from a Finite Population

We consider taking a sample of size n from a finite population of size N. Let X be the random variable of interest and X_i the outcome of the ith observation. Assume that X has a unknown population mean μ and standard deviation σ. Then the sample average is given by

$$\overline{X}_n = \frac{X_1 + \cdots + X_n}{n}.$$

The expected value of the sample average is

$$E\left(\overline{X}_n\right) = \frac{1}{n}E(X_1 + \cdots + X_n) = \frac{1}{n}nE(X_1) = \mu$$

regardless whether sampling is done with or without replacement.

If sampling is done with replacement, then since $\{X_i\}_{i=1}^n$ are independent we have

$$\mathrm{Var}\left(\overline{X}_n\right) = \frac{1}{n^2}\mathrm{Var}\left(X_1 + \cdots + X_n\right) = \frac{1}{n^2}n\,Var(X_1) = \frac{\sigma^2}{n}$$

and thence $Sdv\left(\overline{X}_n\right) = \sigma/\sqrt{n}$. On the other hand, when sampling is done without replacement, then $\{X_i\}_{i=1}^n$ are dependent random variables. Find $\mathrm{Var}\left(\overline{X}_n\right)$ is not as straightforward. Let $S_n = X_1 + \cdots + S_n$. Using (8.1), we find

$$\mathrm{Var}(S_n) = \sum_{i=1}^n \mathrm{Var}(X_i) + 2\sum_{j<k}\mathrm{Cov}(X_j, X_k)$$

$$= n\sigma^2 + (n)(n-1)\,\mathrm{Cov}(X_1, X_2). \tag{8.7}$$

We note that since $\{X_i\}_{i=1}^n$ are exchangeable random variables, the last equality of (8.7) holds. Applying (8.7) with $n = N$ and noting $\mathrm{Var}(S_N) = 0$, it follows that

$$0 = N\sigma^2 + (N)(N-1)\,\mathrm{Cov}(X_1, X_2).$$

Solving the above yields

$$\mathrm{Cov}(X_1, X_2) = -\frac{\sigma^2}{N-1}.$$

Using the above in (8.7) once more gives

$$\mathrm{Var}(S_n) = n\sigma^2 + (n)(n-1)\left(-\frac{\sigma^2}{N-1}\right)$$

$$= n\sigma^2\left(1 - \frac{n-1}{N-n}\right) = n\sigma^2\left(\frac{N-n}{N-1}\right).$$

Thus, we conclude that

$$\mathrm{Var}\left(\overline{X}_n\right) = \mathrm{Var}\left(\frac{S_n}{n}\right) = \frac{1}{n^2}\mathrm{Var}(S_n) = \frac{1}{n}\sigma^2\left(\frac{N-n}{N-1}\right)$$

and

$$Sdv\left(\overline{X}_n\right) = \frac{\sigma}{\sqrt{n}}\sqrt{\frac{N-n}{N-1}},$$

where the term $((N-n)/(N-1))^{1/2}$ is called the *correction factor* for sampling without replacement from a finite population.

We mention in passing that the idea underlying above approach can be applied to derive the variance of a hypergeometric random variable X. The same factor $(N-n)/(N-1)$ has appeared as a part of $\mathrm{Var}(X)$ in Section 5.3.

8.7. Exploring with Mathematica

Illustration 1. The computation of covariance for Example 5 is given below:

```
fn[X_, y_] := 1/6 * Binomial[y, x] * (1/2)^y;

EX = Total[Flatten[Table[x * fn[x, y], {y, 1, 6}, {x, 0, y}]]];

EY = Total[Flatten[Table[y * fn[x, y], {y, 1, 6}, {x, 0, 6}]]];

EXY = Total[Flatten[Table[x * y * fn[x, y], {y, 1, 6}, {x, 0, y}]]];

CovXY = EXY - EX * EY
```

$$\frac{35}{24}$$

☐

Illustration 2. Consider we have the following set of data for the minimum-variance portfolio problem of Example 8.

Stock	Covariance Matrix Σ					$\{R_i\}$
1	2.30	0.93	0.62	0.74	−0.23	15.10
2	0.93	1.40	0.22	0.56	0.26	12.50
3	0.62	0.22	1.80	0.78	−0.27	14.70
4	0.74	0.56	0.78	3.40	−0.56	9.02
5	−0.23	0.26	−0.27	−0.56	2.60	17.68

We use Mathematica's function "QuadraticOptimization" to find the minimum variance portfolio using the above data set. The results are shown below:

```
CovM = {{2.30, 0.93, 0.62, 0.74, -0.23},
    {0.93, 1.40, 0.22, 0.56, 0.26},
    {0.62, 0.22, 1.80, 0.78, -0.27},
    {0.74, 0.56, 0.78, 3.40, -0.56},
    {-0.23, 0.26, -0.27, -0.56, 2.60}};

w := {a, b, c, d, e}

obj = Expand[Flatten[(w.CovM)].w]
```

$2.3\,a^2 + 1.86\,a\,b + 1.4\,b^2 + 1.24\,a\,c + 0.44\,b\,c + 1.8\,c^2 + 1.48\,a\,d + 1.12\,b\,d + 1.56\,c\,d + 3.4\,d^2 - 0.46\,a\,e + 0.52\,b\,e - 0.54\,c\,e - 1.12\,d\,e + 2.6\,e^2$

```
w := {a, b, c, d, e}

obj = Expand[Flatten[(w.CovM)].w]
```

$2.3\,a^2 + 1.86\,a\,b + 1.4\,b^2 + 1.24\,a\,c + 0.44\,b\,c + 1.8\,c^2 + 1.48\,a\,d + 1.12\,b\,d + 1.56\,c\,d + 3.4\,d^2 - 0.46\,a\,e + 0.52\,b\,e - 0.54\,c\,e - 1.12\,d\,e + 2.6\,e^2$

```
w = {a, b, c, d, e} /. QuadraticOptimization[obj,
    {a + b + c + d + e == 1,  15.1 a + 12.5 b + 14.7 c + 9.02 d + 17.68 e == 15.202},
    {a, b, c, d, e}]
```

{0.15836, 0.155337, 0.314268, 0.0379064, 0.334129}

The variance of the minimum variance portfolio given the target expected portfolio return is 15.202:

```
(w.CovM).w
```

0.659161

□

Illustration 3. The following computes the correlation between the daily stock price of Apple and S&P500 Index:

```
x = FinancialData["APPL", "Price", {{2020, 1, 1}, {2021, 2, 28}}]

y = FinancialData["S&P500", "Price", {{2020, 1, 1}, {2021, 2, 28}}]

apple = x["Values"];

sp500 = y["Values"];

Correlation[sp500, apple]

0.898766
```

Illustration 4. In the following, we generate 1000 correlated BVNs from two independent standard normals via a simple linear transformation using Monte-Carlo simulation.

```
SeedRandom[357];

x = RandomVariate[NormalDistribution[0, 1], 1000];

z = RandomVariate[NormalDistribution[0, 1], 1000];

ρ = -0.7; ρ1 = √(1 - ρ²);

y = ρ x + ρ1 z;

Correlation[x, y]

-0.717835
```

Illustration 5. In the following, we generate 1000 correlated BVNs from two independent standard normals via a cosine transformation using Monte-Carlo simulation.

```
SeedRandom[357];

x = RandomVariate[NormalDistribution[0, 1], 1000];

z = RandomVariate[NormalDistribution[0, 1], 1000];

ρ = -0.7; θ = ArcCos[ρ];

w = x Cos[θ] + z Sin[θ];

Correlation[x, w]

-0.717835
```

Illustration 6. The Pearson's 1078 pairs of the father-son heights example. First we plot the joint PDF below.

A plot of the joint PDF of X and Y

$\mu = \{69, 70\}; \Lambda = \{\{2^2, 0.5\,(2^2)\}, \{0.5\,(2^2), 2^2\}\};$

Λ // MatrixForm

rixForm=
$$\begin{pmatrix} 4 & 2. \\ 2. & 4 \end{pmatrix}$$

pXY[x_, y_] = PDF[MultinormalDistribution[μ, Λ], {x, y}];

Plot3D[pXY[x, y], {x, 62, 78}, {y, 62, 78}, PlotRange → All,
  Mesh → None, AxesLabel → {x, y, "f$_{XY}$ (x,y)"}]

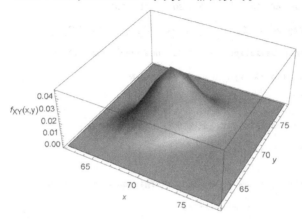

We then present the conditional PDF $f_{Y|X}(y|x = 74)$. In this case $Y|X = 74 \sim N(72.5, \sqrt{2^2(1 - .5^2)})$.

A plot of the conditional PDF f$_{Y|X}$ (y | 74) ~ Normal $\left(72.5, \sqrt{4\,(1 - 0.5^2)}\right)$

= fYgX[y_] = PDF$\left[\text{NormalDistribution}\left[72.5, \sqrt{4\,(1 - 0.5^2)}\,\right], y\right];$

= Plot[fYgX[y], {y, 67, 78}]

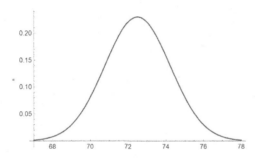

Illustration 7. We do the computation using Approach 1 of Example 15.

$$va = 1 - \frac{1}{4};$$

$$\int_0^\infty \int_0^\infty \frac{1}{2\,\pi\,\sqrt{va}}\,\text{Exp}\Big[-\frac{1}{2\,va}\,\big(x^2 - x\,y + y^2\big)\Big]\,dx\,dy$$

$$\frac{1}{3}$$ □

Illustration 8. Let (X_i, Y_i) denote the scores obtained by student i in the midterm and final exams, respectively. In the illustration, we see that (8.2) is the simple linear regression in action.

```
X = {50, 70, 90, 65, 75, 70, 40, 40, 20, 10, 30};

Y = {65, 30, 48, 28, 40, 60, 55, 30, 25, 12, 25};

data = Transpose[{X, Y}]; μx = Mean[X]; μy = Mean[Y];

σx = StandardDeviation[X]; σy = StandardDeviation[Y];

ρ = Correlation[X, Y] ; b = ρ σy / σx  // N

0.343799

lm = LinearModelFit[data, x, x]   (* From Mathematica *)
```

FittedModel $\Big[\ \boxed{20.4975 + 0.343799\,x}\ \Big]$

```
EYgivenX[xv_] = μy + b (xv - μx); VarYgivenX = (σy)² (1 - ρ²)  // N;

lm[50]   (*    From Mathematica   *)

37.6875

Print["E[Y|X=50]=", EYgivenX[50], "  Var[Y|X=50]= ", VarYgivenX]

E[Y|X=50]=37.6875  Var[Y|X=50]= 209.852

Show[ListPlot[data], Plot[lm[x], {x, 0, 90}]]
```

                                                                              □

Illustration 9. We return to Example 20. Consider a European option on a stock where $S_0 = 40$, $K = 40$, $\sigma = 0.30$ per annum, $r = 0.04$ per annum, $L = 6$ months, and $n = 500$. At the end of the CCR-based computation using (8.4), we compare the result so obtained with that found by Mathematica. While the assumption and approaches used are different, the two results are identical within two decimal places.

```
L = N[6 / 12]; n = 500; Δt = L / n;

dt = Sqrt[Δt]; r = 0.04 * Δt;

u = Exp[0.3 dt]; d = 1 / u;
```

$$p = \frac{1 + r - d}{u - d}; \quad q = \frac{p\,u}{1 + r};$$

```
S0 = 40; K = 40; T = n;
```

$$na = \text{Ceiling}\left[\frac{\text{Log}\left[\frac{K}{S0\,d^T}\right]}{\text{Log}\left[\frac{u}{d}\right]}\right] - 1;$$

```
(S0 (1 - CDF[BinomialDistribution[T, q], na])) - K
```
$$\frac{1}{(1 + r)^T}$$
```
(1 - CDF[BinomialDistribution[T, p], na])
```

3.75449

```
FinancialDerivative[{"European", "Call"}, {"StrikePrice" → 40.00, "Expiration" → 0.5},
   {"InterestRate" → 0.04, "Volatility" → 0.3, "CurrentPrice" → 40, "Dividend" → 0.00}]
```

3.75618

□

Problems

1. Consider a room with 200 people. Let X be the number of days of the year in which there are exactly three persons having the same birthday. Let Y be the number of people having distinct birthdays. (a) Find $E(X)$, and (b) find $E(Y)$.

2. Ten couples are randomly seated at a round table. Let X be the number of couples who are seated next to their partners. Find $E(X)$ and $\text{Var}(X)$.

3. Six people are hunting for partridges in a ranch. When a flock of partridges flies overhead, all fire at the same time. Each independently and randomly chooses the target and with probability of 0.5 of hitting the target. Assume that the number of partridges arriving in a flock follows that of a Poisson distribution with mean 10. Let X denote the number of partridges get hit by the hunters. Find $E(X)$.

4. There are s numbered urns and n indistinguishable balls with $n \geq s$. Each ball can be thrown into any urn with equal probability. Let X_i denote the number of balls being thrown into urn i. (a) Are $\{X_i\}$ independent? (b) To find $P(X_3 = k)$, do we need to be concerned about all other X_is? (c) Find $E(X_i)$ and $\text{Var}(X_i)$. (d) Consider the case with $n = 10$, and $s = 3$. Find $P(X_1 = 3, X_2 = 2, X_5 = 5)$.

5. Let A and B be two events with $P(A) = 0.3$, $P(B) = 0.4$, and $P(A \cup B) = 0.5$. Let $X = I_A$ and $Y = I_B$, where I_A and I_B are the two indicator random variables. Find $\text{Corr}(X, Y)$.

6. You have N boxes labeled Box $1, \ldots,$ Box N. You have k balls. You drop the balls at random into boxes, independent of each other. For each ball the probability that it will land in a particular box is the same for all boxes, namely, $\frac{1}{N}$. Let X_1 be the number of balls in Box 1 and X_N be the number of balls in Box N. Find $\text{Corr}(X_1, X_N)$.

7. A box contains 5 red balls and 8 blue balls. A random sample of size 3 is drawn *without* replacement. Let X be the number of red balls and Y be the number of blue balls selected. Compute (a) $E(X)$; (b) $E(Y)$; (c) $\text{Var}(X)$; (d) $\text{Cov}(X, Y)$.

8. A list of $2n$ numbers has mean μ and variance σ^2. Suppose that n numbers are picked at random from the list. Let A_n be the average of these n numbers, B_n be the average of the remaining n numbers. Find (a) $E(A_n - B_n)$; (b) $\text{Sdv}(A_n - B_n)$

9. Let X denote the score in the midterm exam and Y denote the score in the final exam. Assume $E(X) = 75$, $\text{Var}(X) = 25$, $E(Y) = 80$, and $\text{Var}(Y) = 16$. Assume X and Y follow a bivariate normal distribution with $\rho^2 = 0.75$. What is the probability that a student who obtained an 80 on the midterm will get a higher score on the final?

10. Let $(Z_1, Z_2) \sim \text{BVN}((0, 1), (0, 1), \rho)$. Let $\mu_1, \mu_2, \sigma_1 \geq 0, \sigma_2 \geq 0$ be given constants Define

$$\begin{bmatrix} X_1 \\ X_2 \end{bmatrix} = \begin{bmatrix} \mu_1 \\ \mu_2 \end{bmatrix} + \begin{bmatrix} \sigma_1 & 0 \\ \rho\sigma_2 & \sigma_2\sqrt{1 - \rho^2} \end{bmatrix} \begin{bmatrix} Z_1 \\ Z_2 \end{bmatrix}.$$

11. Assume that $(X, Y) \sim \text{BVN}((0, 1), (0, 1), \rho)$. (a) Let $W = X + 2Y$, Find $f_W(w)$. (b) For $\rho = 0.5$, find $P(W > 3)$.

12. Assume that $(X, Y) \sim \text{BVN}((0, 1), (0, 1), \rho)$. Let $W = X + 2Y$, find the PDF $f_W(w)$ for W.

13. Assume that $(X_1, X_2) \sim \text{BVN}((\mu_1, \sigma_1^2), (\mu_2, \sigma_2^2), \rho)$. (a) Using Mathematica to find $P(X_1 > 0, X_2 > 0)$ in terms of ρ. (b) Compute the probabilities for $\rho \in \{0.9999, 0\}$.

14. Let r_1 and r_2 denote the annual returns of stocks 1 and 2. Assume that $(r_1, r_2) \sim \text{BVN}((0.08, 0.05^2), (0.05, 0.15^2), -0.5)$. (a) What is $P(r_1 > 0.11$ and $r_2 > 0.11)$? (b) What is $P(\bar{r} > 0.11)$, where $\bar{r} = (r_1 + r_2)/2$?

15. Assume that $(X, Y) \sim \text{BVN}((0, 1), (0, 1), \rho)$. Define

$$W = \frac{(Y - \rho X)}{\sqrt{1 - \rho^2}}$$

Find the joint density $f_{X,W}(x, w)$.

16. Assume that (X, Y) follows a standard bivariate normal density with correlation ρ. (a) For $a < b$, find $E(Y | a < X < b)$ as function of a, b, and ρ. (b) Assume that midterm and final scores in a large class follow approximately a bivariate normal distribution. The mean and standard deviation for the midterm scores are 65 and 18, respectively. The mean and standard deviation for the final scores are 60 and 20, respectively. The correlation between the two scores is 0.75. Compute the average final score of students who were above average on the midterm.

17. Let $\mathbf{X} = (X_1, X_2, X_3) \sim \text{MVN}(\boldsymbol{\mu}, \Lambda)$, where $\boldsymbol{\mu} = (1, 1, 1)$ and $\Lambda = I_3$, a three-dimensional identity matrix. This implies that $\{X_i\}_{i=1}^{3}$ are mutually independent. Define

$$\mathbf{Y} \equiv \begin{bmatrix} U \\ V \end{bmatrix} = \begin{bmatrix} 2 & -1 & 1 \\ 1 & 2 & 3 \end{bmatrix} \begin{bmatrix} X_1 \\ X_2 \\ X_3 \end{bmatrix}.$$

Find $f_{V|U}(v|3)$.

18. Assume that $(X, Y) \sim \text{BVN}(2, 4), (2, 4), 0.3)$. Define the following pair of random variables
(i) X and $1 - Y$; (ii) X and $2X + Y$; (iii) X and $-0.3Y$.
(a) For each pair, find the joint density of the corresponding random variables, where we denote the second variable as W. Then each pair becomes (X, W). (b) For each pair, state whether the pair is independent.

19. Let $\{Z_i\}_{i=1}^3$ be independent MVN$((0,0,0), \Lambda)$, where the covariance matrix is a three-dimensional identity matrix. Let

$$
\mathbf{Y} \equiv \begin{bmatrix} Y_1 \\ Y_2 \\ Y_3 \end{bmatrix} = \begin{bmatrix} 1 & 0 & -1 \\ 2 & 1 & -2 \\ -2 & 0 & 3 \end{bmatrix} \begin{bmatrix} Z_1 \\ Z_2 \\ Z_3 \end{bmatrix}.
$$

Find the conditional distribution of Y_2 given $Y_1 + Y_3 = y$.

20. Let X and Y be two random variables. Assume $X \sim N(\mu, \sigma^2)$ and $Y|X = x$ follows $N(x, \tau^2)$. (a) Find $E(Y)$, Var(Y), and Cov(X, Y). (b) Find the joint density of (X, Y). (c) Find the conditional density $X|Y = y$.

21. Let X be the SP500 index and Y be the LIBOR rate of a given day. Assume that $(X, Y) \sim$ BVN$((3250, \sqrt{2500}), (6.12, \sqrt{0.36})), -0.5)$. Find $P(X > 3300|Y < 6.00)$.

22. Consider two random variables X and Y with a joint PDF

$$
f_{X,Y}(x, y) = \begin{cases} \frac{1}{2} & \text{if } x > 0, \ y > 0, \ x + y \leq 1, \\ 0 & \text{otherwise} \end{cases}
$$

(a) Find $f_{Y|X}(y|x)$. (b) Find Var(Y). (c) Find Var$(Y|X = x)$.

23. Let X and Y be two random variables. Assume that $(X, Y) \sim$ BVN$((0, 1), (0, 1), \rho)$. Let $Z = \max(X, Y)$. (a) Find $F_Z(z)$. What interpretation can be drawn by the form of $F_Z(z)$? (b) Find $f_Z(z)$. (c) Find $E(Z)$. (d) Use Mathematica function "Expectation" to numerically computer $E(Z)$ for $\rho \in \{0.2, 0.6, 0.8\}$ and compare the results against the closed-form expression obtained in Part (c).

24. Let X and Y be two random variables. Assume $(X, Y) \sim$ BVN$((0, 1), (0, 1), \rho)$. Find $E(Y|a < X < b)$, where $a < b$.

25. Let X denote the S&P500 index on a given day and Y denote the price of WTI Crude oil per barrel on a given day. Assume that $(X, Y) \sim$ BVN$((4000, 30^2), (55, 5^2), 0.65)$. What is the mean price of WTI crude oil on a given day given the S&P500 index is between 4020 and 4050 on a given day.

26. Let X and Y be two random variables. Show that

$$
E(\max(X, Y)) + E(\min(X, Y)) = E(X) + E(Y).
$$

27. Let X and Y be two independent random variables where $X \sim \text{geom}(p)$ and $Y \sim \text{geom}(q)$. (a) Let $Z = \min(X, Y)$. Find $E(W)$. (b) Find PMF $p_Z(k)$ of Z. (c) Let $W = \max(X, Y)$. Find $E(W)$. (d) Find $E(X|X \leq Y)$.

28. There are s urns and n balls where $n \geq s$. Assume that we randomly place all the balls in the urns, and that each ball has an equal probability of being placed in any urn. What is the probability that no urn is left empty?

29. Avis buys a liability insurance policy from Hertz Mutual of Nebraska. The probability that a claim will be made on this policy is 0.1. When such a claim occurs, the size of the claim follows an exponential distribution with mean $1,000,000. The maximum policy payout is $2,000,000. Find the expected value of the payout Avis would receive.

30. Let X be a random variable with PDF $N(\mu, \sigma^2)$. Let $Y = |X|$. (a) Find the PDF $f_Y(y)$. (b) Find $E(Y)$. (c) Find $E\left[(X - c)^+\right]$, where $c \geq 0$.

31. Let X be the height of a man and Y be the height of his daughter (both in inches). Suppose that the joint probability density function of X and Y is $\text{BVN}((71, 3^2), (60, 2.7^2), 0.45)$. Find the probability that the height of the daughter, of a man who is 70 inches tall, is at least 59 inches.

32. Assume X and Y follow a $\text{BVN}((\mu_1, \sigma_1^2), (\mu_2, \sigma_2^2)\rho))$. For what values of α is the variance of $\alpha X + Y$ minimum?

33. Suppose that X and Y represent SAT mathematics and verbal scores, respectively, of a random individual, and that (X, Y) has a $\text{BVN}((467, 10^2), (423, 10^2), 0.7))$ distribution over the population of individuals taking the SAT. Assume that Choosy University requires that its applicants achieve a total SAT score, $X + Y$, of at least 1100 to be considered for admission.

   (a) Find the distribution of the total SAT score $X + Y$.
   (b) What proportion of all who take the SAT achieve the criterion, $X + Y \geq 1100$, for admission?
   (c) A student who has applied to CU receives a SAT mathematics score of $X = 600$. This student's SAT verbal score is missing on the copy of the report sent by the SAT. Given that $X = 600$, what is the conditional probability that the student's total SAT score $X + Y$ exceeds 1100?

34. Let X be the Dow Jones average of common stocks and Y be the prime lending rate on a randomly chosen day. During a certain period,

X and Y have a $BVN((8500, 40^2), (4.12, 0.3^2), -0.5))$. Find the conditional probability

$$P(X > 8550|Y \le 4.00)$$

that the Dow Jones average exceeds 8550 when the prime lending rate is no greater than 4%.

35. An auto insurance company has found that the number of auto accidents occurring in a given year follows a Poisson distribution with mean 10,000. For each accident, the number of fatalities per accident follows a Poisson distribution with mean 0.1; whereas the number of non-fatal injuries per accident follows a Poisson distribution with mean 2. Also, the correlation between these two types of causalities is 0.5. Let X denote the total number of causalities per year. Find $E(X)$ and $\text{Var}(X)$.

36. Let X and Y be two random variables and assume that (X, Y) follows a BVN with $\mu_X = 2$, $\sigma_X^2 = 3$, $\mu_Y = 3$, σ_Y^2 and $Cov(X, Y) = 1$. (a) Let $Z = X - 2Y$. Find the PDF $f_Z(z)$. (b) Find $f_{Y|X}(y|x)$.

37. An unbiased coin is tossed 50 times. Let E be the event that we flip at least four heads in a row. Find $P(E)$.

38. Let $\{X_i\}_{i=1}^\infty$ be a sequence of independent identically distributed nonnegative random variables with a common mean $E(X)$. Let N be a nonnegative integer-valued random variable such that the event $\{N \ge n\}$ does not depend on $X_n, X_{n+1}, \ldots$, for $n \ge 1$. Show that

$$E\left(\sum_{k=1}^N X_k\right) = E(N)E(X).$$

39. Mississippi Riverboat Authority collaborates with Delta Softdrink Corporation to jointly launch an advertising campaign. Underneath each cap of a Softdrink bottle there is a letter. The possible letters are m, i, s, and p. The probabilities of having these letters under each cap are $\frac{1}{10}$, $\frac{1}{10}$, $\frac{1}{10}$, and $\frac{7}{10}$, respectively. Whoever gets a set of caps that reads "Mississippi" will be awarded a free ticket for a riverboat journey from New Orleans to Chicago. Let N be the number of bottles of Delta Softdrink needed to get a free ticket. Find $E(N)$.

40. There are m balls to be distributed randomly to m urns. Use Poissonization to show that the probability that each box receives exactly one ball is given by $\frac{m!}{m}|$.

Remarks and References

The data for Example 8 involving the minimum-variance portfolio is from Luenberger [5, p. 170]. The exposition on bivariate normal distribution and the related examples is influenced by those appeared in [7]. The two graphs showing rotational symmetry of correlated normals are excerpted form [7, p. 450]. The approach used in finding the various conditional expectations in the craps game (Example 16) appeared in [8]. For more about minimum-variance hedging, one may consult Hull [3], and Luenberger [5]. For the concept of the price of an Arrow–Debreu security, see for example, [4, p. 88]

[1] Cox, J. C., Ross, S. S. and Rubinstein, M., Option pricing: A simplified approach, *J. Financial Econom.*, Vol. 7, pp. 22–64, October, 1979.

[2] Gut, A., *An Intermediate Course in Probability*, Springer, 2009.

[3] Hull, J. C., *Options, Futures, and Other Derivatives*, 10th edn., Pearson, 2018.

[4] Jarrow, R. A., *Continuous-Time Asset Pricing Theory: A Martingale-Based Approach*, Springer, 2018.

[5] Luenberger, D. G., *Investment Science*, 2nd edn., Oxford University Press, 2014.

[6] Mood, A. M. and Graybill, F. A., *Introduction to the Theory of Statistics*, McGraw-Hill, 1963.

[7] Pitman, J., *Probability*, Springer, 1993.

[8] Ross, S. M., *A First Course in Probability*, 10th edn., Pearson, 2019.

Chapter 9

Limit Theorems

9.1. Markov and Chebyshev's Inequalities

Markov's Inequality. X is a *nonnegative* random variable. For any $a > 0$,

$$P(X \geq a) \leq \frac{E(X)}{a}.$$

Proof. Define

$$I = \begin{cases} 1 & \text{if } X \geq a, \\ 0 & \text{otherwise.} \end{cases}$$

Since $a > 0$, we have

$$X \geq a \Longrightarrow \frac{X}{a} \geq 1 \Longrightarrow \frac{X}{a} \geq I.$$

We take expectation of the last expression and conclude

$$E\left(\frac{X}{a}\right) \geq E(I)$$

or

$$P(X \geq a) \leq \frac{E(X)}{a}. \qquad \square$$

Example 1 (Chernoff Bound). Let X be a random variable with a moment generating function $M_X(t)$. Then

$$P(X \geq c) \leq \min_{t>0} e^{-ct} M_X(t) \quad \text{for some constant } c,$$

and

$$P(X \leq c) \leq \min_{t<0} e^{-ct} M_X(t) \quad \text{for some constant } c,$$

where the minimum is over all t for which $M_X(t)$ is finite.

Proof. For $t > 0$, we have

$$P(X \geq c) = P(tX \geq tc) = P\left(e^{tX} \geq e^{tc}\right) = P(Y \geq a).$$

where we define $Y = e^{tX}$ and $a = e^{tc}$. We see that Y is a nonnegative random variable. We apply Markov's inequality. This gives

$$P(X \geq c) \leq \frac{E(e^{tX})}{e^{ct}} = e^{-ct} M_X(t), \quad t > 0.$$

When $t < 0$, we have $P(X \leq c) = P(tX \geq tc)$. It follows that

$$P(X \leq c) \leq \min_{t<0} \frac{E(e^{tX})}{e^{ct}} = e^{-ct} M_X(t) \quad \text{for some constant } c \qquad \square$$

Example 2. Consider a sequence of independent Bernoulli trials. Let X_i denote the outcome of the ith trials and

$$X_i = \begin{cases} +1 & \text{with probability } \dfrac{1}{2}, \\ -1 & \text{with probability } \dfrac{1}{2}. \end{cases}$$

Let $S_n = X_1 + \cdots + X_n$. We want to estimate $P(S_n \geq a)$. To use the Chernoff bound, we need to find the moment generating function of S_n. For X_i, we have

$$M_{X_i}(t) = E[e^{tX}] = e^t \frac{1}{2} + e^{-t} \frac{1}{2} = \frac{1}{2}(e^t + e^{-t}).$$

Now we see

$$e^t + e^{-t} = 2 \left(1 + \frac{1}{2!}t^2 + + \frac{1}{4!}t^4 + \cdots \right)$$

$$= 2 \sum_{n=0}^{\infty} \frac{t^{2n}}{(2n)!}$$

$$\leq 2 \sum_{n=0}^{\infty} \frac{t^{2n}}{n!2^n} \quad (2n)! \geq n!2^n$$

$$= 2 \sum_{n=0}^{\infty} \frac{\left(\frac{t^2}{2} \right)^n}{n!} = 2e^{\frac{t^2}{2}}.$$

Thus

$$M_{X_i}(t) \leq e^{\frac{t^2}{2}} \implies M_{S_n}(t) \leq e^{\frac{nt^2}{2}}.$$

We apply the Chernoff bound and conclude

$$P(S_n \geq a) \leq e^{-at} e^{\frac{nt^2}{2}}.$$

To find the smallest upper bound, we differentiate with respect to the exponent and set the resulting expression to zero.

$$\frac{d}{dt} \exp\left(-at + \frac{nt^2}{2} \right) = \exp\left(-at + \frac{nt^2}{2} \right) \frac{d}{dt} \left(-at + \frac{nt^2}{2} \right)$$

$$= \exp\left(-at + \frac{nt^2}{2} \right) (-a + nt) = 0.$$

Thus we set $t = \frac{a}{n}$, i.e.,

$$P(S_n \geq a) = \exp\left(-\frac{a^2}{n} + \frac{n\left(\frac{a^2}{n^2} \right)}{2} \right)$$

$$= \exp\left(-\frac{a^2}{n} + \frac{a^2}{2n} \right) = \exp\left(-\frac{a^2}{2n} \right).$$

Now $n = 10$, $a = 6$. We have

$$P(S_{10} \geq 6) \leq e^{-\frac{36}{20}} = 0.1653.$$

This can be compared against the exact value found below:

$$P(S_{10} \geq 8) = P(\text{gambler wins at least 8 of the first 10 games})$$

$$= \frac{\binom{10}{8} + \binom{10}{9} + \binom{10}{10}}{2^{10}} = \frac{56}{1024} = 0.0547. \qquad \square$$

Chebyshev's Inequality. X is a a random variable with mean μ and variance σ^2. For any value $k > 0$

$$P(|X - \mu| \geq k) \leq \frac{\sigma^2}{k^2}.$$

Proof. $(X - \mu)^2$ is a nonnegative random variable. Let $a = k^2$. We apply Markov's inequality:

$$P\left\{(X - \mu)^2 \geq k^2\right\} \leq \frac{E[(X - \mu)^2]}{k^2}.$$

Now we see

$$(X - \mu)^2 \geq k^2 \iff |X - \mu| \geq k.$$

Thus we have

$$P(|X - \mu| \geq k) \leq \frac{\sigma^2}{k^2}. \qquad \square$$

Example 3. $X = $ the number of items produced in a factory in a week. Assume X is a random variable with $E(X) = 50$.

(a) $P(X > 75) =?$
    By Markov inequality

$$P(X > 75) \leq \frac{E(X)}{75} = \frac{50}{75} = \frac{2}{3}.$$

(b) If we know that $\text{Var}(X) = 25$, $P(40 < X < 60) =?$

$$P\left\{|X - 50| \geq 10\right\} \leq \frac{\sigma^2}{k^2} = \frac{25}{10^2} = \frac{1}{4}.$$

or

$$1 - P\left\{|X - 50| \geq 10\right\} \geq 1 - \frac{1}{4} \implies P\left\{|X - 50| < 10\right\} \geq \frac{3}{4}.$$

Thus we conclude that

$$P\left\{40 < X < 60\right\} \geq \frac{3}{4}. \qquad \square$$

Example 4. $X \sim U(0, 10)$. We have $E(X) = 5$. $\text{Var}(X) = ?$

$$Y \sim U(0, 1) \implies \text{Var}(Y) = \frac{1}{12},$$

$$X = 10Y, \quad \text{Var}(X) = 10^2 \text{Var}(Y) = \frac{100}{12} = \frac{25}{3}.$$

We apply Chebyshev inequality

$$P\left\{|X - 5| > 4\right\} \leq \frac{25}{3(4)^2} = 0.52.$$

The exact value is

$$P\left\{|X - 5| > 4\right\} = 0.2,$$

i.e.,

$$P\left\{|X - 5| > 4\right\} = 1 - \int_1^9 \frac{1}{10} dx = 1 - \frac{8}{10} = 0.2.$$

The moral of the story: If we know the probability distribution, we can find these probabilities exactly. Chebyshev inequality is mostly for use in theoretical exploration. In the next example, we illustrate one such application. $\qquad \square$

Example 5 (Bernstein Polynomial Approximation of a Continuous Function). Assume f is a real-valued continuous from over $I \equiv [0, 1]$. Then

$$f(x) = \lim_{n \to \infty} P_n(x), \quad x \in I,$$

where

$$P_n(x) = \sum_{k=0}^n f\left(\frac{k}{n}\right)\binom{n}{k} x^k (1 - x)^{n-k}.$$

Proof. The function f is continuous on I and thus uniformly continuous and bounded by M on I. For any $\varepsilon > 0$, we can find a number $\delta(\varepsilon)$ such that if $|y - x| \leq \delta(\varepsilon)$ then $|f(x) - f(y)| \leq \varepsilon$. Let $\{X_i\}$ be a sequence of independent Bernoulli random variables where

$$X_i = \begin{cases} 1 & \text{with probability } x, \\ 0 & \text{with probability } 1 - x. \end{cases}$$

We see $E(X_1) = x$ and $\text{Var}(X_1) = x(1 - x)$. Define $S_n = X_1 + \cdots + X_n$. Then $S_n \sim \text{bino}(n, x)$ with $E(S_n) = nx$ and $\text{Var}(S_n) = nx(1 - x)$. Define $M_n = S_n/n$. It follows that $E(M_n) = x$ and $\text{Var}(M_n) = x(1 - x)/n$. As $x(1 - x) \leq \frac{1}{4}$, if $x \in I$, it follows that $\text{Var}(M_n) \leq \frac{1}{4n}$.

By definition of $P_n(x)$, we note that

$$E[f(M_n)] = \sum_{k=0}^{n} f\left(\frac{k}{n}\right) P(S_n = k) = P_n(x).$$

We now have

$$|P_n(x) - f(x)| = |E[f(M_n)] - f(x)| \leq E[|f(M_n) - f(x)|]$$
$$= E[|f(M_n) - f(x)| \cap \mathbf{1}_A] + E[|f(M_n) - f(x)| \cap \mathbf{1}_{A^c}],$$

where $A = \{\omega \in \Omega \mid |M_n - x| \leq \delta(\varepsilon)\}$. By definition, we know that $E[|f(M_n) - f(x)| \cap \mathbf{1}_A] \leq \varepsilon$. Now since f is bounded by M, we have

$$E[|f(M_n) - f(x)| \, | \, \mathbf{1}_{A^c}] \leq M.$$

We apply Chebyshev's inequality and write

$$P(A^c) = P(|M_n - E(M_n)| \geq \delta(\varepsilon)) \leq \frac{\text{Var}(M_n)}{(\delta(\varepsilon))^2} = \frac{1}{4n(\delta(\varepsilon))^2}.$$

By conditioning on A^c, we obtain

$$E[|f(M_n) - f(x)| \cap \mathbf{1}_{A^c}|] \leq \frac{M}{4n(\delta(\varepsilon))^2}.$$

Combining these results, we conclude

$$|P_n(x) - f(x)| \leq \varepsilon + \frac{M}{4n(\delta(\varepsilon))^2}.$$

As $n \to \infty$, we obtain $|P_n(x) - f(x)| \to 0$ as $\varepsilon > 0$. $\qquad\qquad\square$

One-Sided Chebyshev Inequality. Let X be a random variable with mean 0 and variance σ^2. The one-sided Chebyshev inequality is

$$P(X \geq a) \leq \frac{\sigma^2}{\sigma^2 + a^2}.$$

Proof. Let $b > 0$. We see that

$$X \geq a \iff X + b \geq a + b.$$

Thus

$$P(X \geq a) = P(X + b \geq a + b).$$

Now

$$a + b > 0 \quad \text{and} \quad X + b \geq a + b$$

implies

$$(X + b)^2 \geq (a + b)^2.$$

Considering $Y = X + b$ as a random variable, we apply Markov inequality:

$$P(X \geq a) = P(X + b \geq a + b) \leq \frac{E[(X + b)^2]}{(a + b)^2}.$$

Now

$$E\left[(X + b)^2\right] = E\left[X^2 + 2bX + b^2\right]$$
$$= \text{Var}[X^2] - 2bE(X) + b^2$$
$$= \sigma^2 + b^2.$$

Thus

$$P(X \geq a) \leq \frac{\sigma^2 + b^2}{(a + b)^2}.$$

We minimizing the left side expression with respect to b. The minimum is given by $b^* = \sigma^2/a$. Using this value, we find the upper bound

$$\frac{\sigma^2 + \left(\frac{\sigma^2}{a}\right)^2}{\left(a + \left(\frac{\sigma^2}{a}\right)\right)^2} = \frac{\sigma^2}{\sigma^2 + a^2}. \qquad \square$$

Example 6. Let X be a random variable with $\mu_X = 10$ and $\sigma_X^2 = 5$. We want to find an upper bound for $P(X > 13)$.

$$P(X > 13).$$

Solution. If we use Chebyshev's inequality, we find

$$P(X > 13) = P(X - \mu_X > 3) \leq P(|X - \mu_X| > 3) \leq \frac{\sigma_X^2}{a^2} = \frac{5}{9} = 0.5555.$$

Switching to the one-sided Chebyshev inequality, set $Y = X - \mu_X$ and hence $\mu_Y = 0$ and $\sigma_Y^2 = 5$

$$P(X > 13) = P(Y > 3) \leq \frac{5}{5 + 3^2} = \frac{5}{14} = 0.3571.$$

If we knew $X \sim N(10, 5)$, then the exact value of the probability would be

$$P(X > 13) = P(Z > 0.6) = 0.2742. \qquad \square$$

Example 7. Best-Byte is an on-line merchant selling electronic gadgets. It receives an average of 8700 orders for $E-$ phones per week with a standard deviation of 1200. (a) What is a lower bound for the probability that Best-Byte will receive between 7500 and 10,000 orders next week? (b) What is an upper bound or the probability that 10,000 or more orders will be received next week?

Solution. (a) Let X be the weekly sales of E-phones. We have $\mu_X = 8700$ and $\sigma_X = 1200$. Applying Chebyshev's inequality yields

$$P(|X - 8700| \geq k) \leq \frac{(1200)^2}{k^2}.$$

It is equivalent to

$$1 - P(|X - 8700| \geq k) \geq 1 - \frac{(1200)^2}{k^2}$$

or

$$P(8700 - k \leq X \leq 8700 + k) \geq 1 - \frac{(1200)^2}{k^2}.$$

If we set $k = 1300$, then the LHS of the inequality becomes 7400 whereas the RHS meets the needed bound of 10000. Hence we assert

$$P(7500 < X < 10000) \geq (7400 < X < 10000) \geq 1 - \left(\frac{1200}{1300}\right)^2 = 0.148.$$

So the probability that Best-Byte will receive between 7500 and 10,000 orders next week is at least 0.148.

(b) Let $Y = X - \mu_X$. We apply the one-sided Chebyshev's inequality and obtain

$$P(X \geq 10000) = P(Y \geq 1300) \leq \frac{(1200)^2}{(1200)^2 + (1300)^2} = 0.46.$$

Thus the probability that 10,000 or more orders will be received next week is no more that 0.46. □

9.2. Various Forms of Convergence

Let $X_1, X_2, \ldots$ be random variables. We define several convergence concept in probability theory.

Definition 1. X_n converges almost surely (a.s.) to the random variable X as $n \to \infty$ if and only if

$$P\left(\{\lim_{n \to \infty} |X_n \to X| > \varepsilon\}\right) = 0.$$

We use $X_n \overset{a.s}{\to} X$ to express it.

An equivalent way to state the above is that if $X_n \overset{a.s}{\to} X$, then it means there exists an event $\mathcal{N}$ of zero-probability such that for all $\omega \notin \mathcal{N}$, we have

$$\lim_{n\to\infty} X_n(\omega) = X(\omega),$$

where the limit is in the ordinary sense.

Example 8. Consider a sample space $\Omega = [0,1]$. Define a sample point by $\omega = [a,b]$, namely, an interval of length $b-a$. Assume $P(\omega) = b-a$. We now define a random variable

$$X_n(\omega) = \omega^n, \quad \omega \in \Omega.$$

For any $\omega \in [0,1)$, we see that

$$\lim_{n\to\infty} X_n(\omega) = 0$$

but if $\omega = \{1\}$, we have $\lim_{n\to\infty} X_n(\omega) = 1$. It is clear that

$$\lim_{n\to\infty} X_n(\omega) \neq 0 \quad \text{for all } \omega \in \Omega,$$

i.e., $X_n \nrightarrow 0$.

Since $P(\{1\}) = 0$, we see $\{1\} \in \mathcal{N}$ (the null set defined under Definition 1). We thus conclude that $X_n \overset{a.s.}{\to} X$, where $X(\omega) = 0$. □

Example 9. Let X be a random variable. Assume $\text{Var}(X) = 0$. We now show that $X \overset{a.s.}{\to} E(X) = \mu_X$.

Proof. Let $E_n = \left\{|X - \mu_X| > \frac{1}{n}\right\}$. We see that E_n is a monotone sequence of events. We now invoke the continuity property of probability and write

$$P(\lim_{n\to\infty} E_n) = \lim P(E_n).$$

We apply the Chebyshev's inequality to the right side of the above and conclude

$$P(E_n) = P\left\{|X - \mu_X| > \frac{1}{n}\right\} \leq 0, \quad n \to \infty.$$

Hence $P(\lim_{n\to\infty} E_n) = 0$ and thence $X \overset{a.s}{\to} E(X)$. □

Definition 2. X_n converges in probability to the random variable X as $n \to \infty$ if and only if for all $\varepsilon > 0$

$$\lim_{n\to\infty} P(|X_n - X| > \varepsilon) = 0.$$

We use $X_n \overset{p}{\to} X$ to express it.

We observe that $P(|X - X| > \varepsilon)$ gives a real number. We consider its limit in the ordinary sense (i.e., the usual limit in a sequence of real numbers).

Example 10. Consider a discrete sample space $\Omega = \{0, 1\}$. Let X be a random variable such that its PMF is given by

$$
p_X(x) = \begin{cases} \dfrac{1}{2} & \text{if } x = 0, \\[2mm] \dfrac{1}{2} & \text{if } x = 1. \end{cases}
$$

Define a sequence of random variables

$$
X_n = X + \frac{1}{n}.
$$

Since $|X_n - X| = \frac{1}{n}$, $|X_n - X| \le \varepsilon$ if and only if $\frac{1}{n} \le \varepsilon$, or equivalently, $n \ge \frac{1}{\varepsilon}$, where $\varepsilon > 0$, it follows that

$$
P(|X_n - X| \le \varepsilon) = \begin{cases} \dfrac{1}{2} & \text{if } n < \dfrac{1}{\varepsilon}, \\[2mm] 1 & \text{if } n \ge \dfrac{1}{\varepsilon}. \end{cases}
$$

In other words, we have

$$
\lim_{n \to \infty} P(|X_n - X| > \varepsilon) = \begin{cases} \dfrac{1}{2} & \text{if } n < \dfrac{1}{\varepsilon}, \\[2mm] 0 & \text{if } n \ge \dfrac{1}{\varepsilon}. \end{cases}
$$

Since ε is arbitrary, we conclude that $X_n \xrightarrow{p} X$. □

Example 11. Let $X_n \sim \text{gamma}(\alpha, \lambda)$, where $\alpha = n$ and $\lambda = n$. Then

$$
E(X_n) = \frac{\alpha}{\lambda} = 1 \quad \text{and} \quad \text{Var}(X_n) = \frac{\alpha}{\lambda^2} = \frac{1}{n}.
$$

As $n \to \infty$, $\text{Var}(X_n) \to 0$, we speculate that $X_n \to 1$. We apply Chebyshev's inequality and conclude that for all $\varepsilon > 0$,

$$
P(|X_n - 1| > \varepsilon) \le \frac{\text{Var}(X_n)}{\varepsilon^2} = \frac{1}{n\varepsilon^2} \to 0 \quad \text{as } n \to \infty.
$$

Hence, $X_n \xrightarrow{p} 1$. □

Definition 3. X_n converges in r-mean to the random variable X as $n \to \infty$ if and only if

$$E \, |X_n - X|^r \to 0 \quad \text{as } n \to \infty.$$

We use $X_n \overset{r}{\to} X$ to express it.

For the special case where $r = 2$, convergence in r-mean is known as the *mean square convergence*, and write $X_n \overset{m.s.}{\to} X$. Define the *distance* between two random variables X and Y by

$$d(X, Y) = E \left[(X - Y)^2 \right],$$

where we assume that the above expectation exists. Let X_n be a sequence of random variables. Then $X_n \overset{m.s.}{\to} X$ implies

$$\lim_{n \to \infty} E \left[(X_n - X)^2 \right] = 0.$$

Example 12. Let $\{X_n\}$ be a sequence of discrete random variables with PMF defined by

$$p_{X_n}(x_n) = \begin{cases} \dfrac{1}{n} & \text{if } x_n = n, \\[2mm] 1 - \dfrac{1}{n} & \text{if } x_n = 0, \\[2mm] 0 & \text{otherwise.} \end{cases}$$

Does the mean-square limit exists?

Since

$$\lim_{n \to \infty} P(X_n = 0) = \lim_{n \to \infty} \left(1 - \frac{1}{n} \right) = 1.$$

We expect that $\{X_n\} \to X$, where $X = 0$. Now, we see that

$$E \left[(X_n - X)^2 \right] = E \left[(X_n)^2 \right] = n^2 \frac{1}{n} + (0)^2 \left(1 - \frac{1}{n} \right) = n.$$

As $n \to \infty$, the above expectation does not exists. This implies that the mean-square limit does not exist. $\qquad \square$

Example 13. Let $\{X_n\}$ be a sequence of uncorrelated random variables with $E(X_i) = 0$ and $\text{Var}(X_i) \leq M$ for all i. Define the sample average by

$$M_n = \frac{1}{n}(X_1 + \cdots + X_n).$$

We want to show that $M_n \overset{m.s.}{\to} 0$.

We see that

$$E\left[(M_n - X)^2\right] = \text{Var}(M_n) = \frac{1}{n^2}\text{Var}\left(X_1 + \cdots + X_n\right)$$

$$= \frac{1}{n^2}n\,\text{Var}(X_1) \leq \frac{M}{n} \to 0 \quad \text{as } n \to \infty,$$

where the third equality is due to the fact that $\{X_n\}$ are uncorrelated. Thus we conclude that $M_n \overset{m.s.}{\to} 0$. $\qquad\square$

Definition 4. X_n converges in distribution to the random variable X as $n \to \infty$ if and only if

$$F_{X_n}(x) \to F_X(x) \quad \text{as } n \to \infty \quad \text{for all } x \in C(F_X),$$

where $C(F_X) = \{x : F_X(x) \text{ is continuous at } x\} = $ the continuity set of F_X. We use $X_n \overset{d}{\to} X$ to express convergence in distribution.

Lemma. *Let X and $X_1, X_2, \ldots$ be random variables. The following implications hold as $n \to \infty$:*

$$X_n \overset{a.s.}{\to} X \implies X_n \overset{p}{\to} X \implies X_n \overset{d}{\to} X.$$

In addition, we have

$$X_n \overset{r}{\to} X \implies X_n \overset{p}{\to} X \implies X_n \overset{d}{\to} X.$$

Proof. See references given at the end of the chapter. $\qquad\square$

Example 14. In Chapter 6, we have looked at Cauchy random variables. We now look at a special case where random variable $X > 0$. Then its density is given by

$$f_X(x) = \frac{2}{\pi}\frac{1}{1 + x^2}, \quad x > 0.$$

In Illustration 1, we integrate the density and find

$$F_X(x) = P(X \leq x) = \frac{2}{\pi}\tan^{-1}(x).$$

Consider a sequence of independent random sample $\{X_n\}$ from the common PDF $f_X(x)$. Now we define random variable

$$M_n = \frac{1}{n}\max(X_1, \ldots, X_n).$$

We want to show that $M_n \overset{d}{\to} F_G(x)$, where

$$F_G(x) = e^{-\frac{2}{\pi x}}, \quad x > 0.$$

In other words, M_n converges in distribution to $F_G(x)$.

To establish the above, we observe

$$M_n \le x \iff \frac{1}{n}\max(X_1,\ldots,X_n) \le x \iff \max(X_1,\ldots,X_n) \le nx.$$

The last inequality implies each $\{X_i\} \le nx$. By independent of $\{X_i\}$, we find

$$P(M_n \le x) = P\{X_i \le nx\}^n = \left(\frac{2}{\pi}\tan^{-1}(nx)\right)^n.$$

In Illustration 1, we also verify that

$$\tan^{-1}(x) = \frac{\pi}{2} - \tan^{-1}\left(\frac{1}{x}\right).$$

Hence, we conclude that

$$P(M_n \le x) = \left(1 - \frac{2}{\pi}\tan^{-1}\left(\frac{1}{nx}\right)\right)^n.$$

From the series expansion of $\tan^{-1}(y)$ shown in Illustration 1, we conclude

$$\tan^{-1}\left(\frac{1}{nx}\right) = \frac{1}{nx} - \left(\frac{1}{nx}\right)^3\frac{1}{3} + O\left(\frac{1}{nx}\right)^5.$$

This implies

$$\lim_{n \to \infty} P(M_n \le x) = \lim_{n \to \infty}\left(1 - \frac{2}{\pi}\frac{1}{nx}\right)^n = e^{-\frac{2}{\pi x}}, \quad x > 0.$$

The above establishes $M_n \overset{d}{\to} F_G(x)$. In Illustration 1, we plot the PDF of G and show that $E(G) = 3.49$. $\qquad\square$

9.3. Characteristics Functions

In Section 4.4, we very briefly touched on the notion of the characteristic function (CF) of a random variable. In the subsequent chapters, we made frequent uses of it. In this chapter, we will explore more details about its various important properties. For random variable X, we recall CF is defined by

$$\varphi_X(t) = E\left[e^{itX}\right] = E\left[\cos tX + i\sin tX\right]$$

where $t \in \mathbb{R}$ and $i = \sqrt{-i}$. Since

$$\left|E[e^{itX}]\right| \leq E\left|e^{itX}\right| = E[1] = 1$$

it follows that the characteristic function exists for all $t \in \mathbb{R}$ and all random random variables.

The following is a list of several useful theorems relating to the characteristic function. We state them as we need to use them in the sequel. We skip their proofs but cite direct references. The book by Lukacs [6] focuses exclusively on the subject.

Theorem 1 (Convolution). *Let $\{X_i\}_i^n$ be independent identically distribution random variables with a common characteristic function $\varphi_X(\cdot)$. Let $X = X_1 + \cdots + X_n$. Then*

$$\varphi_X(t) = \prod_{i=1}^{n} \varphi_{X_i}(t) = (\varphi_{X(t)})^n.$$

Proof. See [6, Theorem 3.3.1]. This has been briefly mentioned in Section 6.2. $\qquad\square$

Theorem 2 (Continuity). *Let $X, X_1, \ldots, X_n$ be random variables. Assume that*

$$\varphi_{X_n}(t) \to \varphi_X(t) \quad \text{as } n \to \infty.$$

Then

$$X_n \xrightarrow{d} X \quad \text{as } n \to \infty.$$

Proof. See [6, Theorem 3.6.1]. $\qquad\square$

Theorem 3. Let X be a random variable. If $E\,|X|^n < \infty$ for some $n \in Z^+$, then

(a)

$$\varphi_X^{(i)}(0) = i^k \cdot E[X^k], \quad k = 1, 2, \ldots, n; \tag{9.1}$$

(b)

$$\varphi_X(t) = 1 + \sum_{k=1}^{n} E[X^k] \cdot \frac{(it)^k}{k!} + o\left(|t|^n\right) \quad \text{as } t \to 0. \tag{9.2}$$

Proof. See [6, Section 2.3]. It has been briefly introduced in Section 6.2. □

Digression. Consider two sequences of real numbers $\{a_n\}$ and $\{b_n\}$. We say $\{a_n\}$ is a little-o sequence with respect to $\{b_n\}$, and write $a_n = o(b_n)$ if

$$\frac{a_n}{b_n} \to 0 \quad \text{as } n \to \infty.$$

For a simple illustration, let $a_n = 1/n^2$ and $b = \frac{1}{n}$. Then $a_n = o(b_n)$ because $a_n/b_n = \frac{1}{n} \to 0$ as $n \to 0$.

Theorem 4. *Let X be a random variable and $a, b \in \mathbb{R}$. Define $Y = aX + b$. Then*

$$\varphi_Y(t) = e^{ibt} \cdot \varphi_X(at).$$

Proof. See (6.1) for the derivation. □

Theorem 5. *Let X be a random variable with distribution F and CF φ_X. If F is continuous at a and b and $a < b$, then*

$$F(b) - F(a) = \lim_{T \to \infty} \frac{1}{2\pi} \int_{-T}^{T} \frac{e^{-itb} - e^{-ita}}{-it} \varphi(t)dt. \tag{9.3}$$

Proof. This is the main inversion theorem (see [6, Theorem 3.2.1]). □

Theorem 6. *Let X be a continuous random variable with CF φ_X. Then its PDF is given by*

$$f(x) = \frac{1}{2\pi} \int_{-\infty}^{\infty} e^{-itx} \varphi(t)dt.$$

Proof. See [6, Theorem 3.2.2]. □

Theorem 7. *Let X be a discrete random variable with CF φ_X. Then its PMF is given by*

$$p_X(x) = \lim_{T \to \infty} \frac{1}{2T} \int_{-T}^{T} e^{-itx} \varphi(t)dt. \tag{9.4}$$

Proof. See [6, Theorem 3.2.3]. □

In summary, every random variable has a unique CF and the CF characterizes the distribution uniquely.

Example 15. This example illustrates an application of Theorems 2 and 3. Let X be a binomial random variable with parameters (n, p) and Y be a binomial random variable with parameters (n, p). Assume $X \perp Y$. Define $Z = X + Y$. What is the distribution for Z?

Solution. We see that $X = X_1 + \cdots + X_n$ where

$$P(X_i = 1) = p, \quad P(X_i = 0) = 1 - p \equiv q.$$

Thus

$$\varphi_{X_i}(t) = q + pe^{it}.$$

Since $\{X_i\}_{i=1}^n$ are independent and identically distributed random variables, we apply Theorem 2 and obtain

$$\varphi_X(t) = \left(q + pe^{it}\right)^n.$$

Similarly, we have

$$\varphi_Y(t) = \left(q + pe^{it}\right)^m.$$

Since $Z = X + Y$ and $X \perp Y$, we apply Theorem 2 once more and conclude that

$$\varphi_Z(t) = \left(q + pe^{it}\right)^{n+m}.$$

Since the above is the characteristic function for a binomial random variable with parameters $n + m$ and p, we conclude $Z \sim$ binomial$(n + m, p)$.

In Illustration 2, we demonstrate the use of Theorem 7 where (9.5) inverts the CF of Z with $p = 0.6, n = 3$, and $m = 5$ into the PGF of bino$(8, 0.6)$. □

Example 16. Let X be a uniform random variable over (a,b) with $a < b$. We want to find the characteristic function of X.

Solution. Using the identity $e^{itx} = \cos tx + i \sin tx$, we write

$$\varphi(t) = \int_a^b e^{itx} \frac{1}{b-a} dx$$

$$= \frac{1}{b-a} \int_a^b (\cos(tx + i \sin tx) dx$$

$$= \frac{1}{b-a} \left[\frac{1}{t} \sin tx - i \frac{1}{t} \cos tx \right]_a^b$$

$$= \frac{1}{(b-a)} \cdot \frac{1}{t} (\sin bt - \sin at - i \cos bt + i \cos at)$$

$$= \frac{1}{it(b-a)} (i \sin bt - i \sin at + \cos bt - \cos at)$$

$$= \frac{e^{itb} - e^{ita}}{it(b-a)}.$$

When $a = 0$ and $b = 1$, we have $X \sim U(0,1)$. The characteristic function of X specializes to

$$\varphi_X(t) = \frac{e^{it} - 1}{it}$$

and when $a = 0$ and $b = 2$, we have $X \sim U(0,2)$. The characteristic function of X specializes to

$$\varphi_X(t) = \frac{e^{2it} - 1}{2it}.$$

Using Theorem 5, we show an example in Illustration 3 that uses (9.3) to compute $F(5) - F(2)$ for $X \sim U(2,12)$. While trivial, it demonstrates that (9.3) indeed produces the correct answer. $\square$

Example 17. Let X be a Bernoulli random variable with $P(X = 1) = \frac{1}{2}$ and $P(X = 0) = \frac{1}{2}$. Let $Y \sim U(0,1)$. Assume $X \perp Y$. Define $Z = X + Y$. What is f_Z?

Solution. We find the respective characteristic functions as follows:

$$\varphi_X(t) = e^{it \cdot 1} \frac{1}{2} + e^{it \cdot 0} \frac{1}{2} = \frac{1}{2} \left(e^{it} + 1 \right)$$

and

$$\varphi_Y(t) = \frac{e^{it} - 1}{it}.$$

So we apply the convolution theorem and write

$$\varphi_Z(t) = \frac{e^{2it} - 1}{2it}.$$

Using the result obtained from Example 16, we conclude $Z \sim U(0,2)$. In Illustration 4, we use (9.4) to invert the CF and plot the resulting density function. $\square$

Example 18. Assume $X \sim N(0,1)$. Then

$$\varphi_X(t) = \int_{-\infty}^{\infty} e^{itx} \frac{1}{\sqrt{2\pi}} e^{-\frac{1}{2}x^2} dx$$

$$= e^{-\frac{t^2}{2}} \int_{-\infty}^{\infty} \frac{1}{\sqrt{2\pi}} \exp\left(-\frac{1}{2}(x - it)^2\right) dx = e^{-\frac{t^2}{2}}.$$

When $Y \sim N(\mu, \sigma^2)$, then $Y = \mu + \sigma X$, where $X \sim N(0,1)$. Using Theorem 4, we find

$$\varphi_Y(t) = e^{it\mu} e^{-\frac{(\sigma t)^2}{2}} = \exp\left(it\mu - \frac{\sigma^2 t^2}{2}\right).$$
$\square$

Example 19 (Compound Poisson). For $t > 0$, we let $N(t)$ denote a Poisson random variable with PMF:

$$P(N(t) = n) = e^{-\lambda t} \frac{(\lambda t)^n}{n!}, \quad n = 0, 1, \ldots.$$

$\{N(t)\}_{t \geq 0}$ is the number of arrivals in an interval of length t where the inter-arrival times of successful arrivals follow an exponential distribution with rate λ. Assume that arrival i generate a random revenue of Z_i where Z_i follows a common characteristic function $\varphi_Z(t)$. We also assume $\{Z_i\}$ are mutually independent and $\{Z_i\} \perp \{N(t)\}$. Let

$$X(t) = \sum_{i=1}^{N(t)} Z_i.$$

Then $X(t)$ represents the total revenue received by time t. (a) What is the characteristic function $\varphi_{N(t)}(s)$ of $N(t)$? (b) What is the characteristic function $\varphi_{X(t)}(s)$ of $X(t)$? (c) For $Z \sim N(10, 2^2)$ and $\lambda = 2$ and $t = 2$, what are $E(X(t))$, and $\text{Var}(X(2))$?

Solution. (a) We first derive the characteristic function of $N(t)$

$$\varphi_{N(t)}(s) = E\left[e^{isN(t)}\right] = \sum_{n=0}^{\infty} e^{isn} \cdot e^{-\lambda t}\frac{(\lambda t)^n}{n!} = e^{-\lambda t}\sum_{n=0}^{\infty}\frac{(\lambda t e^{is})^n}{n!}$$

$$= e^{-\lambda t}\left(e^{\lambda t e^{is}}\right) = e^{-\lambda t(1-e^{is})}.$$

(b) For the characteristic functions of $X(t)$, we use the law of total probability by conditioning on $N(t)$. We find

$$\varphi_{X(t)}(s) = E[e^{osX(t)}] = E_{N(t)}\left[E\left[\exp\left(is\sum_{i=1}^{N(t)}Z_i\right)\right]\Bigg|N(t)\right]$$

$$= E_{N(t)}\left[E\left(\exp\left(is\sum_{i=1}^{N(t)}Z_i\right)\right)\right] \qquad \text{by } \{Z_i\} \perp \{N(t)\}$$

$$= E_{N(t)}\left[(\varphi_Z(s)^n]\right] \qquad \text{by mutual independent of } \{Z_i\}$$

$$= \sum_{n=0}^{\infty}(\varphi_Z(s))^n\, e^{-\lambda t}\frac{(\lambda t)^n}{n!} = e^{-\lambda t}\sum_{n=0}^{\infty}\frac{(\lambda t\varphi_Z(s))^n}{n!}$$

$$= e^{-\lambda t}e^{\lambda t\varphi_Z(s)} = e^{\lambda t(\varphi_Z(s)-1)}.$$

(c) For the numerical example, we condition on $N(2)$ and write

$$E[X(2)] = E_{N(2)}\left[E[X(2)|N_2]\right] = \sum_{n=0}^{\infty}e^{-4}\frac{4^n}{n!}(nE(Z)) = e^{-4}\sum_{n=0}^{\infty}\frac{4^n}{n!}(10n) = 40.$$

In computing $\mathrm{Var}(X(2))$, we use (8.4) and write its first part:

$$E_{N(2)}\left[\mathrm{Var}[X(2)|N(2)]\right] = \sum_{n=0}^{\infty}e^{-4}\frac{4^n}{n!}(n\mathrm{Var}(Z)) = \sum_{n=0}^{\infty}e^{-4}\frac{4^n}{n!}(4n) = 16.$$

For the second part of (8.4), we have

$$\mathrm{Var}[E(X(2))|N(2)] = \mathrm{Var}[10N(2)] = 100^2\,\mathrm{Var}[N(2)] = 100^2(4) = 400.$$

Combining the two terms, we obtain

$$\mathrm{Var}[X(2)] = 416.$$

We use the result from Example 18 to find the characteristic functions for Z:

$$\varphi_Z(s) = \exp[i(10s) - 2s^2].$$

In Illustration 5, we verify the aforementioned numerical results. □

9.4. Weak Law of and Strong Law of Large Numbers

Theorem 8. (The Weak Law of Large Numbers). *Let $\{X_i\}_{i=1}^n$ be a sequence of independent and identically distributed random variables each with mean $E(X_i) = \mu$ and finite variance σ^2. Then for any $\varepsilon > 0$,*

$$P\left\{\left|\frac{X_1 + \cdots + X_n}{n} - \mu\right| \geq \varepsilon\right\} \to 0 \quad as \ n \to \infty.$$

Proof. We see

$$E\left[\frac{X_1 + \cdots + X_n}{n}\right] = \mu, \quad \mathrm{Var}\left(\frac{X_1 + \cdots + X_n}{n}\right) = \frac{\sigma^2}{n}.$$

We now apply Chebyshev's inequality

$$P\left\{\left|\frac{X_1 + \cdots + X_n}{n} - \mu\right| \geq \varepsilon\right\} \leq \frac{\sigma^2}{n\varepsilon^2} \to 0 \quad as \ n \to \infty. \qquad □$$

The weak law of large numbers deals with the *convergence in probability* whereas the strong law of large number deals with *almost surely convergence*. The latter implies the former but not vice versa. We shall skip the proof of the next important theorem as it is involved and lengthy. References will be given at the end of the chapter.

Theorem 9. (The Strong Law of Large Numbers). *Let $\{X_i\}_{i=1}^n$ be a sequence of independent and identically distributed random variables each with mean $E(X_i) = \mu$ and $\mathrm{Var}(X_i) < \infty$. Then*

$$P\left\{\lim_{n\to\infty} \frac{X_1 + \cdots + X_n}{n} = \mu\right\} = 1.$$

Example 20. Let $\{X_i\}$ be a sequence of independent and identically distributed random variables with $\mathrm{Var}(X_i) < \infty$ and a common distribution function F. Assume that F is unknown and we would like to estimate F. Define

$$Y_j(x) = 1_{\{X_j \leq x\}}$$

and for a fixed x, define

$$F_n(x) = \frac{1}{n} \sum_{j=1}^{n} Y_j(x).$$

We see $\{Y_j\}$ are independent and identically distributed random variables with a finite variance. $F_n(x)$ is called the empirical distribution function. By the strong law of large numbers, we have

$$\lim_{n \to \infty} F_n(x) = \lim_{n \to \infty} \frac{1}{n} \sum_{j=1}^{n} Y_j(x) = E(Y_1(x)).$$

Now

$$E(Y_1(x)) = E\left(1_{\{X_1 \leq x\}}\right) = P(X_1 \leq x) = F(x)$$

So we conclude

$$\lim_{n \to \infty} F_n(x) \stackrel{a.s.}{=} F(x)$$

A numerical example is given in Illustration 6.   □

Example 21 (Kelly's Betting Formula). Consider the betting on a sequence of independent trials. Each bet paying out f times your stake with probability p. Let T_k denote the payoff associated with the kth bet. Then

$$T_k = \begin{cases} f - 1 & \text{if the } k\text{th bet is a win} \\ -1 & \text{if the } k\text{th bet is a loss} \end{cases}$$

It follow that $E(T_k) = p(f-1) - q = pf - 1$. Assume $E(T_k) > 0$ for each k. As an example, $f = 3$ and $p = 0.4$. Thus $p \cdot f = 1.2 > 1$. Thus the game is in player's favor. On the other hand, if $f = 2$ and $p = 0.5$, then $p \cdot f = 1$ and the game is even. Using the strong law of large number, Kelly shows that the optimal betting strategy, expressed in term of the fraction α of your current capital to be placed in the next bet is

$$\alpha^* = \frac{pf - 1}{f - 1}.$$

Solution: Let X_0 denote the player's initial capital. Let X_n denote the capital after n bets. Let R_n denote the payoff if the nth bet is a win. Then

$$X_n = (1 - \alpha)X_{n-1} + \alpha X_{n-1} R_n.$$

We see that the first term on the right side is the untouched portion after the nth play and the second term on the right side is the gain resulting from the nth playoff after the bet αX_{n-1} is placed. Since $X_n = (1 - \alpha + \alpha R_n)X_{n-1}$, we have

$$X_n = (1 - \alpha + \alpha R_1) \cdots (1 - \alpha + \alpha R_n)V_0$$

by repeated substitutions. We take the logarithm of the above and rewrite the last expression as

$$\ln X_n - \ln X_0 = \ln(1 - \alpha + \alpha R_1) + \cdots + \ln(1 - \alpha + \alpha R_n)$$

or

$$\ln\left(\frac{X_n}{X_0}\right) = \ln(l - \alpha + \alpha R_1) + \cdots + \ln(1 - \alpha + \alpha R_n)$$

Define

$$G_n \equiv \frac{1}{n} \ln\left(\frac{V_n}{V_0}\right) = \frac{1}{n}(\ln(1 - \alpha + \alpha R_1) + \cdots + \ln(1 - \alpha + \alpha R_n)).$$

Therefore, we have

$$nG_n = \ln\left(\frac{X_n}{X_0}\right) \quad \text{or} \quad X_n = X_0 e^{nG_n}, \tag{9.5}$$

where we call G_n the growth factor.

It is obvious that $\{1 - a + aR_k\}_{k=1}^{n}$ are i.i.d. random variables. We apply strong law of large numbers and conclude

$$\lim_{n \to \infty} G_n \overset{a.s.}{=} p \ln(1 - \alpha + \alpha f) + (1 - p) \ln(1 - \alpha)$$
$$\equiv g(\alpha)$$

We call $g(\alpha)$ the long term growth rate. As shown in Illustration 7, the maximum long term growth rate occurs at

$$\alpha^* = \frac{pf - 1}{f - 1}$$

We now derive the asymptotic rate of return for this optimal betting strategy. Define

$$V_n = (1 + \gamma_n)^n V_0. \tag{9.6}$$

We equate (9.5) and (9.6) and obtain

$$e^{G_n} = 1 + \gamma_n \implies \gamma_n = e^{G_n} - 1$$

Thus

$$\lim_{n \to \infty} \gamma_n = e^{g(\alpha)} - 1 \quad \text{with probability 1}$$

and

$$e^{g(\alpha)} = e^{p \ln(1 - \alpha + \alpha f) + (1 - p) \ln(1 - \alpha)}$$
$$= (1 - \alpha + \alpha f)^p (1 - \alpha)^{1-p}.$$

So we conclude that

$$\lim_{n \to \infty} \gamma_n = (1 - \alpha + \alpha f)^p (1 - \alpha)^{1-p} - 1.$$

With $p = 0.4$ and $f = 3$, we find $\alpha^* = 0.1$ and $\lim_{n \to \infty} \gamma_n = 0.0098$. $\quad\square$

Example 22. Assume that $\{U_i\}_{i=1}^n$ are independent and identically distributed $U(0,1)$ random variables. Define the geometric mean by

$$G_n = (U_1 \cdots U_n)^{\frac{1}{n}}.$$

We want to find an approximation for G_n when $n \to \infty$. Taking the log yields

$$\log G_n = \frac{1}{n} \log (U_1 \cdots U_n)$$

$$= \frac{1}{n} (\log(U_1) + \cdots + \log(U_n)).$$

By the strong law of large numbers, we conclude that

$$\lim_{n \to \infty} \log G_n \overset{a.s.}{\to} E[\log(U_1)].$$

Now

$$E[\log(U_1)] = \int_0^1 \log(u)du = -1 \implies \log G_n \overset{a.s.}{\to} -1.$$

Thus, for large n, we have $G_n \overset{a.s.}{\to} e^{-1}$. $\qquad\square$

9.5. Central Limit Theorem

Theorem 10 (The Central Limit Theorem). *Let $\{X_i\}$ be independent and identically distributed random variables with mean μ and variance σ^2. Let*

$$S_n = X_1 + \cdots + X_n.$$

Then $E(S_n) = n\mu$ and $\text{Var}(S_n) = n\sigma^2$. Then we have

$$S_n \overset{d}{\to} N(n\mu, n\sigma^2) \quad \text{as } n \to \infty$$

or, equivalently,

$$\frac{S_n - n\mu}{\sigma\sqrt{n}} \overset{d}{\to} N(0,1) \quad \text{as } n \to \infty.$$

Proof. We prove this theorem for the case where $\mu = 1$ and $\sigma^2 = 1$. This means we want to show

$$\frac{S_n}{\sqrt{n}} \overset{d}{\to} N(0,1) \quad \text{as } n \to \infty.$$

Since there is one-to-one correspondence between $F_X(x)$ and $\varphi_X(t)$, It is equivalent to show that $\varphi_{\frac{S_n}{\sqrt{n}}}(t) \to e^{-\frac{t^2}{2}}$ as $n \to \infty$. We note

$$\varphi_{\frac{S_n}{\sqrt{n}}}(t) = \varphi_{S_n}\left(\frac{t}{\sqrt{n}}\right) \qquad \text{recall } \varphi_{aX}(t) = E(e^{i(aX)t}) = E(e^{iX(at)}) = \varphi_X(at)$$

$$= \left(\varphi_{X_1}\left(\frac{t}{\sqrt{n}}\right)\right)^n \qquad \text{convolution}$$

$$= \left(1 - \frac{t^2}{2n} + o\left(\frac{t^2}{n}\right)\right)^n,$$

which implies $\varphi_{\frac{S_n}{\sqrt{n}}}(t) \to e^{-\frac{t^2}{2}}$ as $n \to \infty$.

The last assertion uses an important consequence of Theorem 3 about characteristic function: when $E(X) = 0$ and $\text{Var}(X) = \sigma^2$, the characteristic function has the property that

$$\varphi_X(t) = 1 - \frac{1}{2}t^2\sigma^2 + o(t^2) \quad \text{as } t \to 0. \qquad \square$$

Theorem 11 (Central Limit Theorem (CLT) for Independent Random variables). *Let $\{X_i\}$ be independent random variables with*

$$E(X_i) = \mu_i, \quad \text{Var}(X_i) = \sigma_i^2.$$

Define

$$Y = X_1 + \cdots + X_n.$$

Then

$$E(Y) = \sum_{i=1}^n \mu_i, \quad \text{Var}(Y) = \sum_{i=1}^n \sigma_i^2.$$

Under some regularity conditions, we have

$$P\left(\frac{Y - E(Y)}{\sqrt{\text{Var}(Y)}} \le a\right) \xrightarrow{d} \Phi(a) \quad \text{as } n \to \infty.$$

The above convergence is in distribution.

One of the most celebrated examples in the applications of the Central Limit theorem is its use in approximating a binomial random variable.

Example 23. Let $\{X_i\}$ be a sequence of independent Bernoulli trials with $P(X_i = 1) = p$ and $P(X_i = 0) = q = 1 - p$. Let $X = X_1 + \cdots + X_n$. Then we know $X \sim$ Binomial(n, p). $E(X) = np$ and Var$(X) = npq$. By CLT, we have

$$\frac{X - np}{\sqrt{npq}} \xrightarrow{d} N(0, 1).$$

Thus we formally establish our earlier assertion. □

9.6. Other Inequalities

Definition. A twice differentiable real-valued function $f(x)$ is said to be convex if $f''(x) \geq 0$ for all x. It is said to be concave if $f''(x) \leq 0$.

Theorem 12 (Jensen's inequality). *Assume that $f(x)$ is a convex function. If X is a random variable, then*

$$E[f(X)] \geq f(E(X))$$

provided the expectation exists and is finite.

Proof. We expand $f(x)$ in a Taylor series about $\mu = E(X)$. Thus

$$f(x) = f(\mu) + f'(\mu)(x - \mu) + \frac{f''(\xi)(x - \mu)^2}{2},$$

where $\xi \in (x, \mu)$. Since $f''(\xi) \geq 0$, we conclude

$$f(x) \geq f(\mu) + f'(\mu)(x - \mu).$$

Thus

$$f(X) \geq f(\mu) + f'(\mu)(X - \mu).$$

We take the expectations on both sides of the above

$$E[f(X)] \geq f(E(X)) + f'(\mu)(E(X) - \mu) = f(E(X)). \qquad \square$$

Example 24. $X \sim$ expo$(\frac{1}{4})$, i.e., $E(X) = 4$.

Let $h(x) = x^2$ denote the cost of producing x units of X. Assume X denote the amount produced per day. We see that h is a convex function.

$$E[h(x)] = \int_0^\infty h(x) f_X(x) dx = \int_0^\infty x^2 \left(\frac{1}{4} e^{-\frac{x}{4}}\right) dx = 32$$

while $h(E(X)) = h(4) = 16$. We see

$$E[h(x)] > h(E(X)).$$

Using an incorrect approach yields an underestimate of the expected cost.

Let $g(x) = \sqrt{x}$ denote the revenue resulting from selling X units of X. We see that g is a concave function.

$$E[g(x)] = \int_0^\infty g(x) f_X(x) dx = \int_0^\infty \sqrt{x} \left(\frac{1}{4} e^{-\frac{x}{4}} \right) dx = \sqrt{\pi} = 1.772$$

while $g(E(X)) = \sqrt{4} = 2$. We see

$$E[g(X)] > g(E(X)).$$

Using the simple-minded approach yields an overestimate of the expected revenue. □

Example 25. Let X denote the price of a stock when the call options expires. Assume $X \sim N(6, 1)$. Assume that the strike price of the call option is 5. The payoff function of the call option is $g(X) = (X - 5)^+$. We note that f is a convex function. Now

$$g(E(X)) = g(6) = (6 - 5)^+ = 1.$$

Now

$$
\begin{aligned}
E[g(X)] &= \int_{-\infty}^\infty g(x) f_X(x) dx \\
&= \int_{-\infty}^\infty (x - 5)^+ \frac{1}{\sqrt{2\pi}} e^{-\frac{1}{2}(x-6)^2} dx \\
&= \int_5^\infty (x - 5)^+ \frac{1}{\sqrt{2\pi}} e^{-\frac{1}{2}(x-6)^2} dx \\
&= 1.0833.
\end{aligned}
$$

We see that $E[g(X)] > g(E(X))$. □

Theorem 13 (Gibb's Inequality). Let $\{p_i\}_{i=1}^n$ and $\{q_i\}_{i=1}^n$ be two PMF. Assume $p_i > 0$ for all i. Then

$$-\sum_{i=1}^n p_i \log p_i \leq -\sum_{i=1}^n p_i \log q_i$$

Proof. It is well-known that, for $z > 0$, we have

$$\log z \leq z - 1.$$

Let $z_i = q_i/p_i$ for all i. Then

$$\log\left(\frac{q_i}{p_i}\right) \leq \frac{q_i}{p_i} - 1, \quad i = 1, \ldots, n.$$

We multiply the ith inequality by p_i and sum over all i. After simplifications, it yields the needed result. □

Gibb's inequality is used in the context information and coding theories. If random variable X assumes the values $\{x_i\}_{i=1}^n$ with respective probabilities $\{p_i\}_{i=1}^n$. Define

$$H(X) = -\sum_{i=1}^n p_i \log p_i.$$

Then $H(X)$ is known in the information theory as the *entropy* of the random variable X. $H(X)$ represents the average amount of surprise we receive upon learning the value of X. It can also be considered the amount of *uncertainty* that exists relating to X. The following example shed some lights along these lines of interpretation.

Example 26. Assume that random variables X and Y follow PMFs:

$$p_X(x) = 0.2, \quad x = 1, \ldots, 5.$$

and

$$p_Y(y) = \begin{cases} 0.8 & \text{if } x = 1, \\ 0.05 & \text{if } x = 2, 3, 4, 5.. \end{cases}$$

Simple calculations yield $H(X) = 1.6094$ and $H(Y) = 0.7777$. This implies that X contains more uncertainty than Y. As a matter of fact, it can be shown that the entropy is maximized when a random variable assumes a uniform PMF. □

9.7.   Exploring with Mathematica

Illustration 1. In Example 14, we present the PDF of a one-sided Cauchy random variable. We integrate the PDF to construct the CDF and plot it below:

```
Fx[x_] = Integrate[ 2/(π(1+τ²)), {τ, 0, x}, Assumptions -> Re[x] > 0 && Im[x] == 0]
```

$$\frac{2 \, \text{ArcTan}[x]}{\pi}$$

```
Plot[Fx[x], {x, 0, 10}, AxesLabel → {"x", "fₓ(x)"}, AxesOrigin → {0, 0},
  AxesStyle → {Black, Medium}]
```

```
Equal[ArcTan[x] == π/2 - ArcTan[1/x] ]
```

True

```
Series[ArcTan[y], {y, 0, 4}]
```

$$y - \frac{y^3}{3} + O[y]^5$$

In the example, we show that the maximum of a random sample converges to F_G shown below. We also compute $E(G)$ using the fact that

$$E(G) = \int_0^\infty P(G > x)\,dx$$

for a nonnegative random variable G by numerical integration.

```
Gx[x_] := Exp[- 2/(π x)];      the CDF of G|
```

```
Plot[Gx'[x], {x, 0, 10}, PlotRange → All, AxesLabel → {"x", "gx(x)"}]
```

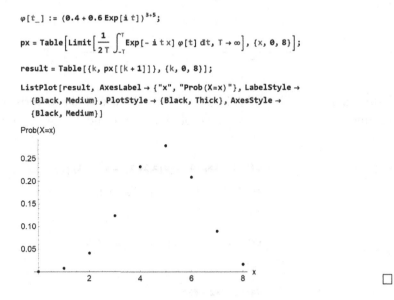

```
EG = NIntegrate[1 - Exp[- 2/(π x)], {x, 0, 100}]
```

3.49041 ☐

Illustration 2. Consider Example 15. Here we show the inversion of a CF of a discrete random variable:

```
φ[t_] := (0.4 + 0.6 Exp[i t])^(3+5);

px = Table[Limit[1/(2T) ∫_{-T}^{T} Exp[- i t x] φ[t] dt, T → ∞], {x, 0, 8}];

result = Table[{k, px[[k + 1]]}, {k, 0, 8}];

ListPlot[result, AxesLabel → {"x", "Prob(X=x)"}, LabelStyle →
    {Black, Medium}, PlotStyle → {Black, Thick}, AxesStyle →
    {Black, Medium}]
```

☐

Illustration 3. Based on the result given in Example 15, we demonstrate the use of (9.4) to do compute the probability related to the uniform density.

$$\varphi[t\_] := \frac{\text{Exp}[i\,12\,t] - \text{Exp}[i\,2\,t]}{i\,10\,t};$$

$$\text{Limit}\left[\frac{1}{2\,\pi}\int_{-T}^{T}\frac{\text{Exp}[-i\,5\,t] - \text{Exp}[-i\,2\,t]}{-i\,t}\,\varphi[t]\,dt,\,T \to \infty\right]$$

$$\frac{3}{10}$$

□

Illustration 4. The CF inversion for Example 17:

$$f[x\_] = \frac{1}{2\,\pi}\int_{-\infty}^{\infty}\text{Exp}[-i\,t\,x]\,\varphi[t]\,dt$$

```
Plot[f[x], {x, -2, 3}, AxesLabel → {"x", "f(x)"},
  AxesStyle → {Black, Medium}, LabelStyle → {Black, Medium},
  PlotStyle → {Black, Medium}]
```

□

Illustration 5. The following is to show the use of characteristic function to produce the mean and variance of the compound Poisson random variable $X(t)$.

$$\varphi[s\_] = \text{Exp}\left[4\left(\text{Exp}\left[i\,10\,s - 2\,s^2\right] - 1\right)\right];$$

$$\text{EX} = (i)^{-1}\varphi'[s]\,/.\,s \to 0\,//\,N$$

40.

$$\text{EX2} = (i)^{-2}\varphi''[s]\,/.\,s \to 0\,//\,N$$

2016.

$$\text{VarX} = \text{EX2} - \text{EX}^2$$

416.

□

Illustration 6. We construct an empirical PDF based on observed sample of 100 to illustrate the procedure stated in Example 20.

```
SeedRandom[123412]; nInterval = 60;

nSample = 100; cx = Table[0.1*i, {i, 1, nInterval}];

x = RandomVariate[NormalDistribution[4, 1], nSample];

CF = Table[N[Total[Table[If[x[[i]] < cx[[j]], 1, 0],
        {i, 1, nSample}]/nSample]] , {j, 1, nInterval}];

ListPlot[CF]
```

Illustration 7. For Example 21, we find the optimal betting fraction using Mathematica and show that it is the unique maximum as $g''(\alpha^*) < 0$.

```
g[α_, p_] = p Log[1 - α + α f] + (1 - p) Log[1 - α];

α* = α /. Solve[D[g[α, p], {α, 1}] == 0, {α}]
```

$$\left\{ \frac{-1 + f\,p}{-1 + f} \right\}$$

```
D2[α_] = D[g[α, p], {α, 2}];

D2[α*]
```

$$\left\{ -\frac{1 - p}{\left(1 - \frac{-1+f\,p}{-1+f}\right)^2} - \frac{(-1 + f)^2\, p}{\left(1 - \frac{-1+f\,p}{-1+f} + \frac{f\,(-1+f\,p)}{-1+f}\right)^2} \right\}$$

Problems

1. Let $\{X_i\}$ be independent and identically distributed random variables with

$$f_X(x) = \frac{\alpha}{x^{\alpha+1}}, \quad x > 1, \ \alpha > 0.$$

Define

$$Y_n = n^{-\frac{1}{\alpha}} \max_{1 \le k \le n} \{X_k\}, \quad n \ge 1.$$

(a) What is the random variable X? (b) We note that $Y_n \xrightarrow{d} F_Z$. Find F_Z.

2. Let $\{X_i\}$ be independent and identically distributed random variables with

$$f_X(x) = \frac{1}{2}(1+x)e^{-x}, \quad x > 0.$$

Define

$$Y_n = \min_{1 \le k \le n} \{X_k\}, \quad n \ge 1.$$

We note that $n \cdot Y_n \xrightarrow{d} F_Z$. Find F_Z.

3. Use Chebyshev's inequality to find how many times a fair coin must be tossed so that

$$P(0.4 < \overline{X} < 0.6) \ge 0.90,$$

where $Y \equiv \overline{X} = (X_1 + \cdots + X_n)/n$.

4. The insurance company Mutual of Wyoming has 30,000 policy holders. The amount claimed yearly by a policy holder per year has an expected value of $250 and a standard deviation of $500. What is the approximate probability that the total amount of claims in the next year will be larger than 7.6 million dollars?

5. Casino Gomez just opens its door and offer the following promotional roulette game to its first 1000 gamblers. For each turn of the roulette wheel, the payoffs X_i to the ith gambler who bets on red are as follows:

$$X_i = \begin{cases} -5 & \text{if the gambler loses,} \\ 10 & \text{otherwise,} \end{cases}$$

where the probability of winning (on red) is $\frac{18}{37}$. What is the approximate probability that Gomez will lose at least $2500 in this promotional campaign?

6. Consider a coin is to be tosses repeatedly. Each toss will yield heads with probability P. Let $X = X_1 + \cdots + X_n$. It follows that $X \sim \text{bino}(n, P)$.

We are uncertain about the parameter P. We now assume $P \sim U(0,1)$, (a) What is the PMF of X? (b) Given $X = k$, what is the PDF for P?

7. Let $\{X_n\}$ be a sequence of independent and identically distributed random variables with a common PDF $U(0,1)$. Define

$$Y_n = \min(X_1, \ldots, X_n)$$

and

$$Z_n = \frac{Y_n}{n}.$$

Show that

$$Z_n \xrightarrow{d} W,$$

where $W \sim \text{expo}(1)$.

9. Consider the sequence of random variables $\{X_i\}_{i=1}^n$. Assume they are independent and with a common PDF $U(0,1)$. Define

$$Y_n = \prod_{i=1}^n X_i.$$

Now

$$\log Y_n = \sum_{i=1}^n \log X_i.$$

As $\{\log X_i\}_{i=1}^n$ are independent random variables with a common PDF, by the Central Limit Theorem, $\log Y$ follows a normal distribution. (a) Find the $E(\log Y)$ and $\text{Var}(\log Y)$. (b) Compute the approximate value for

$$P\left(\frac{1}{3^{100}} < Y_{100} < \frac{1}{2^{100}}\right).$$

10. Suppose Channel 35 wants to take a public opinion poll for a proposed property tax hike. Let p be the percent of people who are for the tax hike. Assume X_i is the response of the ith respondent of the poll. Let

$$X_i = \begin{cases} 1 & \text{if the } i\text{th respondent is for the tax hike,} \\ 0 & \text{otherwise.} \end{cases}$$

Then

$$Y_n = \frac{X_1 + \cdots + X_n}{n} = \text{the fraction of the respondents who are}$$

for the tax hike.

How large should the sample size n be so that the unknown parameter p is within an error of at most 0.05 within a 95% confidence interval. In other words, we want to determine the smallest value of n such that

$$P\left(Y_n - 0.05 < p < Y_n + 0.05\right) \geq 0.95 \qquad (*)$$

Here, we assume $\{X_i\}_{i=1}^{n}$ are independent random variables and each follows a Bernoulli distribution with parameter p.

11. Let $\{X_i\}_i^n$ be a sequence of independent random variables with a common PDF $U(0,1)$. Define

$$P_n = (X_1 X_2 \cdots X_n)^{\frac{1}{n}}$$

Show that $P_n \overset{a.s.}{\to} \frac{1}{e}$ as $n \to \infty$.

12. Let $\{X_i\}$ be a sequence of random variables and X be a random variable to which $\{X_i\}$ converge. Assume that

$$\sum_{n=1}^{\infty} P\left(|X_n - X| > \varepsilon\right) < \infty. \qquad (*)$$

Show that $X_n \overset{a.s}{\to} X$.

13. For $i = 1, \ldots, n$, we let X_i be a random variable defined by

$$X_i = \begin{cases} 1 & \text{with probability } p_i, \\ 0 & \text{otherwise.} \end{cases}$$

We can view X_i as a generalized Bernoulli random variable. Assume that $\{X_i\}$ are independent random variables. Let $X = X_1 + \cdots + X_n$.

Define

$$\mu \equiv E(X) = \sum_{i=1}^{n} E(X_i) = \sum_{i=1}^{n} p_i$$

(a) Using the Markov inequality to establish an upper bound $P(X \geq (1 + \delta)\mu)$, where $\delta > 0$. (b) Using the Markov inequality to establish an upper bound $P(X \leq (1 - \delta)\mu)$, where $\delta > 0$. (c) Compute the two bounds for $\mu = 1$ and $\delta = 0.2$.

14. For $i = 1, \ldots, n$, we let X_i be a random variable defined by

$$X_i = \begin{cases} 1 & \text{with probability } p_i, \\ 0 & \text{otherwise.} \end{cases}$$

We can view X_i as a generalized Bernoulli random variable. Assume that $\{X_i\}$ are independent random variables. Prove that

$$X_n \xrightarrow{a.s.} 0$$

if and only if

$$\sum_{i=1}^{\infty} p_i < \infty.$$

15. (a) Apply the Central Limit Theorem to establish the following

$$\lim_{n \to \infty} e^{-n} \sum_{k=0}^{n} \frac{n^k}{k!} = \frac{1}{2}.$$

(b) Use Mathematica to compute the left-side terms for $n \leq 500$.

16. Assume that $X \sim \text{pois}(\lambda t)$ random variable. (a) Find the characteristic function $\varphi_X(s)$ of X. (b) Use the characteristic functions to find the probability generating function $p_X^g(z)$. (c) Define $Y = (X - \lambda t)/\sqrt{\lambda t}$. Find the characteristic function $\varphi_Y(s)$ of Y. (d) Apply Theorem 2 of this chapter to show that $Y \xrightarrow{d} N(0, 1)$.

17. Let $\{\varepsilon_i\}_{i=1}^n$ be a sequence of independent and identically distributed $N(0,1)$ random variables. Define

$$X_n = \varepsilon_1 + \cdots + \varepsilon_n.$$

(a) Are $\{X_i\}_{i=1}^n$ independent normal random variables? (b) Define

$$Y = \frac{X_n}{n\sqrt{n}}.$$

What is the characteristic function $\varphi_Y(s)$ of Y. (c) As $n \to \infty$, what can you say about the limiting distribution for Y?

18. Let $\{\varepsilon_i\}_{i=0}^\infty$ be a sequence of independent and identically distributed $N(0,1)$ random variables. Define

$$X_{n+1} = \begin{cases} \varepsilon_0 & \text{if } n = 0, \\ aX_n + \varepsilon_{n+1} & \text{if } n = 1, 2, \ldots. \end{cases}$$

(a) Consider the case when $a < 1$. Show that $X_n \xrightarrow{d} Y$ and identify the PDF of Y. (b) Consider the case when $a > 1$. Show that $\frac{X_n}{a^n} \xrightarrow{m.s.} W$ and identify the constant W.

19. Let $X_1, X_2, \ldots$ be a sequence of random variables such that

$$P\left(X_n = \frac{k}{n}\right) = \frac{1}{n}, \qquad k = 1, 2, \ldots, n.$$

Find the limit distribution of X_n as $n \to \infty$.

20. Let $X_1, X_2, \ldots$ be independent and identically distributed random variables with expectation 1 and finite variance σ^2. Let

$$S_n = X_1 + \cdots + X_n, \qquad n = 1, 2, \ldots.$$

Define random variable $V_n = \sqrt{S_n} - \sqrt{n}$, for $n = 1, 2, \ldots$. Does V_n converge to a limiting distribution as $n \to \infty$? If yes, show why this is so and state the limiting distribution; if not, justify your assertion.

21. Let $\{X_i\}_{i=1}^\infty$ be a sequence of independent random variable with a common PDF $U(0,1)$.

(a) Prove that

$$\max_{1 \le k \le n} X_k \xrightarrow{p} 1 \quad \text{as } n \to \infty.$$

(b) Prove that

$$\min_{1 \le k \le n} X_k \xrightarrow{p} 0 \quad \text{as } n \to \infty.$$

22. Let $\{X_n\}_{n=1}^{\infty}$ be a sequence of independently random variables with a commo PDF

$$f(x) = \begin{cases} e^{-(x-a)} & \text{if } x > a, \\ 0 & \text{otherwise.} \end{cases}$$

Define

$$Y_n = \min_{1 \le k \le n} \{X_k\}.$$

Prove that

$$Y_n \xrightarrow{p} a \quad \text{as } n \to \infty.$$

23. Let $\{X_i\}_{i=1}^{n}$ be a sequence of independently distributed random variables with a common mean μ and variance $\sigma^2 < \infty$. In statistics, the sample mean and variance are defined by

$$\overline{X}_n = \frac{1}{n} \sum_{i=1}^{n} X_i, \quad s_n^2 = \frac{1}{n-1} \sum_{i=1}^{n} (X_k - \overline{X}_n)^2.$$

(a) Show that $E(s_n^2) = \sigma^2$. (b) Show that $s_n^2 \xrightarrow{p} \sigma^2$ as $n \to \infty$.

24. Let $\{X_i\}_{i=1}^{\infty}$ be a sequence of independently distributed random variables with a common PDF Pareto(2,1). Define

$$Y_n = \min_{1 \le k \le n} X_k.$$

(a) Show that

$$Y_n \to 1 \quad \text{as } n \to \infty.$$

(b) Show that $n(Y_n - 1) \xrightarrow{d} W$ as $n \to \infty$. What is the PDF of W?

25. Let $\{X_i\}_{i=1}^n$ be a sequence of independently distributed random variables with a common PDF $C(0,1)$ (c.f., Section 6.7, Cauchy random variables). Define

$$S_n = \sum_{k=1}^n X_k.$$

(a) Show that S_n/n follows $C(0,1)$. (b) Define

$$Z_n \equiv \left(\frac{1}{n}\right) \sum_{k=1}^n \frac{S_k}{k}.$$

Show that Z_n follows $C(0,1)$.

26. Let $X \sim N(0,1)$ and $Y = XZ$, where

$$Z = \begin{cases} 1 & \text{with probability } \dfrac{1}{2}, \\ -1 & \text{with probability } \dfrac{1}{2} \end{cases}$$

and $Z \perp X$. (a) Find the CDF $F_Y(y)$ of Y. (b) Are $X \perp Y$? (c) Find the correlation coefficient of X and Y. (d) Are X and Y jointly normal?

27. Let $X \sim U(-n,n)$. (a) Find the characteristic functions $\varphi_X(t)$ for X. (b) For $n = 2$, find $E(X)$ and $\text{Var}(X)$. (c) For $n = 2$, by inverting the characteristic functions to find $f_X(1)$.

28. Let $\{X_i\}_i^n$ be a sequence of independent and identically distributed random variables with a common PDF $N(0,\sigma^2)$. Let

$$Y = \frac{X_1 + \cdots + X_n}{\sqrt{n}}.$$

Find $F_Y(y)$.

29. Let $X \sim \text{bino}(n,p)$. (a) Find the characteristic function $\varphi_X(t)$ for X. (b) Set $\lambda = np$. Show that $X \overset{d}{\to} \text{pois}(\lambda)$.

30. Let $X \sim \text{geom}(p)$. (a) Find the characteristic function $\varphi_X(t)$ for X. (b) Let $Z \sim \text{expo}(\lambda)$. Find the characteristic function $\varphi_Z(t)$ for Z. (c) Set $\lambda = np$. Define $Y = \frac{X}{n}$. Show that $Y \overset{d}{\to} \text{expo}(\lambda)$

31. Let X be discrete random variable where

$$P(X_i = x_i) = p_i \quad i = 1, \ldots, n$$

and $p_1 + \cdots + p_n = 1$. Using the Lagrange multiplier approach to show that the discrete uniform distribution over the set of integers $\{1, \ldots, n\}$ maximize the entropy.

32. Let X be discrete random variable over the set of positive integers $\{1, 2, \ldots\}$ with $p_1 + \cdots + p_\infty = 1$. Given we know that $E(X) = \mu$, where $\mu > 1$, use the Lagrange multiplier approach to show that under the maximum entropy paradigm we have $X \sim geom(p)$, where $p = \frac{1}{\mu}$.

33. Let X be a discrete positive-valued random variable whose values are $\{x_1, \ldots, x_n\}$, where $x_1 < x_2 < \cdots < x_n$. Assume $P(X = x_i) = p_i$, $i = 1, \ldots, n$ and that $E(X) = \mu$. Find the maximum entropy distribution for $\{p_i\}$.

34. Rosie is the owner of Casino Carabra in Las Vegas. She receives a batch of six-sided dice from her supplier. Rosie selects one die and rolled it for 500 times and computes its average. The average is found to be 3.4. Rosie wants to find the maximum entropy estimates of $\{p_i\}$, $i = 1, \ldots, 6$, where p_i is the probability that a roll of the die will show a face value of i. (a) Can you help? (b) What would be the estimates for $\{p_i\}$, if Rosie finds $\mu = 3.5$?

35. Arron owns a gas station. Each week he observes the sales for the three types of gasoline: regular, premium, and super. Typically, they were 2.5, 3.4, and 3.8 thousands of dollars. Last week, Arron found the average sales was 3.2 thousands of dollars. What are the maximum entropy estimates for $\{p_i\}$, where i corresponds to the three types of gasoline.

36. Let $f(x) = \sin(3x)$. Write a Mathematica code to do the Bernstein polynomial approximation to compute $f(x)$ for $x \in \{0.1, 0.2, \ldots, 0.9\}$ for $n = 20$. Compare the approximations against $\sin(x)$ found at the indicated values.

37. An eighteen-wheeler is waiting at the shipping dock ready for loading. Assume that items to be loaded whose weights in pounds $\{X_i\}$ are random variables. Assume that $\{X_i\}$ are independently distributed following a common density $U(5, 40)$. What is the maximum number of items to be loaded into the truck so that the probability is at least 95% that the total weight is no more than 8000 pounds?

38. We toss a fair die 100 times. Let X_i be the result from the ith toss and $S = X_1 + \cdots + X_{100}$. (a) State the characteristic function $\varphi_S(t)$. (b) Use $\varphi_S(t)$ to find $E(S)$ and $\mathrm{Var}(S)$. (c) Use Chebyshev inequality to find $P(|S - 350| > 25)$. (d) Use the central limit theorem to find the approximate probability that $P(|S - 350| > 25)$.

39. Casino Gomez just opens its door and offer the following promotional roulette game to its first 1000 gamblers. For each turn of the roulette wheel, the payoffs X_i to the ith gambler who bets on red are as follows:

$$X_i = \begin{cases} -5 & \text{if the gambler loses} \\ 10 & \text{otherwise} \end{cases}$$

where the probability of winning (on red) is $\frac{18}{37}$. What is the approximate probability that Gomez will lose at least \$2,500 in this promotional campaign?

40. Consider the sequence of random variables $\{X_i\}_{i=1}^n$. Assume they are independent and with a common PDF $U(0, 1)$. Define

$$Y_n = \prod_{i=1}^n X_i$$

Now

$$\log Y_n = \sum_{i=1}^n \log X_i.$$

As $\{\log X_i\}_{i=1}^n$ are independent random variables with a common PDF, by the central limit theorem $\log Y$ follows a normal distribution, (a) Find the $E(\log Y_n)$ and $\mathrm{Var}(\log Y_n)$. (b) Compute the approximate value for

$$P\left(\frac{1}{3^{100}} < Y_{100} < \frac{1}{2^{100}}\right).$$

41. Suppose Channel 35 wants to take a public opinion poll for a proposed property tax hike. Let p be the percent of people who is for the tax hike.

Assume X_i is the response of the ith respondent of the poll. Let

$$X_i = \begin{cases} 1 & \text{if the } i\text{th respondent is for the tax hike} \\ 0 & \text{otherwise} \end{cases}$$

Then

$$Y_n = \frac{X_1 + \cdots + X_n}{n}$$

$$= \text{the fraction of the respondents who are for the tax hike.}$$

How large should the sample size n be so that the unknown parameter p is within an error of at most 0.05 within a 95% confidence interval. In other words, we want to determine the smallest value of n such that

$$P(Y_n - 0.05 < p < Y_n + 0.05) \geq 0.95$$

Here we assume $\{X_i\}_{i=1}^n$ are independent random variables and each follows a Bernoulli distribution with parameter p (Hint: $p(1-p) \leq \frac{1}{4}$).

42. Consider that there are n independent events $\{A_1, \ldots, A_n\}$. For $i = 1, \ldots, n$, let p_i be the probability that event i will occur. Define

$$I_i = \begin{cases} 1 & \text{if } A_i \text{ occurs} \\ 0 & \text{otherwise} \end{cases}$$

and

$$X = \sum_{k=1}^n I_k$$

(a) Find expression for computing $E(X)$ and $\text{Var}(X)$. (b) Let $n = 4$ and $\{p_1, \ldots, p_4\} = \{0.5, 0.3, 0.7, 0.6\}$. Compute $E(X)$ and $\text{Var}(X)$.

43. **Another Version of Kelly's Formula.** Consider a sequence of independent coin tosses. Let p be the probability that a single toss results a win. Define

$$T_k = \begin{cases} +1 & \text{if the } k\text{th toss is a win} \\ -1 & \text{if the } k\text{th toss is a loss} \end{cases}$$

Thus $p = P(T_k = +1)$. Assume $p > \frac{1}{2}$. Let B_k be the bet size made by the gambler before the kth toss. Then the payoff to the gambler is $B_k T_k$

and its mean is

$$E(B_k T_k) = pE(B_k) - qE(B_k) = (p - q)E(B_k)$$

Let X_n denote the size of the gambler's capital after n bets. We consider the betting strategy

$$B_i = f X_{i-1}$$

where $0 < f < 1$.

(a) Let S denote the number of wins in n bets and $F = n - S$. State an expression for X_n given X_0. (b) Use SLLN to derive the expected gain rate function $E[G_n(f)]$ where

$$E[G_n(f)] = \log \left(\frac{X_n}{X_0} \right)^{\frac{1}{n}} \quad \text{with } n \to \infty$$

(c) Find the optimal betting strategy f^* when $p = 0.53$. (e) For what value n we would expect the gambler's capital will be doubled?

44. **Another Version of Kelly's Formula.** Consider a sequence of independent coin tosses. Let p be the probability that a single toss results a win. Define the payoff for the kth bet as follows

$$T_k = \begin{cases} b & \text{if the } k\text{th toss is a win} \\ -1 & \text{if the } k\text{th toss is a loss} \end{cases}$$

Thus $p = P(\text{the } k\text{th play is a win})$. Assume that the expected gain at a single bet $E(T_k) = bp - q > 0$. Let B_k be the bet size made by the gambler before the $k$the toss. Let X_{k-1} denote the size of the gambler's capital before the $k$the bet. We consider the betting strategy

$$B_k = f X_{k-1}$$

where $0 < f < 1$. Let H_k denote the outcome of the $k$the bet. (a) Using SLLN, express $E(X_n)$ as a function of X_0 and n, p, and b for large n. (b) Let S denote the number of wins in n bets and $F = n - S$. State an expression for X_n given X_0. (c) Use SLLN to derive the expected gain rate function $E[G_n(f)]$ where

$$E[G_n(f)] = \log \left(\frac{X_n}{X_0} \right)^{\frac{1}{n}} \quad \text{with} \quad n \to \infty$$

(d) Find the optimal betting strategy f^* when $p = 0.53$. (e) For what value n we would expect the gambler's capital will be doubled?

45. Let $g : [0, \infty)$ be strictly increasing and nonnegative. Show that

$$P(|X| \geq a) \leq \frac{E[g(x)]}{g(a)} \quad a > 0$$

46. Let X be a non-negative random variable with finite mean μ and variance σ^2. Assume that $b > 0$.

Define the function $g(x)$ as follows:

$$g(x) = \frac{((x - \mu)b + \sigma)^2}{\sigma^2(1 + b^2)^2}.$$

(a) Show that $\{X \geq \mu + b\sigma\} \Rightarrow \{g(X) > 1\}$.
(b) Use Markov inequality to establish

$$P(X \geq \mu + b\sigma) \leq \frac{1}{1 + b^2}.$$

(c) In Example 4, we found that the random variable $X \sim unif(0, 10)$ had $\mu = 5$ and $\sigma^2 = \frac{25}{3}$. Use Part (b) to obtain an upper bound for $P(X \geq 9)$.
(d) What is the exact probability $P(X \geq 1)$?

47. Assume $X \sim unif(-\frac{1}{2}, \frac{1}{2})$ and $Y \sim unif(-\frac{1}{2}, \frac{1}{2})$. Suppose $X \perp Y$. Define random variable $Z = X + Y$.

(a) Find the characteristic function $\varphi_X(u)$.
(b) Find the characteristic function $\varphi_Z(u)$.
(c) Use Mathematica to invert $\varphi_Z(u)$ to find the PDF to Z (note when invoking Mathematica function "Inverse Fourier Transform", use the option Fourier Parameter $\leq \{1, 1\}$).
(d) Use Mathematics to plot $f_Z(z)$ over the interval $[-1, 1]$.

48. Let $\{X_n\}_{n=1}^{\infty}$ be a sequence of i.i.d. random variables with PMF given by

$$P(X_n = 1) = \frac{1}{2} \quad P(X_n = -1) = \frac{1}{2}$$

Let $S_n = X_1 + \cdots + X_n$. Define $W_n = \frac{1}{n} S_n$. Show that $W_x \xrightarrow{P} 0$.

Remarks and References

For expositions on the various concepts of convergence, see [1] or [4]. For proofs of the strong law of large numbers, excellent references can be found in many sources, e.g., Chung [2], Jacd and Protter [5], and Fristedt and Gray [3].

[1] Brémaud, P., *An Introduction to Probabilistic Modeling*, Springer, 1988.
[2] Chung, K. L., *A Course in Probability Theory*, Academic Press, 1974.
[3] Fristedt, B. and Gray, L., *A Modern Approach to Probability Theory*, Birkhäuser, 1997.
[4] Gut, A., *An Intermediate Course in Probability*, 2nd edn., Springer, 2009.
[5] Jacod, J. and Protter, P., *Probability Essentials*, Springer, 2004.
[6] Lukacs, E., *Characteristics Functions*, Griffin, 1970.

A Terse Introduction to Mathematica

This introduction is written for use under Microsoft® Windows 10

Tutorial 1:

Assume that you have just launched Mathematica (by clicking its icon)

Select New Document

>click □ on the top right to use the full screen

To adjust the screen font size (if you would like to enlarge it): at the bottom right side corner, you see "100%"; click ▲ and select the font size of your choice (e.g., 200%). The screen fonts immediately change their sizes.

Type "1+1". Then **hold the shift key and click "enter"**. You will see "2" at the next line.

Move the cursor to the top of the screen and select "Palettes". There is a panel of choices. Select "Other". Choose "Basic Math Input"

A symbol panel will appear at the upper right-hand side of your screen. Select

$$\sum_{\square=\square}^{\square} \blacksquare$$

Fill in the squares so that it will read

$$\sum_{i=1}^{10} i$$

Then **hold the shift key and click "enter"** and you will see "55" at the next line.

Now that you have completed the first encounter with Mathematica. You want to leave.

At the moment, the "Basic Math Input" panel is still open. To close it, move the cursor to the upper-right corner of the screen and click the red exit button (with X sign)

To leave Mathematica, if you want to save your notebook, then choose

> File> SaveAs followed by typing the filename (e.g., XYZ) of the notebook in a directory of your choice and save the notebook. Then click the red exit bottom (with X sign) on the upper right-hand-side corner of the screen to exit.

Tutorial 2: How to write up a report for a homework assignment.

Launch Mathematica to open an existing notebook or create a new notebook

If you want to type text, do the following >Format >Style > Text, (equivalently, you just do Alt-7) and click the mouse, then start typing your text. You may need math symbols. Then do the following >Palettes > Other>Basic Math Input (this will give a set of symbols to see, subscript, etc. at the bottom of the little panel of symbols), or >Palettes > Other>Basic Typesetting (this gives you some more, including our friendly binomial coefficient). Once you are done with these panels, you can use "X" to remove it from the screen, or you can just keep them there.

When you have completed your text and would like to do a little computation using Mathematica, you just click the mouse once (right after you are done with the text). A line will be drawn by Mathematica indicating you can make your next move. You do >Format >Style > Input, (equivalently, you just do Alt-9). You then enter your Mathematica commands. Once done, do "shift" + "enter", it will give the answer on the next line with the answer printed there. You then give another mouse click, a new line appears. Now you are ready to make the next move (e.g., writing some more text? or do some more calculations?)

When you are done with your assignment, save it as a notebook (e.g., HW1.nb).

Four Simple Examples

Example 1. It computes the outcomes in tossing two dice. The first two lines illustrate the use of conditional statements. "Table" is a command to compute the values a function given its arguments. "Total" computes the sum of the elements of the list.

$$p[j\_] := \frac{j-1}{36} \ /; \ j \le 7$$

$$p[j\_] := p[14 - j] \ /; \ 8 \le j \le 12$$

```
Table[p[j], {j, 2, 12}]
```

$$\left\{ \frac{1}{36}, \ \frac{1}{18}, \ \frac{1}{12}, \ \frac{1}{9}, \ \frac{5}{36}, \ \frac{1}{6}, \ \frac{5}{36}, \ \frac{1}{9}, \ \frac{1}{12}, \ \frac{1}{18}, \ \frac{1}{36} \right\}$$

```
Total[Table[p[j], {j, 2, 12}]]
```

1

Alternatively:

```
Sum[p[j], {j, 2, 12}]
```

1 □

Example 2. The first line retrieves the output λ_2 for later use. If you just do "NSolve" (without the preceding syntaxes), then the output will be in the form of a list. You cannot use it for further work directly. The term "Reals" only picks up the real solution. In the third line, the term "Quiet" is used to suppress outputting of warning messages.

$$\lambda_2 = \lambda_2 \ /. \ \text{NSolve}\left[\left\{\sum_{i=1}^{6} i \ \text{Exp}[\lambda_2 \ i] = 3.4 \sum_{i=1}^{6} \text{Exp}[\lambda_2 \ i]\right\}, \ \{\lambda_2\}, \ \text{Reals}\right]$$

{-0.0343106}

$$\lambda_1 = \lambda_1 \ /. \ \text{Quiet}\left[\text{NSolve}\left[\left\{\text{Exp}[\lambda_1 - 1] = \frac{1}{\sum_{i=1}^{6} \text{Exp}[\lambda_2 \ i]}\right\}, \ \{\lambda_1\}\right]\right]$$

{-0.673389}

```
Table[Exp[λ₁ - 1 + λ₂ i], {i, 1, 6}]
```

{{0.181282}, {0.175168}, {0.16926}, {0.163551}, {0.158035}, {0.152704}}

There are many other examples shown as a part of illustrations at the end of each chapter. Readers are encouraged to review them and explore other usages of Mathematica. The "Help" icon at the top of the screen provides valuable help when clicked. □

Example 3. The following shows how to input a function with multiple right-hand sides. Note that the first three lines (the narratives) are typed using the text input mode (Alt-7) and the rest is done with Mathematica (Alt-9, i.e., the default setting).

How to type the following: first type f[x_]:=, then do ESC pw ESC, then do CTRL enter as many times as you want to enter the number of lines you need. "True" means "otherwise".

$$f[x\_] := \begin{bmatrix} 1 & 0 < x < 1 \\ 0.5 & 3 < x < 4 \\ 0 & \text{True} \end{bmatrix}$$

`Plot[f[x], {x, 0, 5}, PlotRange → All]`

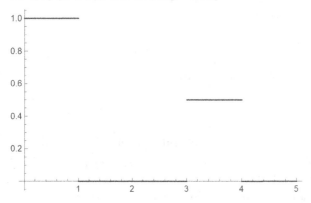

Example 4. Retrieving results from a Mathematica run requires some attention. The following example demonstrates a few alternatives to do this. Assume that we are solving the following system linear equations for x and y

$$ax + by = 1$$
$$bx + cy = 0.$$

The first line of the Mathematica output solves the problem with the output shown as a list. The third line assign the result of variable x for future uses.

The fifth line gives the results in a matrix form. We see that the variable XY is an 1×2 matrix. We can retrieve the components of XY as we normal do in linear algebra.

```
Solve[{a x - b y == 1, b x + a y == 0}, {x, y}]
```

$$\left\{ \left\{ x \to \frac{a}{a^2 + b^2}, y \to -\frac{b}{a^2 + b^2} \right\} \right\}$$

```
x = x /. Solve[{a x - b y == 1, b x + a y == 0}, {x, y}]
```

$$\left\{ \frac{a}{a^2 + b^2} \right\}$$

```
XY = {x, y} /. Solve[{a x - b y == 1, b x + a y == 0}, {x, y}];

XY // MatrixForm
```

xForm=

$$\left(\begin{array}{cc} \frac{a}{a^2+b^2} & -\frac{b}{a^2+b^2} \end{array} \right)$$

```
Dimensions[XY]
```

$\{1, 2\}$

```
z = XY[[1, 1]]
```

$$\frac{a}{a^2 + b^2}$$

```
w = XY[[1, 2]]
```

$$-\frac{b}{a^2 + b^2}$$

□

Answers to Odd-Numbered Problems

Chapter 1

1. 2880

3. $4!(3!)^2(4!)^2 = 497\,664$

5. (a) 56   (b) 28

7. (a) $\binom{20}{5,5,5,5}$ (b) $\binom{20}{5,5,5,5}/4!$

9. $\binom{12}{4,3,5}$

11. $\binom{12}{4,4,4}$

13. (a) 8!   (b) 1152   (c) 2880   (d) 1152

15. (a) 220   (b) 572

17. 70560

19. (a) 36   (b) 26

21. (a) $n(n-1)$ (b) $n(n-1)(n!/2)$

23. (a) 8   (b) 6

25. (a) 4!   (b) $2 \times 4! \times 4!$

27. (a) $\binom{6}{4}4!$ (b) $\binom{6}{3}\binom{4}{3}3!$

29. (a) 1001   (b) 126

31. (a) 6^{15} (b) $6^{14} \times 3$ (c) 4^{15}

Chapter 2

1. (a) 0.4   (b) 0.1

3. (a) $\frac{1}{128}$ (b) $\frac{7}{128}$ (c) $\frac{21}{128}$ (d) $\frac{35}{128}$

5. (a) $\frac{1}{n}$ (b) $\left(\frac{n-1}{n}\right)^{k-1}\left(\frac{1}{n}\right)$

7. (a) $\frac{1}{12} \le P(EF) \le \frac{1}{3}$ (b) $\frac{3}{4} \le P(E \cup F) \le 1$

9. (a) $\frac{15}{216}$ (b) $\frac{5}{72}$

11. 0.91091

13. 0.85616

15. $\frac{20}{27}$

17. (a) 0.0886   (b) 0.2456

19. (a) $\frac{28}{115}$ (b) $\frac{21}{230}$

21. (a) 0.851682   (b) 0.85625

23. 0.4654

27. (a) $\frac{3}{10}$ (b) $\frac{1}{5}$ (c) $\frac{1}{10}$

29. $\dfrac{\binom{m}{k-1}\binom{n}{r-k}\binom{n-r+1}{1}}{\binom{m+n}{r-1}\binom{m+n-r+1}{1}}$

31. (a) 0.1055   (b) $\frac{4}{13}$

33. 0.34985

35. 0.36608

37. (a) $\sum_{m=n-i}^{2n-(i+j)-1} \left(\frac{1}{2}\right)^m \binom{m-1}{n-i-1}$ (b) $P(E_1) = \frac{3}{4}$, $P(E_2) = \frac{11}{16}$

39. (a) $\binom{2N-k}{N} \left(\frac{1}{2}\right)^{2N-k}$, $\quad k = 0, 1, \ldots, N$

(b)

```
p[k_, BN_] = Binomial[2 BN - k, BN] (0.5)^(2 BN-k);

Total[Table[p[k, BN], {k, 0, BN}]]

1.

ListPlot[Table[{k, p[k, BN]}, {k, 0, BN}]]
```

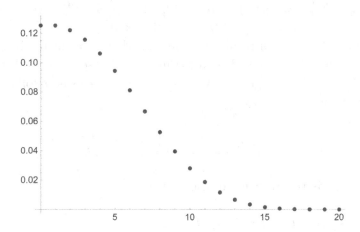

Chapter 3

1. (a) $\frac{1}{2}$ (b) $\frac{1}{10}$

3. $\frac{2}{3}$

5. (a) 0.25455 (b) 0.057

7. (a) $\frac{2}{7}$ (b) $\frac{7}{8}$

9. $\frac{1}{6}$

11. (a) $\frac{1}{66}$ (b) $\frac{3}{5}$ (c) $\frac{13}{30}$

13. 0.26239

15. $\frac{9}{108}$

17. $\frac{2}{9}$

19. (a) $\frac{5}{36}$ (b) $\frac{1}{210}$

21. $\frac{4}{11}$

23. 0.16923

25. 0.69591

27. (a) $p_n = \sum_{k=1}^{6} p_{n-k}\left(\frac{1}{6}\right)$, $n = 1, 2, \ldots$;   specifically, $p_n = p_{n-1}\left(\frac{7}{6}\right)$,
$n = 1, \ldots, 6$; $p_0 = 1$,   and $p_j = 0$ if $j < 0$
(b) Since the value of resulting from a single roll of the die is $(1 + \cdots +$
$6)/6 = 3.5$,   the expected increase in the total score per roll is $\frac{1}{3.5} =$
$0.285\,7 \approx p_n$ for a large n.   Another way to see this is as follows: $P(A)$
can be approximated by

$$\lim_{n \to \infty} \sum_{i=1}^{n} \frac{P(1\{X_i = 1\}}{n} \approx \frac{1}{3.5}$$

where X_i is the outcome of the ith   roll

29. (a) $P_n = p(1 - P_{n-1}) + (1 - p)P_{n-1}$

31. The posterior probabolieties for Professor Abel, Braz, and Clarence are
0.90391, 0.090361, and 0.00602, respectively

33. (a) $b/(r + b)$ (b) $b/(r + b)$

Chapter 4

1. $p_X(x) = (x - 1)/36$, $x = 2, 3, ..., 7$; $p_X(x) = (13 - (x - 1))/36$, $c = 8, 9, ..., 12$

3. (b) $F_X(x) = 1 - (3/7)^x$, $x = 1, 2, ...$ (c) $P(X > k) = (3/7)^k$,
$k = 1, 2, , ...$ (d) $E(X) = 7/4$.   (e) $P(X = \text{even}) = 0.3$

```
p[x_] = 4 (3^(x-1)) / 7^x   // N; sEven = 0; sOdd = 0;

Table[If[EvenQ[x], sEven = sEven + p[x], 0], {x, 1, 30}];

Table[If[OddQ[x], sOdd = sOdd + p[x], 0], {x, 1, 70}];

Print["P(x is even) = ", sEven, "    P(X is odd) = ", sOdd]

P(x is even) = 0.3    P(X is odd) = 0.7
```

5. (a) $\frac{1}{2}$ (b) $\frac{1}{2}$ (c) $\frac{1}{3}$

7. (a) 3295 (b) 0.305

9. (a) 3 (b) $F_X(x) = 0,\ x \le 0;\quad F_X(x) = x^3,\ 0 < x < 1;\quad F_X(x) = 1,$ if $x \ge 1$ (c) $\frac{3}{4}$ (d) $\frac{3}{80}$

11.

$$\text{pgf}[z\_,\ p\_] := \frac{p\,z}{1 - (1 - p)\,z};$$

CoefficientList[Series[pgf[z, 0.6], {z, 0, 6}], z] // N

$\{0.,\ 0.6,\ 0.24,\ 0.096,\ 0.0384,\ 0.01536,\ 0.006144\}$

The above shows that the PMF

x	$p_X(x)$
0	0
1	0.6
2	$(0.4)(0.6) = 0.24$
3	$(0.4)^2(0.6) = 0.096$
4	$(0.4)^3(0.6) = 0.038\,4$
5	$(0.4)^4(0.6) = 0.015\,36$
6	$(0.4)^5(0.6) = 0.006144$

13. (a) $P(X = n) = q^{n-1}p,\quad n = 1, 2, \ldots$ (b) $E(X) = 1/p$ (c) $P(X \ge k) = q^{k-1},\quad k = 1, 2, \ldots$

15. $F_Y(y) = 0,\ y \le 0,\ F_Y(y) = y,\ 0 < y < 1,\ F_Y(y) = 1,\ y \ge 1$

17. (a) $\frac{1}{2}$ (b) $\frac{1}{6}$ (c) $\frac{1}{4}$ (d) $\frac{1}{2}$ (e) 0 (f) $\frac{1}{3}$ (g) 0

19. (a) $p_X^g(z) = 2/(2 - z)$

(b) The following series is the expansion of the PGF up to order 6:

$$\text{pgf}[z\_] := \frac{z}{2 - z};$$

Series[pgf[z], {z, 0, 6}]

$$\frac{z}{2} + \frac{z^2}{4} + \frac{z^3}{8} + \frac{z^4}{16} + \frac{z^5}{32} + \frac{z^6}{64} + O[z]^7$$

So it implies that $p_X(x) = 1/2^x,\quad x = 1, 2, \ldots$

21. Let X_i = the gain from each of the five cases shown below:

Case	Outcome of each cases	X_i	Probability
1	10	10	$\left(\frac{18}{38}\right) = 0.473\,68$
2	$-10, 10, 10$	10	$\left(\frac{20}{38}\right)\left(\frac{18}{38}\right)^2 = 0.118\,09$
3	$-10, 10, -10$	-10	$\left(\frac{20}{38}\right)^2\left(\frac{18}{38}\right) = 0.131\,21$
4	$-10, -10, 10$	-10	$\left(\frac{20}{38}\right)^2\left(\frac{18}{38}\right) = 0.13121$
5	$-10, -10, -10$	-30	$\left(\frac{20}{38}\right)^3 = 0.145\,79$

$P(X > 0) = P(\text{Case } 1) + P(\text{Case } 2) = 0.47368 + 0.11809 = 0.591\,77$
(b) $E(X) = (10)(0.59177) + (-10)(0.13121)(2) + (-30)(0.14579) = -1.0802$

(c) It depends on your personal risk preference

23. (a) $\frac{1}{2}$

(b) $f_Y(y) = \begin{cases} \frac{1}{a}e^{-\left(\frac{y-b}{a}\right)}, & y > b \\ \frac{1}{a}e^{-\left(\frac{b-y}{a}\right)}, & y < b \end{cases}$

25.   1.3889,   25.015,   0.62193,   1.8667   (excess kurtosis $= -1.1333$)

27.

$$f_Y(y) = \begin{cases} \frac{2}{9\sqrt{y}}, & 0 < y < 1 \\ \frac{1}{9\sqrt{y}}(\sqrt{y}+1), & 1 < y < 4 \\ 0, & y \geq 4 \end{cases}$$

29. (b) $p_X^g(z) = -\log(1-z) + z^{-1}\log(1-z) + 1$

31. **Proof.**

$$E(X^2) = \int_0^\infty x^2 f_X(x)dx = \int_0^\infty \left(\int_0^x (2y)dy\right) f_X(x)dx$$
$$= \int_0^\infty 2y\left(\int_y^\infty f_X(x)\right)dy = \int_0^\infty 2yP(Y > y)dy$$

□

Chapter 5

1. (a) 0.24451   (b) 0.23884   (c) $p_C(x) = q^{x-1}p, \ x = 1, 2, \ldots,$ where $p = 1/13$

3. 0.27156

5. (a) 0.9502   (b) 0.0658   (c) 0.0037

7. 0.0592

9. (a) $\varphi_X(t) = \exp(\lambda(e^{it} - 1))$ (b) $p_X^g(z) = \exp(\lambda(z - 1))$ (c) same as (b)

11. (a) $p_X^g(z) = \frac{1}{2}(1/(1 - z) + 1/(1 - Qz))$ (b) $P(X = n) = \frac{1}{2}(1 - (q - p)^n), \ n = 0, 1, \ldots$

13. (a) $p_X^g(z) = \frac{1}{6}(z^1 + \cdots + z^6)$. Thus $p_{Z_n}^g(z) = \frac{1}{6^n}(z^1 + \cdots + z^6)^n$.
    (b) We pick the coefficients associated with the polynomial expansion. They give the PMF of S_5. We plot the PMF below:

$$\mathtt{pgf[z\_] = \frac{1}{6^5}\left(z^1 + z^2 + z^3 + z^4 + z^5 + z^6\right)^5;}$$

$$\mathtt{x = Flatten\left[Coefficient\left[Expand[pgf[z]], \{Table[z^i, \{i, 1, 30\}]\}\right]\right] \ // \ N;}$$

$$\mathtt{ListPlot[x]}$$

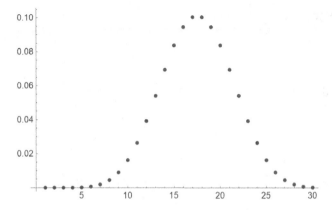

15. (a) Let $(x, y) =$ (number of heads, number of tails). The scenrios $(15,0)$, $(15,1)$, ..., $(15,4)$ all meet the stipulation. We see

$$P(15,0) = \binom{14}{14}(0.65)^{15}$$

$$P(15,1) = \binom{15}{14}(0.65)^{15}(0.35)$$

$$P(15,2) = \binom{16}{14}(0.65)^{15}(0.35)^2$$

$$P(15,3) = \binom{17}{14}(0.65)^{15}(0.35)^3$$

$$P(15,4) = \binom{18}{14}(0.65)^{15}(0.35)^4$$

or

$$\sum_{n=15}^{19} \binom{n-1}{14}(0.65)^{15}(0.35)^{n-15} = 0.15$$

(b) Note that MATHEMATICA uses the alternative version of the negative binomial random variable, namely the number of failures needed to obtained n successes for the first time. But the parameters state the same, i.e., $n = 15$, and $p = 0.65$

```
negb[x_] = PDF[NegativeBinomialDistribution[15, 0.65], x];

Total[Table[negb[x], {x, 0, 4}]]

0.149996
```

17. (a) 108.33   (b) 3741.1

19. (a)

$$p_X(x) = \frac{\binom{1}{1}\binom{x-1}{k-1}}{\binom{n}{k}}, \qquad x = k, k+1, \ldots, n$$

(b) $F_X(x) = \binom{x}{k} / \binom{n}{k}, \quad x = k, k+1, \ldots, n$

(c)

```
EX = FullSimplify[∑(x=k to n) x Binomial[(x-1), (k-1)] / Binomial[n, k]];

EX2 = FullSimplify[∑(x=k to n) x² Binomial[(x-1), (k-1)] / Binomial[n, k]];

VarX = EX2 - EX²;

mean[n_, k_] = k (1+n) / (1+k);   variance[n_, k_] = - k² (1+n)² / (1+k)² + k (1+n) (k+n+kn) / ((1+k) (2+k));

m = mean[10, 5] // N;   v = variance[10, 5] // N;

Print["   EX = ", m, "    VarX = ", v]
    EX = 9.16667    VarX = 1.09127
```

21. (a) $p_X^g(z) = z/(6 - (z^2 + \cdots + z^6))$ (b) $E(X) = 21$ and $\mathrm{Var}(X) = 490$
(c)

```
h[z_] = z / (6 - (z² + z³ + z⁴ + z⁵ + z⁶));

EX = h'[1];

EX2mX = h''[1];

VarX = EX2mX + EX - EX²;

Print["  E(X)  = ", EX, "    Var(X) = ", VarX]
    E(X)  = 21    Var(X) = 490

pmf = Flatten[Table[Coefficient[Series[h[z], {z, 0, 80}], {z^i}], {i, 1, 80}] // N];

ListPlot[pmf]
```

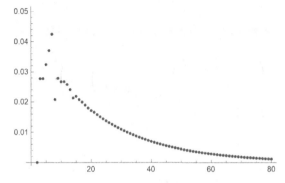

23. (a) $p_X(x) = qp^{k-1}$, $k = 1, 2, \ldots, 6$, $p_X(x) = q^6$, $x = 7$, where $p = 0.1$
and $q = 0.9$ (b) $E(Y) = 5.61 \times 10^6$ and $Sdv(Y) = 5.8507 \times 10^6$

25. (a) $P(X = k) = q^k/k$, $k = 1, 2, \ldots$ (b) $E(X) = \text{Var}(X) = e - 1$

```
Simplify[Series[Log[1/(1 - q t)], {t, 0, 6}]]
```

$$q t + \frac{q^2 t^2}{2} + \frac{q^3 t^3}{3} + \frac{q^4 t^4}{4} + \frac{q^5 t^5}{5} + \frac{q^6 t^6}{6} + O[t]^7$$

```
Gxt = Log[1/(1 - q t)];
```

```
EX = Simplify[D[Gxt, {t, 1}] /. {t → 1, q → 1 - e^-1}]
```

$-1 + e$

```
a = Simplify[D[Gxt, {t, 2}] /. {t → 1, q → 1 - e^-1}]
```

$(-1 + e)^2$

```
EX2 = a + EX;
```

```
VarX = Simplify[EX2 - EX^2]
```

$-1 + e$

27. (a)

$$P(N = n) = \left(1 - \left(\frac{5}{6}\right)^n\right)^{10} - \left(1 - \left(\frac{5}{6}\right)^{n-1}\right)^{10}, \qquad n = 1, 2, \ldots, n$$

(b) $T \sim \text{negb}\left(10, \frac{1}{6}\right)$

(c)

$$P(N = k, L = l) = \binom{10}{l}\left(1 - \left(\frac{5}{6}\right)^{k-1}\right)^{10-l}\left(\frac{1}{6}\left(\frac{5}{6}\right)^{k-1}\right)^{l},$$

$$k \geq 2, l = 1, \ldots, 10$$

and

$$P(N = 1, L = 10) = \left(\frac{1}{6}\right)^{10} \quad \text{and}$$

$$P(N = 1, L = k) = 0, \quad k = 1, \ldots, 9.$$

29. Let E be the event that Wayne made 7 successful shot out of 10. Then

$$P(E) = \sum_{i=1}^{3} P(E|\theta_i)P(\theta_i)$$

$$= \binom{10}{7}(0.2)^3(0.8)^7(0.25) + \binom{10}{7}(0.5)^3(0.5)^7(0.6)$$

$$+ \binom{10}{7}(0.75)^3(0.25)^7(0.2)$$

$$= 0.12126$$

The posterior probabilities on θ are

$$P(\theta_1|E) = \frac{\binom{10}{7}(0.2)^3(0.8)^7(0.25)}{0.12126} = 0.41507$$

$$P(\theta_2|E) = \frac{\binom{10}{7}(0.5)^3(0.5)^7(0.6)}{0.12126} = 0.57985$$

$$P(\theta_3|E) = \frac{\binom{10}{7}(0.75)^3(0.25)^7(0.2)}{0.12126} = 0.00509$$

$$\sum_{i=1}^{3} P(\theta_i|E) = 0.41507 + 0.57985 + 0.00509 = 1.0$$

31. **Proof.** Define

$$x_n = P(X \text{ is even})$$

For $n = 1$

$$x_1 = 1 - p = \frac{1}{2}\left(1 + (1 - 2p)^1\right) = 1 - p$$

Assume the formula is true for $n - 1$. Then

$$x_n = \{\text{the } (n-1)\text{th trial is even}\}(1 - p) + \{\text{the } (n-1)\text{th trial is odd}\}p$$

$$= x_n(1 - p) + (1 - x_n)p$$

$$= (1 - p)\left(\frac{1}{2}(1 + (1 - 2p)^{n-1})\right) + p\left(1 - \frac{1}{2}(1 + (1 - 2p)^{n-1})\right)$$

$$= \frac{1}{2}(1 + (1 - 2p)^n)$$

$\square$

33. We see that

$$P(\{X = 0\} = 1 - e^{-\lambda}$$

Thus

$$P(Y = k) = P(\{X = k\} \cap \{X \neq 0\}|\{X \neq 0\})$$

$$= \frac{\left(\frac{e^{-\lambda}\lambda^x}{k!}\right)}{1 - e^{-\lambda}} \qquad k = 1, 2, \dots, n \qquad \Box$$

Chapter 6

1. (a) $\frac{4}{9}$

 (b)

$$F_X(x) = \begin{cases} 0 & \text{if } x \leq 2 \\ \frac{1}{18}x^2 + \frac{1}{6}x - \frac{5}{9} & \text{if } 2 < x < 4 \\ 1 & \text{if } x \geq 4 \end{cases}$$

 (c) 3.0741

3.

$$f_Y(y) = \begin{cases} \sqrt{\frac{2}{\pi}}e^{-\frac{1}{2}y^2} & \text{if } Y > 0 \\ 0 & \text{otherwise} \end{cases}$$

5. $f_Y(y) = e^{-y}$, $y > 0$; and 0, otherwise

7. (a) $Y \sim \text{geom}(1 - e^\alpha)$ (b) $f_W(w) = \alpha e^{-\alpha}/(1 - e^{-\alpha})$, $0 < w < 1$ (c) Yes, $Y \perp W$

9. (a) 0.6065 (b) 0.49659

11. (a) 0 (b) $\frac{2}{\lambda^2}$. The following does the computation:

: `Integrate[`$\frac{\lambda\,x}{2}$` Exp[-λ Abs[x]], {x, -∞, ∞}, Assumptions → Re[λ] > 0]`

0

: `Integrate[`$\frac{\lambda\,x^2}{2}$` Exp[-λ Abs[x]], {x, -∞, ∞}, Assumptions → Re[λ] > 0]`

$\frac{2}{\lambda^2}$

(c) It is a double exponential random variable with

$$f_X(x) = \begin{cases} \frac{1}{2}e^{-\lambda x} & \text{if } x > 0 \\ \frac{1}{2}e^{\lambda x} & \text{if } x < 0 \end{cases}$$

13. (a) $F_X(x) = 1 - e^{-(x-\ln(x+1))}$, $x > 0$ (b) $P(X > s + t | X > t) = \frac{1+t+s}{1+t}e^{-s}$

15. 2.273

17. 0.0211

19. (a) $E(X) = 50$ and $\text{Var}(X) = 500$ (b) 0.94138   (c) Per MATHE-MATICA syntax, the second parameter is entered as $1/\lambda$:

`CDF[GammaDistribution[5, 10], 90] - CDF[GammaDistribution[5, 10], 10] // N`

`0.941377`

21. (a) 0.3028   (b) $E(X) = 8.23$ (c) 0.3009

(a)

`1 - CDF[WeibullDistribution[2.396, 9.2851], 10]`

`0.302858`

(b)

`Mean[WeibullDistribution[2.396, 9.2851]]`

`8.23082`

(c)

`CDF[WeibullDistribution[2.396, 9.2851], 20] - CDF[WeibullDistribution[2.396, 9.2851], 10]`

`0.300998`

23. Y follows a lognormal distribution with parameters $(\mu_1 + \mu_2, \sqrt{\sigma_1^2 + \sigma_2^2})$

25. (a) $\varphi_Y(t) = \frac{1}{(1-2it)^{\frac{n}{2}}}$ (b) $E(Y) = n$ and $\text{Var}(Y) = 2n$

```
φ[t_] = ———————————————
               1
         ———————————————
                      n
         (1 - 2 i t) 2
```

```
(1 - 2 i t) -n/2
```

```
        1
EX = ——— φ'[0]
        i
```

```
n
```

```
         1
EX2 = ——— φ''[0];
        i²
```

$$\text{VarX} = \text{Simplify}\left[\text{EX2} - \text{EX}^2\right]$$

```
2 n
```

27. The following MATHEMATICA output answers (a)–(c)

```
          Exp[-x]
f[x_] = ——————————————;
        (1 + Exp[-x])²
```

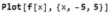

```
1
```

$$\text{EX} = \int_{-\infty}^{\infty} x\, f[x]\, dx$$

```
0
```

$$\text{EX2} = \int_{-\infty}^{\infty} x^2\, f[x]\, dx$$

```
π²
——
 3
```

```
VarX = EX2 - EX²
```

```
π²
——
 3
```

```
Plot[f[x], {x, -5, 5}]
```

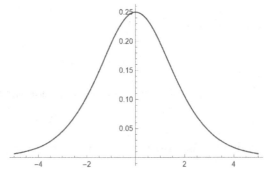

29. (a)

```
a = 10; b = 0.7;
```

$$f[x\_] := \frac{b\,x^{b-1}}{a^b};$$

$$\int_0^{10} f[x]\,dx$$

```
1.
```

```
Plot[f[x], {x, 0, 10}, AxesLabel → {"x", "f(x)"}]
```

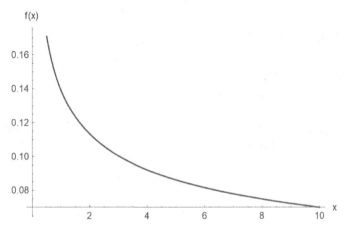

(b) The following MATHEMATICA output gives $F_X(x) = a^{-b}x^b$ for $0 < x < a$:

$$\texttt{cdf}[x\_] = \int_0^x \frac{b\,\tau^{b-1}}{a^b}\,d\tau$$

$$\boxed{a^{-b}\,x^b \quad \text{if} \quad \text{Re}[b] > 0}$$

It follows that

$$P(Y \le y) = P\left(\frac{1}{X} \le y\right) = P\left(X \ge \frac{1}{y}\right) = 1 - F_X\left(\frac{1}{y}\right)$$

$$= 1 - (a^{-b})\left(\frac{1}{y}\right)^b = 1 - (a\,y)^{-b}, \qquad y > \frac{1}{a}$$

$$f_Y(y) = \frac{d}{dy}\left(1 - (a\,y)^{-b}\right) = \frac{b}{y(ay)^b}, \qquad y > \frac{1}{a}$$

(c) $P(Z \le z) = P\left(\frac{X}{a} \le z\right) = P(X \le az) = F_X(az) = a^{-b}\,(az)^b = z^b, \quad 0 < z < 1$

Hence

$$f_Z(z) = \frac{d}{dz}z^b = bz^{b-1}, \qquad 0 < z < 1$$

We can write the above as

$$f_Z(x) = \frac{1}{B(b,1)}z^{b-1}(1-z)^{1-1} = \frac{\Gamma(b+1)}{\Gamma(b)\Gamma(1)}z^{b-1} = z^{b-1}, \qquad 0 < z < 1$$

Thus $Z \sim \text{Beta}(b,1)$.

(d) From Part (c), we easily obtain

$$E(X) = \frac{b}{b+1}, \qquad E(X^2) = \frac{(b+1)b}{(b+2)(b+1)} = \frac{b}{b+2}$$

31. $\left(\frac{b}{c}\right)^\lambda$

33.

$$f_X(x) = \begin{cases} \dfrac{1}{x\log(10)} & 1 < x < 10 \\ 0 & \text{otherwise} \end{cases}$$

35. See the MATHEMATICA output. (b) We see that a microprocessor with indicated parameters of the failure rate function has a high failure risk initially and the risk then diminishes as it ages. (c) We see that

$$F_X(t) = 1 - \exp\left(-\int_0^t \lambda(s)ds\right)$$
$$= 1 - (1 + t^\alpha)^{-\beta}$$

We then obtain PDF with

$$f_X(t) = \frac{\alpha\beta t^{\alpha-1}}{(1+t^\alpha)^{\beta+1}}$$

(d) A plot of the PDF $f_X(\cdot)$ is also shown below

$$\lambda[t\_, \beta\_, \alpha\_] = \frac{\beta \alpha t^{\alpha-1}}{1 + t^{\alpha}};$$

`Plot[λ[t, 3, 1.5], {t, 0, 30}, PlotRange → All]`

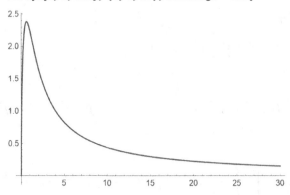

$$Ft[t\_] = 1 - Exp\left[-\int_0^t \lambda[s, \beta, \alpha] \, ds\right]$$

$$\boxed{1 - (1 + t^{\alpha})^{-\beta} \quad \text{if} \quad Re[\alpha] > 0}$$

`f[t_] = Ft'[t]`

$$\boxed{t^{-1+\alpha} (1 + t^{\alpha})^{-1-\beta} \alpha \beta \quad \text{if} \quad Re[\alpha] > 0}$$

$$ff[t\_, \beta\_, \alpha\_] = \frac{\alpha \beta t^{\alpha-1}}{(1 + t^{\alpha})^{\beta+1}};$$

$$\int_0^{\infty} ff[t, 3, 1.5] \, dt$$

1.

`Plot[ff[t, 3, 1.5], {t, 0, 30}, PlotRange → All]`

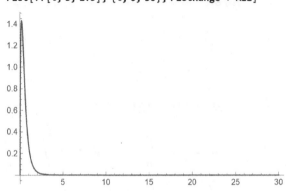

Chapter 7

1. $(1 - 2d)^3$

3. $P_Y(y) = \frac{1}{n+1}$, $y = 0, 1, \ldots, n$

5. $P(R) = \frac{1}{m+1}$

7. (a) $\frac{1}{4}$ (b) $\frac{1}{2}$ (c) $\frac{1}{2}$ (d) $\frac{3}{4}$ (e) $\frac{1}{8}$

9. $1 - \left(\frac{2}{e} - \frac{1}{e^2}\right)^3$

11. (a) $c = \frac{1}{8}$ (b) $f_X(x) = \frac{1}{4}e^{-x}(1+x)$ and $f_Y(y) = \frac{1}{6}e^{-y}y^3$

    (a)

$$\text{Solve}\left[\int_0^\infty \int_{-y}^y c\,(y^2 - x^2)\, \text{Exp}[-y]\, dx\, dy == 1,\, c\right]$$

$$\left\{\left\{c \to \frac{1}{8}\right\}\right\}$$

    (b) The range for fx: x< |y|; the range for fy: y > 0

$$\text{fx}[x\_] = \frac{1}{8}\int_x^\infty (y^2 - x^2)\, \text{Exp}[-y]\, dy$$

$$\frac{1}{4}\,e^{-x}\,(1 + x)$$

$$\text{fy}[y\_] = \int_{-y}^y \frac{1}{8}\,(y^2 - x^2)\, \text{Exp}[-y]\, dx$$

$$\frac{1}{6}\,e^{-y}\,y^3$$

13. (a) Let $Z = X - Y$. We see that $|Z| < a$, or $-a < X - Y < a$. Thus

$$F_Z(z) = P(Z < z) = \frac{1}{L^2}\left(\int_0^z \int_0^{x+z} dy dx + \int_z^{L-z}\int_{x-z}^{x+z} dy dx\right.$$
$$\left. + \int_{L-z}^L \int_{x-z}^L dy dx\right) = \frac{z}{L}\left(2 - \frac{z}{L}\right)$$

(b)

$$f_Z(z) = \frac{d}{dz}\frac{z}{L}\left(2 - \frac{z}{L}\right) = \frac{2}{L^2}(L - z), \qquad 0 < z < L$$

$$\int_0^L \frac{2}{L^2}(L - z)dz = 1$$

(c)

$$E(Z) = \int_0^L P(Z > z)dz = \int_0^L \left(1 - \frac{z}{L}\left(2 - \frac{z}{L}\right)\right)dz = \frac{L}{3}$$

15. We see that

$$U = g_1(X, Y) = X + Y$$
$$V = g_2(X, Y) = X - Y$$

Thus

$$\frac{\partial g_1}{\partial X} = 1, \qquad \frac{\partial g_1}{\partial Y} = 1, \qquad \frac{\partial g_2}{\partial X} = 1, \qquad \frac{\partial g_2}{\partial Y} = -1$$

and

$$J(X, Y) = \begin{vmatrix} 1 & 1 \\ 1 & -1 \end{vmatrix}^{-1} = |-2|^{-1} = \frac{1}{2}$$

Since $X = (U + V)/2$ and $Y = (U - V)$, we have

$$f_{U,V}(u, v) = f_{X,Y}\left(\frac{u + v}{2}, \frac{u - v}{2}\right)\frac{1}{2}$$

$$= \frac{1}{\sqrt{2\pi}}e^{-\frac{1}{2}\left(\frac{u+v}{2}\right)^2}\frac{1}{\sqrt{2\pi}}e^{-\frac{1}{2}\left(\frac{u-v}{2}\right)^2} \times \frac{1}{2}$$

$$= \frac{1}{2\pi}e^{-\frac{1}{2^2}(u^2+v^2)} \cdot \frac{1}{2}$$

$$= \frac{1}{\sqrt{2\pi \cdot 2}}e^{-\frac{1}{2\cdot 2}u^2}\frac{1}{\sqrt{2\pi \cdot 2}}e^{-\frac{1}{2\cdot 2}v^2},$$

$$-\infty < u < \infty \quad \text{and} \quad -\infty < v < \infty.$$

Thus $U \sim N(0, 2)$ and $V \sim N(0, 2)$ — results also could have been obtained via convolution. Also we conclude that $U \perp V$ as the two densities factor

17. We first define

$$Z = (2Y)^{\frac{1}{2}}$$

We see that

$$F_Z(z) = P(Z \le z) = P\left((2Y)^{\frac{1}{2}} \le z\right)$$

$$= P(2Y \le z^2) = F_Y\left(\frac{z^2}{2}\right) = 1 - e^{-\frac{z^2}{2}}$$

Thus

$$f_Z(z) = z e^{-\frac{z^2}{2}}, \qquad z > 0$$

Now $W = X \cdot Z$, where X has the PDF

$$f_X(x) = \frac{1}{\sqrt{2\pi}} e^{-\frac{1}{2}x^2}, \qquad -\infty < x < \infty$$

When $x > 0$, we have $w > 0$ and when $x < 0$, we have $z < 0$. By symmetry, we first consider the case when $x > 0$. For $W > 0$, $W = X \times Z$ and both are positive. So we use Mellin transform to get the joint density below

```
f[s_] = MellinTransform[ 1/√(2 π) Exp[- 1/2 x²], x, s];

g[s_] = MellinTransform[z Exp[- z²/2], z, s];

h[s_] = f[s] × g[s];

InverseMellinTransform[h[s], s, w]
```

$$\frac{e^{-w}}{2}$$

Thus

$$f_W(w) = \frac{1}{2} e^{-w}, \qquad w > 0$$

By symmetry, we have

$$f_W(w) = \frac{1}{2} e^{-|w|}, \qquad -\infty < w < \infty$$

The above equation is known as the Laplace distribution

19. 0.44

21. (a) Yes   (b) $E(X) = \frac{1}{2}$, $E(Y) = \frac{2}{3}$ (c) Var$(X) = \frac{1}{20}$,   Var$(Y) = \frac{1}{18}$

23. (a) 0.0296   (b) 0.0063

25. $\frac{203}{256}$

27. $E(X) = 3.5$

29. (a) $E(X) = \frac{1}{3}$, Var$(X) = \frac{1}{18}$ (b) $E(X|Y = \frac{1}{3}) = \frac{1}{3}$,   Var$(X|Y = \frac{1}{3}) = \frac{1}{27}$

31. (a) No   (b) $P(X + Y > 3) = 0.7333$

(c) $f_{U,V}(u, v) = \dfrac{1}{16}u + \dfrac{3}{80}v$ if $(u, v) \in D$

where

$$D = \{(u, v) | 0 < u + v, u + v < 2, 2 < u - v, u - v < 10\}$$

In other words, D is the region enclosed by the four vertices $\{(1, -1), (5, -5), (6, -4), (2, 0)\}$

33. Let E be the event that the largest of the three is greater than the sum of the other two. The needed probability is given by

$$P(E) = P(X_1 > X_2 + X_3) + P(X_2 > X_1 + X_3) + P(X_3 > X_1 + X_2)$$

$$= 3P(X_1 > X_2 + X_3) \qquad \text{by symmetry}$$

So we have

$$P(E) = 3 \int_0^1 \int_0^{1-x_3} \int_{x_2+x_3}^1 dx_1 dx_2 dx_3$$

$$= 3 \int_0^1 \int_0^{1-x_3} (1 - x_2 - x_3) dx_2 dx_3 = 3 \int_0^1 \frac{(1 - x_3)^2}{2} dx_3 = \frac{1}{2}$$

35. (a) $f_Y(y) = \sqrt{\frac{2}{\pi}} e^{-\frac{y^2}{2}}$, $y > 0$, and $f_Y(y) = 0$, otherwise   (b) $X|Y \sim$ expo(y)   (c) $P(X < 1|Y = 1) = 0.63212$

$$\texttt{f[x\_, y\_] :=} \sqrt{\frac{2}{\pi}} \ y \ \texttt{Exp[- x y]} \times \texttt{Exp}\left[-\frac{y^2}{2}\right]$$

$$\texttt{Integrate[f[x, y], \{x, 0, \infty\}, Assumptions} \to \texttt{y > 0]}$$

$$e^{-\frac{y^2}{2}} \sqrt{\frac{2}{\pi}}$$

$$\int_0^\infty \frac{1}{\sqrt{2\pi}} \ \texttt{Exp}\left[-\frac{y^2}{2}\right] \texttt{dy}$$

$$\frac{1}{2}$$

37. (a) $c = 1$
   (b) $f_Y(y) = \frac{1}{6} e^{-y} y^3$, $y > 0$, $f_X(x) = xe^{-x}$, $x > 0$

(a)

$$\texttt{Solve}\left[\int_0^\infty \int_x^\infty \texttt{c x (y - x) Exp[-y] dy dx == 1, \{c\}}\right]$$

$$\{\{c \to 1\}\}$$

(b)

$$\texttt{fY[y\_] =} \int_0^y \texttt{x (y - x) Exp[-y] dx}$$

$$\frac{1}{6} e^{-y} y^3$$

$$\texttt{fX[x\_] =} \int_x^\infty \texttt{x (y - x) Exp[-y] dy}$$

$$e^{-x} x$$

(c) $f_{Y|X}(y|x) = e^{x-y}(y-x)$, $y > x$, $f_{X|Y}(x|y) = \frac{6x(y-x)}{y^3}$, $y > x$

(c)

$$\text{fYgX}[x\_] = \frac{x\,(y-x)\,\text{Exp}[-y]}{e^{-x}\,x}$$

$e^{x-y}\,(-x+y)$

$$\text{fXGy}[y\_] = \frac{x\,(y-x)\,\text{Exp}[-y]}{\frac{1}{6}\,e^{-y}\,y^3}$$

$$\frac{6\,x\,(-x+y)}{y^3}$$

(d)

$$\text{EXgy} = \int_0^y x\,\frac{x\,(y-x)\,\text{Exp}[-y]}{\frac{1}{6}\,e^{-y}\,y^3}\,dx$$

$$\frac{y}{2}$$

$$\text{EYgX} = \int_x^\infty y\,e^{x-y}\,(-x+y)\,dy$$

$2 + x$

(d) $E(X|Y) = \frac{Y}{2}$ and $E(Y|X) = X + 2$

41. Proof.

$$\varphi_{\log X}\left(\frac{s-1}{i}\right) = \varphi_Z\left(\frac{s-1}{i}\right) \qquad \text{define } Z = \log X$$

$$= E\left[e^{i\left(\frac{s-i}{i}\right)Z}\right] \qquad \text{by definition of } \varphi_Z$$

$$= E\left[e^{(s-1)Z}\right]$$

$$= E\left[e^{(s-1)\log X}\right]$$

$$= E\left[e^{\log(X)^{s-1}}\right]$$

$$= E\left[X^{s-1}\right] = \mathcal{M}_X(s) \qquad \text{by definition of } \mathcal{M}_X(s)$$

$\square$

Chapter 8

1. (a) 5.74   (b) 154.14

3. 2.61181

5. 0.35635

7. (a) 1.1538   (b) 1.8462   (c) 0.59172   (d) −0.59172

9. 0.9583

11. (a)

$$f_W(w) = \frac{1}{\sqrt{2\pi(5+4\rho)}} e^{-\frac{1}{2(5+4\rho)}w^2}, \qquad -\infty < w < \infty$$

(b) 0.12842

13. (a) We see that

$$P(X_1 > 0, X_2 > 0) = P\left(\frac{X_1 - \mu_1}{\sigma_1} > 0, \frac{X_2 - \mu_2}{\sigma_2} > 0\right)$$

$$\equiv P(Y > 0, Z > 0)$$

where $(Y, Z) \sim \text{BVN}((0,1),(0,1),\rho)$. We use MATHEMATICA to do the integration. This yields

```
f[y_, z_] = 1/(2 π √(1-ρ²)) Exp[- 1/(2 (1-ρ²)) (y² - 2 ρ y z + z²)];

Integrate[f[y, z], {y, 0, ∞}, {z, 0, ∞}, Assumptions → Re[ρ²] ≤ 1 && Re[ρ/√(2 - 2ρ²)] > 0]
```

$$\frac{\pi + 2\,\text{ArcTan}\left[\sqrt{-\frac{\rho^2}{-1+\rho^2}}\right]}{4\,\pi}$$

Thus we see that

$$P(X_1 > 0, X_2 > 0) = \frac{\pi + 2\tan^{-1}\left(\frac{\rho}{\sqrt{1-\rho^2}}\right)}{4\pi} = \frac{1}{4} + \frac{1}{2\pi}\tan^{-1}\left(\frac{\rho}{\sqrt{1-\rho^2}}\right)$$

(b) We see when $\rho = 0.9999$, $P(X_1 > 0, X_2 > 0) \approx P(X_1 > 0) \approx 0.5$. On the other hand, when $\rho = 0$, $X_1 \perp X_2$. Thus $P(X_1 > 0, X_2 > 0) = P(X_1 > 0)P(X_2) = (0.5)(0.5) = 0.25$

15. Since W is a linear transformation of two normal, we know that W follows a normal distribution. We see that

$$E(W) = \frac{1}{\sqrt{1-\rho^2}} E(W - \rho X) = 0$$

and

$$\text{Var}(W) = \frac{1}{1-\rho^2} \text{Var}\,(Y - \rho X)$$

$$= \frac{1}{1-\rho^2}\,(\text{Var}(Y) + \text{Var}(\rho X) - 2\text{Cov}(Y, \rho X))$$

$$= \frac{1}{1-\rho^2}(1 + \rho^2 - 2\rho^2) = 1$$

Finally, we have

$$\text{Cov}(X, W) = \text{Cov}\left(X, \frac{(Y - \rho X)}{\sqrt{1-\rho^2}}\right) = \frac{1}{\sqrt{1-\rho^2}}\text{Cov}(X, (Y - \rho X))$$

$$= \frac{1}{\sqrt{1-\rho^2}}\,(\text{Cov}(X, Y) - \text{Cov}(X, \rho X))$$

$$= \frac{1}{\sqrt{1-\rho^2}}\,(\text{Cov}(X, Y) - \rho\text{Cov}(X, X))$$

$$= \frac{1}{\sqrt{1-\rho^2}}(\rho - \rho) = 0$$

This implies that $X \perp W = 0$ and $X \sim N(0, 1)$ and $W \sim N(0, 1)$

17. $V|U = 3 \sim N(6.5, \sqrt{12.5})$

19. $(Y_2|Y_1 + Y_3 = x) \sim N(-12.x, 5)$

21. We see that

$$P(X > 3300|Y < 6.00) = \frac{P(X > 3300 \text{ and } Y < 6.00)}{P(Y < 6.00)}$$

We note that $Y \sim N(6.12, 0.6)$. The covariance matrix is

$$\Lambda = \begin{bmatrix} 2500 & -15.0 \\ -15.0 & 0.36 \end{bmatrix}$$

```
f[x_, y_] := PDF[MultinormalDistribution[{3250, 6.12},
    {{2500, -15}, {-15, 0.36}}], {x, y}];

num = NIntegrate[f[x, y], {x, 3300, ∞}, {y, -∞, 6.00}]

0.116586
```

```
Clear[x]

dno = NIntegrate[PDF[NormalDistribution[6.12, 0.6], x], {x, -∞, 6.0}]

0.42074
```

```
Answer = num/dno
```

```
0.277096
```

Thus we find that

$$P(X > 3300|Y < 6.00) = 0.277096$$

23. (a) We see that $F_Z(z) = P(Z \le z) = P(X \le z, Y \le z)$. Thus

$$F_Z(z) = \int_{-\infty}^{z} \int_{-\infty}^{z} \frac{1}{2\pi\sqrt{1-\rho^2}} \exp\left(-\frac{1}{2(1-\rho^2)}(x^2 - 2\rho xy + y^2)\right) dx dy$$

To complete the squares, we see that

$$x^2 - 2\rho xy + y^2 = x^2 - 2\rho xy + (\rho y)^2 - (\rho y)^2 + y^2$$
$$= (x - \rho y)^2 + y^2(1 - \rho^2)$$

Thus

$$F_Z(z) = \int_{-\infty}^{z} \frac{1}{\sqrt{2\pi}} e^{-\frac{y^2}{2}} \int_{-\infty}^{z} \frac{1}{\sqrt{2\pi(1-\rho^2)}} \exp\left(-\frac{1}{2(1-\rho^2)}(x - \rho y)^2\right)$$
$$= \int_{-\infty}^{z} \frac{1}{\sqrt{2\pi}} e^{-\frac{y^2}{2}} \Phi\left(\frac{z - \rho y}{\sqrt{1-\rho^2}}\right) dy$$

We see that

$$F_Z(z) = \int_{-\infty}^{z} P(X \le z|Y = y) f_Y(y) dy$$

Now by (8.2) and (8.3), we know

$$E(X|Y = y) = \mu_X + \frac{\rho \sigma_X}{\sigma_Y}(y - \mu_Y) = \rho y, \qquad \text{Var}(X|Y = y) = 1 - \rho^2$$

Thus $X|Y = y \sim N(\rho y, \sqrt{1 - \rho^2})$. This means

$$P(X \leq z) = \Phi\left(\frac{z - \rho y}{\sqrt{1 - \rho^2}}\right)$$

(b) We find

$$f_Z(z) = \frac{d}{dz} F_Z(z)$$

$$= \frac{d}{dz} \int_{-\infty}^{z} \int_{-\infty}^{z} \frac{1}{2\pi\sqrt{1 - \rho^2}} \exp\left(-\frac{1}{2(1 - \rho^2)}\left((x - \rho y)^2 + y^2(1 - \rho^2)\right)\right) dx dy$$

$$= \frac{d}{dz} \int_{-\infty}^{z} \int_{-\infty}^{z} \frac{1}{2\pi\sqrt{1 - \rho^2}} \exp\left(-\frac{y^2}{2} - \frac{(x - \rho y)^2}{2(1 - \rho^2)}\right) dx dy$$

$$= \frac{d}{dz} \int_{-\infty}^{z} f(z, x) dy$$

where

$$f(z, x) = \int_{-\infty}^{z} \frac{1}{2\pi\sqrt{1 - \rho^2}} \exp\left(-\frac{y^2}{2} - \frac{(x - \rho y)^2}{2(1 - \rho^2)}\right) dx$$

We need to apply Leibnitz's rule

$$f_Z(z) = \frac{d}{dz} \int_{-\infty}^{z} f(z, y) dy = \int_{-\infty}^{z} \frac{\partial f(z, y)}{\partial z} + f(z, z)\frac{dz}{dz} = f(z, z)$$

$$+ \int_{-\infty}^{z} \frac{\partial f(z, y)}{\partial z}$$

Now

$$f(z, z) = \frac{1}{\sqrt{2\pi}} e^{-\frac{y^2}{2}} \Phi\left(\frac{z - \rho y}{\sqrt{1 - \rho^2}}\right)\bigg|_{y=z} = \frac{1}{\sqrt{2\pi}} e^{-\frac{z^2}{2}} \Phi\left(\frac{z - \rho z}{\sqrt{1 - \rho^2}}\right)$$

$$= \frac{1}{\sqrt{2\pi}} e^{-\frac{z^2}{2}} \Phi\left(\frac{(1 - \rho)z}{\sqrt{1 - \rho^2}}\right)$$

and

$$\frac{\partial f(z, y)}{\partial z} = \frac{\partial}{\partial z} \int_{-\infty}^{z} \frac{1}{2\pi\sqrt{1 - \rho^2}} \exp\left(-\frac{y^2}{2} - \frac{(x - \rho y)^2}{2(1 - \rho^2)}\right) dx$$

$$= \frac{1}{2\pi\sqrt{1 - \rho^2}} \exp\left(-\frac{z^2}{2} - \frac{(z - \rho y)^2}{2(1 - \rho^2)}\right)$$

Hence we conclude that

$$
f_Z(z) = \frac{1}{\sqrt{2\pi}} e^{-\frac{z^2}{2}} \Phi\left(\frac{(1-\rho)z}{\sqrt{1-\rho^2}}\right)
$$
$$
+ \int_{-\infty}^{z} \frac{1}{2\pi\sqrt{1-\rho^2}} \exp\left(-\frac{z^2}{2} - \frac{(z-\rho y)^2}{2(1-\rho^2)}\right) dy
$$
$$
= \frac{1}{\sqrt{2\pi}} e^{-\frac{z^2}{2}} \Phi\left(\frac{(1-\rho)z}{\sqrt{1-\rho^2}}\right)
$$
$$
+ \int_{-\infty}^{z} \frac{1}{2\pi\sqrt{1-\rho^2}} \exp\left(-\frac{z^2 - 2\rho xy + y^2}{2(1-\rho^2)}\right) dy
$$
$$
= \frac{1}{\sqrt{2\pi}} e^{-\frac{z^2}{2}} \Phi\left(\frac{(1-\rho)z}{\sqrt{1-\rho^2}}\right)
$$
$$
+ \frac{1}{\sqrt{2\pi}} e^{-\frac{z^2}{2}} \int_{-\infty}^{z} \frac{1}{2\pi\sqrt{1-\rho^2}} \left(-\frac{1}{2}\left(\frac{y-\rho z}{\sqrt{1-\rho^2}}\right)^2 dy\right)
$$
$$
= \frac{2}{\sqrt{2\pi}} e^{-\frac{z^2}{2}} \Phi\left(\frac{(1-\rho)z}{\sqrt{1-\rho^2}}\right) \quad \text{by a change of variable}
$$

(c) We see that

$$
E(Z) = \int_{-\infty}^{\infty} \frac{2}{\sqrt{2\pi}} z \times e^{-\frac{z^2}{2}} \Phi\left(\frac{(1-\rho)z}{\sqrt{1-\rho^2}}\right) dz
$$
$$
= \sqrt{\frac{2}{\pi}} \int_{-\infty}^{\infty} z \times e^{-\frac{z^2}{2}} \Phi\left(\frac{(1-\rho)z}{\sqrt{1-\rho^2}}\right) dz
$$
$$
= -\sqrt{\frac{2}{\pi}} \int_{-\infty}^{\infty} \Phi\left(\frac{(1-\rho)z}{\sqrt{1-\rho^2}}\right) d\left(e^{-\frac{z^2}{2}}\right)
$$
$$
= \sqrt{\frac{2}{\pi}} \int_{-\infty}^{\infty} d\Phi\left(\frac{(1-\rho)z}{\sqrt{1-\rho^2}}\right)
$$
$$
= \sqrt{\frac{2}{\pi}} \int_{-\infty}^{\infty} \frac{1-\rho}{\sqrt{2\pi(1-\rho^2)}} \exp\left(-\frac{z^2}{2} - \frac{(1-\rho)^2 z^2}{2(1-\rho^2)}\right) dz
$$
$$
= \sqrt{\frac{2}{\pi}} \sqrt{\frac{1-\rho}{2}} = \sqrt{\frac{1-\rho}{\pi}}
$$

$$= \int_{-\infty}^{\infty} \frac{1-\rho}{\sqrt{2\pi\left(1-\rho^2\right)}} \, \mathrm{Exp}\left[-\frac{z^2}{2} - \frac{\left(1-\rho^2\right)z^2}{2\left(1-\rho^2\right)}\right] dz$$

$$\frac{1-\rho}{\sqrt{2-2\rho^2}}$$

(c) We use MATHEMATICA's function "Expectation" to do the numerical experiment with this problem

pdf = ProbabilityDistribution$\left[\frac{1}{2\pi\sqrt{1-\rho^2}} \, \mathrm{Exp}\left[-\frac{1}{2\left(1-\rho^2\right)}\left(x^2 - 2\rho xy + y^2\right)\right],\right.$

$\left.\{x, -\infty, \infty\}, \{y, -\infty, \infty\}\right]$;

soln = Expectation[Max[x, y], {x, y} ≈ pdf];

soln /. $\rho \rightarrow$ {0.2, 0.6, 0.8}

{0.504627, 0.356825, 0.252313}

$\sqrt{\frac{1-\rho}{\pi}}$ /. $\rho \rightarrow$ {0.2, 0.6, 0.8}

{0.504627, 0.356825, 0.252313}

25. We convert all in terms of the standardize unit. Thus

$$(4020 < X < 4050) \qquad \Longleftrightarrow \qquad \frac{4020 - 4000}{30} < Z < \frac{4050 - 4000}{30}$$

or

$$0.6667 < Z < 1.6667$$

Thus $\Phi(1.6667) - \Phi(0.6667) = 0.204695$

CDF[NormalDistribution[0, 1], 1.6667] - CDF[NormalDistribution[0, 1], 0.6667]

0.204695

Thus the expected price in standardized unit is

$$E(Y|a < X < b) = \frac{0.65}{\sqrt{2\pi}\,(0.204695)}\left[e^{-\frac{1}{2}(0.6667)^2} - e^{-\frac{1}{2}(1.6667)^2}\right]$$

$$= \frac{0.65}{\sqrt{2(3.1416)}\,(0.204695)}(0.80072 - 0.24934) = 0.31602$$

Thus we convert it back to the original unit

$$\frac{Y - \mu_Y}{\sigma_Y} = 0.31602, \quad E(Y|4020 < X < 4050) = 55 + 5(0.31602) = 56.58$$

27. (a) $E(Z) = \frac{1}{p+q-pq}$ (b) $p_Z(k) = [(1-q)(1-p)]^{k-1}(p+q-pq)$, $k = 1, 2, \ldots$ (c) $E(W) = \frac{1}{p} + \frac{1}{q} - \frac{1}{p+q-pq}$ (d) $E(X|X \leq Y) = \frac{1}{p+q-pq}$

29. 864,660

31. 0.5987

33. (a) $W \sim N(890, 340)$ (b) 0   (c) 0.98

35. (a) 21000   (b) 69224

37. Let $P_k = P\{E^c$ in k tosses$\}$. For $k = 0, 1, \ldots, 3$, it follows that $P_k = 1$. Consider the following diagram depicting the initial four tosses leading to the "system-restart"

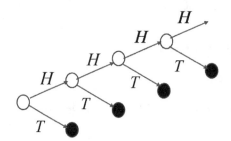

The system-restart points are shown in the above graph as the solid dots. We follow on the system-restart points and state the following recursion:

$$P_k = \sum_{i=1}^{4} P_{k-i} \left(\frac{1}{2}\right)^i$$

```
Clear[p]

Table[p[k_] = 1, {k, 0, 3}];

Table[p[k_] = (∑_{i=1}^{4} p[k - i] (0.5)^i), {k, 4, 50}];

1 - p[50]

0.951844
```

We see that $P(E) = 1 - P(E^c) = 0.9518$. In other words, if we flip a fair coin 50 times, the probability of at least four heads showing sccessively is 95.18%

39. Let E_i be the event that the last ball enters urn i. The Possionization approach calls for the probability that the rate each ball enters an urn being $\lambda p = (1)\frac{1}{m} = \frac{1}{m}$. Mimicking the approach used in Example 23, conditioning on the time of occurrence of the event E_i, we have

$$P(E_i) = \int_0^\infty \left(\frac{t}{m}e^{-\frac{1}{m}t}\right)^{m-1} \frac{1}{m}e^{-\frac{t}{m}} dt = \frac{1}{m^m} \int_0^m t^{m-1}e^{-t}dt$$

$$= \frac{1}{m^m}\Gamma(m) = \frac{(m-1)!}{m^m}$$

We remark that

$$\left(\frac{t}{m}e^{-\frac{1}{m}t}\right)^{m-1}$$

is the joint probability that exactly one Poisson event occurs before time t for the sub-Poisson process associated with $k = 1, \ldots, m$ and $k \neq i$. The occurrence time T_i associated with E_i follows $\text{expo}(\frac{1}{m})$, i.e.,

$$f_{T_i}(t) = \frac{1}{m}e^{-\frac{t}{m}}$$

So we conclude that

$$P(\text{each box contains exactly one ball}) = \sum_{i=1}^m P((E_i) = \frac{m!}{m^m}$$

Chapter 9

1. (a) Pareto(λ, a) (b) $F_Z(x) = e^{x^{-\alpha}}$ $x > 1$.

3. 250

5. 0.1963

7. We observe

$$
\begin{aligned}
P(Y_n \le x) &= P\left(\min(X_1, \ldots, X_n) \le x\right) \\
&= 1 - P\left(\min(X_1, \ldots, X_n) > x\right) \\
&= 1 - P\left(X_1 > x, \ldots, X_n > x\right) \\
&= 1 - (1 - x)^n
\end{aligned}
$$

Thus

$$
P(nZ_n \le x) = P\left(Y_n \le n\right) = 1 - (1 - x)^n
$$

or

$$
P\left(Z_n \le \frac{x}{n}\right) = 1 - \left(1 - \frac{x}{n}\right)^n \to 1 - e^{-x} \quad \text{as} \ \ n \to \infty
$$

So we conclude that $Z_n \xrightarrow{d} W$, where $W \sim \text{expo}(1)$

9. (a) $\log(Y_n) \approx N(-n, n)$ (b) 0.836858

11. We note that

$$
\ln P_n = \frac{1}{n} \sum_{i=1}^{n} \ln X_i
$$

and

$$
E(\ln X_i) = \int_0^1 \ln x \, dx = -1
$$

By the strong law of large numbers, we have

$$
P\left(\omega : \lim_{n \to \infty} \ln\left(P_n(\omega)\right) = -1\right) = 1
$$

This implies that

$$
P\left(\omega : \lim_{n \to \infty} P_n(\omega) = e^{-1}\right) = 1
$$

Thus we conclude that $P_n \xrightarrow{a.s.} \frac{1}{e}$ as $n \to \infty$

13. The moment generating function for X_i is

$$M_{X_i}(t) = E\left(e^{tX_i}\right) = p_i e^t + (1 - p_i) = 1 + p_i(e^t - 1) \le e^{p_i(e^t - 1)}$$

as $1 + y \le e^y$ for any y. Since X is the convolution of $\{X_i\}_{i=1}^n$, it follows that

$$M_X(t) = \prod_{i=1}^n M_{X_i}(t) \le \prod_{i=1}^n e^{p_i(e^t - 1)} = \exp\left(\sum_{i=1}^n p_i(e^t - 1)\right)$$
$$= e^{(e^t - 1)\mu}$$

(a) Let $a = (1 + \delta)\mu$. The Markov inequality states

$$P(X \ge a) \le \frac{E(X)}{a}$$

For any $t > 0$, it follows that

$$P(X \ge (1 + \delta)\mu)) = P(e^{tX} \ge e^{t(1+\delta)\mu})$$
$$\le \frac{M_X(t)}{e^{t(1+\delta)\mu}} \qquad \text{by Markov inequality}$$
$$\le \frac{e^{(e^t - 1)\mu}}{e^{t(1+\delta)\mu}}$$

For any $\delta > 0$, we set $t = \ln(1 + \delta)$. Then $t > 0$. Thus we have

$$P(X \ge (1 + \delta)\mu)) \le \left(\frac{e^\delta}{(1 + \delta)^{(1+\delta)}}\right)^\mu$$

(b) Consider

$$P(X \le (1 - \delta)\mu)$$

We multiply both sides by t, where $t < 0$ and then exponentiate both sides. This gives

$$P(X \le (1 - \delta)\mu) = P\left(e^{tX} \ge e^{t(1-\delta)\mu}\right)$$
$$\le \frac{M_X(t)}{e^{t(1-\delta)\mu}} \quad \text{by Markov inequality}$$
$$\le \frac{e^{(e^t - 1)\mu}}{e^{t(1-\delta)\mu}}$$

For any $\delta > 0$, we set $t = \ln(1 - \delta)$. Then $t < 0$. Thus we have

$$P(X \geq (1 + \delta)\mu)) = \left(\frac{e^{-\delta}}{(1 - \delta)^{(1-\delta)}} \right)^{\mu}$$

(c) With $\delta = 0.2$, $\mu = 1$, we have

$$P(X > 1.2) \leq \frac{e^{0.2}}{(1 + 0.2)^{(1.2)}} = 0.98139$$

and

$$P(X < 0.8) \leq \frac{e^{-0.2}}{(1 - 0.2)^{0.8}} = 0.97874$$

15. (a) Let $\{X_n\}$ be a sequence of i.i.d. random variable where $X_i \sim \text{pois}(1)$. Let $Y = X_1 + \cdots + X_n$. Then $Y \sim \text{pois}(n)$ and

$$P(Y \geq n) = \sum_{k=n}^{\infty} \frac{n^k}{k!} e^{-n} = P\left(\frac{Y - n}{\sqrt{n}} \geq 0 \right) = 1 - \Phi(0) \quad \text{by CLT}$$

$$= \frac{1}{2}$$

(b)

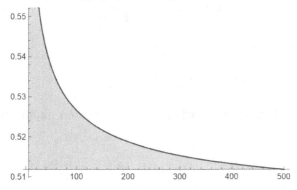

```
p[n_] := N[Exp[-n] * Total[Table[ (n^k)/(Factorial[k]) , {k, 0, n}]]];

DiscretePlot[p[n], {n, 5, 500}]
```

17. (a) Is is clear that $X_2 = X_1 + \varepsilon_2$ and $X_2 \perp X_1$. We can show the independence successively. Hence we conclude that $\{X_i\}_{i=1}^{n}$ are mutually independent.

(b) Applying Theorem 1, we have

$$\varphi_{X_n}(u) = \exp\left(-\frac{1}{2}\left(u^2 + 2u^2 + 3u^2 + \cdots\right)\right)$$

$$= \exp\left(-\frac{1}{2}u^2(1^2 + \cdots + n^2)\right)$$

Now $Y = n^{-\frac{3}{2}} X_n$. Thus

$$\varphi_Y(u) = \varphi_{X_n}\left(n^{-\frac{3}{2}}u\right) = \exp\left(-\frac{1}{2}(n^{-\frac{3}{2}}u)^2\right)(1^2 + \cdots + n^2)$$

$$= \exp\left(-\frac{1}{2}u^2\left(\frac{1}{n^3}(1^2 + \cdots + n^2)\right)\right)$$

$$\text{Limit}\left[\text{Exp}\left[-\frac{1}{2}u^2 \ \frac{1}{n^3}\sum_{i=1}^{n}i^2\right], \ n \to \infty\right]$$

$$e^{-\frac{u^2}{6}}$$

The above equation shows that

$$\varphi_Y(u) \to e^{-\frac{u^2}{6}} = \exp\left(-\frac{1}{2}\left(\frac{1}{3}\right)u^2\right)$$

We recall that if $W \sim N(\mu, \sigma^2)$, then

$$\varphi_W(u) = \exp\left(iu\mu - \frac{1}{2}\sigma^2 u^2\right)$$

Using Theorem 2, we conclude that

$$Y \xrightarrow{d} W$$

where $W \sim N(0, \sigma^2) = N(0, \frac{1}{3})$.

19. We first obtain the characteristic function for $X_{(n)}$:

$$\varphi_{X_n}(t) = E[e^{itX_{(n)}}] = \sum_{k=1}^{n} e^{\frac{itk}{n}} \frac{1}{n} = \frac{1}{n} \sum_{k=1}^{n} e^{\frac{itk}{n}}$$

$$= \frac{1}{n} \sum_{k=1}^{n} \left(e^{\frac{it}{n}} \right)^k \qquad (*)$$

As

$$\sum_{k=1}^{n} a^k = a \sum_{k=0}^{n-1} a^k = a \frac{1 - a^n}{1 - a}$$

$(*)$ can be stated as

$$\varphi_{X_n}(t) = \frac{1}{n} e^{\frac{it}{n}} \frac{1 - e^{\frac{it}{n}(n)}}{1 - e^{\frac{it}{n}}} = \frac{1}{n} e^{\frac{it}{n}} \frac{1 - e^{it}}{1 - e^{\frac{it}{n}}}$$

Since

$$\lim_{n \to \infty} n \left(1 - e^{\frac{it}{n}} \right) = -it, \qquad \lim_{n \to \infty} e^{\frac{it}{n}} = 1$$

we have

$$\lim_{n \to \infty} \varphi_{X_n}(t) = \frac{e^{it} - 1}{it}$$

The right-hand side is the characteristic function of $U(0,1)$. By Theorem 2, $X_n \xrightarrow{d} U(0,1)$

21. (a) Define

$$Y \equiv X_{(n)} = \max_{1 \le k \le n} X_k$$

When $X \in U(0,1)$, we have $F_X(x) = x$ and

$$f_Y(x) = n \cdot x^{n-1} \cdot 1 = n \cdot x^{n-1}(1-x)^{1-1}, \qquad 0 < x < 1$$

Hence $Y \in \beta(n, 1)$ and we have

$$m \equiv E[Y] = \frac{n}{n+1}, \qquad \text{Var}[Y] = \frac{n}{(n+1)^2(n+2)}$$

We want to show that

$$P(|Y - 1| > \varepsilon) \to 0 \qquad \text{as } n \to \infty$$

Note

$$|Y - 1| = |Y - m + m - 1| \leq |Y - m| + |m - 1|$$

Now

$$|Y - m| \leq \frac{\varepsilon}{2}, \qquad |m - 1| \leq \frac{\varepsilon}{2}$$

imply

$$|Y - 1| \leq \varepsilon$$

Hence

$$\{|Y - 1| > \varepsilon\}$$

implies

$$\left\{|Y - m| > \frac{\varepsilon}{2}\right\} \cup \left\{|m - 1| > \frac{\varepsilon}{2}\right\}$$

So we conclude

$$P\left(\{|Y - 1| > \varepsilon\}\right) \leq P\left(\left\{|Y - m| > \frac{\varepsilon}{2}\right\}\right) + P\left(\left\{|m - 1| > \frac{\varepsilon}{2}\right\}\right)$$

We apply Chebyshev's inequality to the first term on the right-hand side

$$P\left(\left\{|Y - m| > \frac{\varepsilon}{2}\right\}\right) \leq \frac{\text{Var}(Y)}{\left(\frac{\varepsilon}{2}\right)^2} = \frac{4n}{\varepsilon^2 (n+1)^2 (n+2)} \to 0 \qquad \text{as } n \to \infty$$

The second term on the right-hand side becomes

$$P\left(\left\{|m - 1| > \frac{\varepsilon}{2}\right\}\right) = \left|\frac{n}{n+1} - 1\right| = \left|\frac{1}{n+1}\right| \to 0 \qquad \text{as } n \to \infty$$

It is now clear that

$$P\left(\{|Y - 1| > \varepsilon\}\right) \to 0 \qquad \text{as } n \to \infty.$$

Hence

$$\max_{1 \leq k \leq n} X_k \xrightarrow{p} 1 \qquad \text{as } n \to \infty.$$

(b) Define

$$Y := X_{(1)} = \min_{1 \leq k \leq n} X_k$$

Again, when $X \in U(0,1)$, we have $F_X(x) = x$ and

$$f_Y(x) = n \cdot (1 - x)^{n-1} \cdot 1 = n \cdot x^{1-1}(1 - x)^{n-1}, \qquad 0 < x < 1$$

Hence $Y \in \beta(1, n)$ and we have

$$m \equiv E[Y] = \frac{1}{n+1}, \qquad \text{Var}[Y] = \frac{n}{(n+1)^2(n+2)}.$$

We want to show that

$$P(|Y| > \varepsilon) \to 0 \qquad \text{as } n \to \infty$$

Since $0 < Y < 1$, the above result is equivalent to show that

$$P(Y > \varepsilon) \to 0 \qquad \text{as } n \to \infty$$

We apply Markov's inequality:

$$P(Y > \varepsilon) \le \frac{E[Y]}{\varepsilon} = \frac{1}{\varepsilon(n+1)} \to 0 \qquad \text{as } n \to \infty$$

for any $\varepsilon > 0$. So we conclude

$$\min_{1 \le k \le n} X_k \xrightarrow{p} 0 \qquad \text{as } n \to \infty$$

23. (a) We see that

$$s_n^2 = \frac{1}{n-1} \sum_{i=1}^{n} (X_i - \overline{X}_n)^2 = \frac{1}{n-1} \sum_{i=1}^{n} (X_i - 2X_i\overline{X}_n + \overline{X}_n^2)$$

$$= \frac{1}{n-1} \left(\left(\sum_{i=1}^{n} X_i^2 \right) - 2n\overline{X}_n^2 + n\overline{X}_n^2 \right) = \frac{1}{n-1} \left(\left(\sum_{i=1}^{n} X_i^2 \right) - n\overline{X}_n^2 \right)$$

Now

$$\text{Var } X_i = EX_i^2 - (EX_i)^2 \implies EX_i^2 = \sigma^2 + \mu^2$$

and

$$\text{Var}\overline{X}_n = E\overline{X}_n^2 - (E\overline{X}_n)^2 \implies E\overline{X}_n^2 = \text{Var}\overline{X}_n + (E\overline{X}_n)^2$$

We see

$$\text{Var } \overline{X}_n = \frac{1}{n^2} (n \,\text{Var}(X_k)) = \frac{1}{n}\sigma^2 \qquad \text{and} \qquad E\overline{X}_n = \sum_{i=1}^{n} \frac{1}{n} X_i = \mu$$

and so

$$E\overline{X}_n^2 = \frac{1}{n}\sigma^2 + \mu^2$$

We take expectation of s_n^2 and conclude that

$$E[s_n^2] = \frac{1}{n-1}\left(n\left(\sigma^2 + \mu^2\right) - n\left(\frac{1}{n}\sigma^2 + \mu^2\right)\right) = \sigma^2$$

(b) Define $U_k = (X_k - \overline{X}_n)^2$. We see that $\{U_k\}$ are i.i.d with finite expectation σ^2. We now apply WLLN and conclude

$$\frac{1}{n}\sum_{k=1}^{n} U_k = \frac{1}{n}\sum_{k=1}^{n}(X_k - \overline{X}_n)^2 \overset{p}{\to} EU_k = \sigma^2$$

Now

$$s_n^2 = \frac{1}{n-1}\sum_{k=1}^{n}(X_k - \overline{X}_n)^2$$

$$= \left(\frac{n}{n-1}\right)\left(\frac{1}{n}\sum_{k=1}^{n}(X_k - \overline{X}_n)^2\right)$$

The first term tends to 1 as $n \to \infty$. The second term tends to σ^2 with probability one. Therefore, we conclude that $s_n^2 \overset{p}{\to} \sigma^2$ as $n \to \infty$

25. (a) $\{X_i\}$ are i.i.d $C(0,1)$. The characteristic function of X_i is given by (cf., Chapter 6.9)

$$\varphi_X(t) = e^{-|t|}$$

Since $\{X_i\}$ are i.i.d, we apply Theorem 1 and find the characteristic function for S_n as

$$\varphi_{S_n}(t) = (\varphi_X(t))^n = e^{-|t|n}$$

Let

$$Z = \frac{S_n}{n}$$

We apply Theorem 4 with $a = \frac{1}{n}$ and obtain

$$\varphi_Z(t) = \varphi_{S_n}\left(\frac{1}{n}t\right) = e^{-\left|\frac{t}{n}\right|n} = e^{-|t|}$$

By the uniqueness of the characteristic function, we conclude that $Z \in C(0,1)$

(b) We observe

$$\left(\frac{1}{n}\right) \sum_{k=1}^{n} \frac{S_k}{k} = \left(\frac{1}{n}\right) \left(\frac{X_1}{1} + \frac{X_1 + X_2}{2} + \cdots + \frac{X_1 + \cdots + X_n}{n}\right)$$

$$= \left(\frac{1}{n}\right) \left[\left(1 + \frac{1}{2} + \cdots + \frac{1}{n}\right) X_1\right.$$

$$+ \left(\frac{1}{2} + \cdots + \frac{1}{n}\right) X_2$$

$$+ \left(\frac{1}{3} + \cdots + \frac{1}{n}\right) X_2$$

$$\left. + \cdots + \left(\frac{1}{n}\right) X_n\right]$$

$$= \left(\frac{1}{n}\right) (a_1 X_1 + \cdots + a_n X_n)$$

where we define

$$a_1 = \left(1 + \frac{1}{2} + \cdots + \frac{1}{n}\right)$$

$$a_2 = \left(\frac{1}{2} + \cdots + \frac{1}{n}\right)$$

$$\vdots$$

$$a_n = \left(\frac{1}{n}\right)$$

It is easy to verify that

$$\sum_{k=1}^{n} a_k = n \qquad (*)$$

Define

$$Z = a_1 X_1 + \cdots + a_n X_n$$

and $Y_i = a_i X_i$. Thus $Z = Y_1 + \cdots + Y_n$. We see that $\{Y_k\}_{k=1}^{n}$ are independent random variables. Thus, the characteristic function of Z is given by

$$\varphi_Z(t) = \prod_{k=1}^{n} \varphi_{Y_k}(t) \qquad \text{by independence of } \{Y_k\}$$

$$= \prod_{k=1}^{n} \varphi_{X_k}(a_i t) \qquad \text{by Theorem 4}$$

$$= \prod_{k=1}^{n} e^{-|a_k t|}$$

$$= \prod_{k=1}^{n} e^{-|t|a_k} \qquad \text{as } a_k > 0 \qquad \forall k$$

$$= e^{-|t|(a_1 + \cdots + a_k)}$$

$$= e^{-n|t|} \qquad \text{by (*)}$$

Let

$$W = \left(\frac{1}{n}\right) \sum_{k=1}^{n} \frac{S_k}{k} = \frac{Z}{n}$$

We conclude

$$\varphi_W(t) = \varphi_Z\left(\frac{1}{n}t\right) = e^{-n\left|\frac{t}{n}\right|} = e^{-|t|}$$

By the uniqueness of the characteristic function, we conclude that $W \in C(0, 1)$

27. (a) Using the result from Example 16 with $a = -n$ and $b = n$, after simplification by MATHEMATICA we find

$$\varphi_X(t) = \frac{e^{it(n)} - e^{it(-n)}}{it(n - (-n))} = \frac{\sin(nt)}{nt}$$

(b) For $n = 2$, we find

$$E(X) = 0, \qquad \text{Var}(X) = \frac{4}{3}$$

(c) We use

$$f_X(x) = \frac{1}{2\pi} \int_{-\infty}^{\infty} e^{-itx} \varphi_X(x) dx$$

to find $f_X(1)$. The following gives $f_X(1) = 0.25$ — as expected.

```
f[t_, n_] = FullSimplify[ (Exp[i t n] - Exp[i t (-n)]) / (2 i t n) ]
```

$$\frac{Sin[n t]}{n t}$$

```
1
- Limit[D[f[t, 2], {t, 1}], t → 0]
i
```

0

```
1
-- Limit[D[f[t, 2], {t, 2}], t → 0]
i²
```

$$\frac{4}{3}$$

```
         1
f1 = ---- Limit[∫_{-T}^{T} Exp[-i t (1)] f[t, 2] dt, T → ∞]
        2 π
```

$$\frac{1}{4}$$

29. (a) Since $X = X_1 + \cdots + X_n$, where $\{X_i\}$ are i.i.d bern(p), we have

$$X_i = \begin{cases} 1 & \text{with probability } p \\ 0 & \text{with probability } 1 - p \end{cases}$$

and

$$\varphi_{X_i}(t) = E(e^{itX_i}) = e^{it}p + 1 - p = 1 - p(1 - e^{it})$$

where $q = 1 - p$. Thus

$$\varphi_X(t) = (1 - p(1 - e^{it}))^n$$

(b) Let $p = \frac{\lambda}{n}$, the above equation becomes

$$\varphi_X(t) = \left(1 - \frac{\lambda}{n}(1 - e^{it})\right)^n$$

```
        λ                    n
Limit[ (1 - - (1 - Exp[i t]))  , n → ∞, Assumptions → {λ > 0}]
        n
```

$$e^{(-1+e^{it})\lambda}$$

Since $\varphi_X(t) \to \varphi_Y(t)$, where $Y \sim$ pois(λ), Theorem 2 implies $X \xrightarrow{d}$ pois(λ)   □

31. We want to maximize

$$-\sum_{i=1}^{n} p_i \log p_i$$

subject to

$$\sum_{i=1}^{n} p_i = 1 \qquad\qquad (1)$$

We write Lagrange function

$$L(p_1, \ldots, p_n, \lambda) = -\sum_{i=1}^{n} p_i \log p_i + \lambda \left(\sum_{i=1}^{n} p_i - 1 \right)$$

We differentiate the above with respect to $p_1, \ldots, p_n$ and find

$$-1 - \log p_i + \lambda = 0 \qquad\qquad i = 1, \ldots, n \qquad\qquad (2)$$

Equation (2) implies $p_i = e^{\lambda-1}$. Now using (1), we find

$$\sum_{i=1}^{n} e^{\lambda-1} = 1 \implies \sum_{i=1}^{n} c = 1$$

Thus $c = \frac{1}{n}$ and we conclude $p_i = \frac{1}{n}$. □

33. We mimic the development of Problem 32. We write Lagrange function

$$L(p_1, \ldots, p_n, \lambda_1, \lambda_2) = -\sum_{i=1}^{n} p_i \log p_i + \lambda_1 \left(\sum_{i=1}^{n} p_i - 1 \right) + \lambda_2 \left(\sum_{i=1}^{n} p_i x_i - \mu \right)$$

where λ_1 and λ_2 are the two Lagrange multipliers. Setting $\partial L/\partial p_i = 0$ yields

$$-1 - \log p_i + \lambda_1 + \lambda_2 x_i = 0, \qquad\qquad i = 1, \ldots, n$$

The above equation implies that

$$p_i = e^{\lambda_1 - 1 + \lambda_2 x_i}, \qquad\qquad i = 1, \ldots, n \qquad\qquad (1)$$

Using the constraint $p_1 + \cdots + p_n = 1$, we have

$$e^{\lambda_1 - 1} \sum_{i=1}^{n} e^{\lambda_2 x_i} = 1, \qquad e^{\lambda_1 - 1} = \frac{1}{\sum_{i=1}^{n} e^{\lambda_2 x_i}} \qquad\qquad (2)$$

Substituting the above equations into (1) yields

$$p_i = \frac{e^{\lambda_2 x_i}}{\sum_{i=1}^{n} e^{\lambda_2 x_i}}, \qquad i = 1, \ldots, n \tag{3}$$

We multiply the ith equation of (3) by x_i and sum over all i. This gives

$$\mu = \frac{1}{\sum_{i=1}^{n} e^{\lambda_2 x_i}} \sum_{i=1}^{n} x_i e^{\lambda_2 x_i}$$

or

$$\sum_{i=1}^{n} x_i e^{\lambda_2 x_i} = \mu \sum_{i=1}^{n} e^{\lambda_2 x_i} \tag{4}$$

We can solve (4) numerically for λ_2. Once λ_2 is found, we solve (2) numerically for λ_1. Finally (1) gives the maximum entropy estimates for $\{p_i\}$ □

35. The following expression gives the answers to the problem:

```
x = {2.5, 3.4, 3.8};
```

$$\lambda_2 = \lambda_2 \ /. \ \text{NSolve}\left[\left\{\sum_{i=1}^{3} x[\![i]\!] \ \text{Exp}[\lambda_2 \ x[\![i]\!]] == 3.2 \sum_{i=1}^{3} \text{Exp}[\lambda_2 \ x[\![i]\!]]\right\}, \{\lambda_2\}, \text{Reals}\right]$$

```
{-0.111433}
```

$$\lambda_1 = \lambda_1 \ /. \ \text{Quiet}\left[\text{NSolve}\left[\left\{\text{Exp}[\lambda_1 - 1] == \frac{1}{\sum_{i=1}^{3} \text{Exp}[\lambda_2 \ x[\![i]\!]]}\right\}, \{\lambda_1\}\right]\right]$$

```
{0.259836}
```

```
Table[Exp[λ₁ - 1 + λ₂ x[[i]]], {i, 1, 3}]
```
$$\text{Table}[\text{Exp}[\lambda_1 - 1 + \lambda_2 \ x[\![i]\!]], \{i, 1, 3\}]$$

```
{{0.361048}, {0.326595}, {0.312357}}
```

Since the mean tilts to the lower end, the estimates of $\{p_i\}$ give a larger weight to $i = 1$.

37. Since $X_i \sim U(5, 40)$, we find

$$E(X_i) = \int_5^{40} \frac{x}{35} dx = 22.5 \qquad E(X_i)^2 = \int_5^{40} \frac{x^2}{35} dx = 608.33$$

$$\text{Var}(X_i) = 608.33 - (22.5)^2 = 102.08$$

Let $Y_n = X_1 + \cdots + X_n$. Thus $E(Y_n) = nE(X_1) = 22.5n$ and $\text{Var}(Y_n) = n\text{Var}(X_1) = 102.08n$. We require that $P(Y_n < 8000) \geq 0.95$. We apply

CLT and require

$$P\left(Z < \frac{8000 - 22.5n}{\sqrt{102.08n}}\right) \approx 0.95$$

This implies, theta

$$\frac{8000 - 22.5n}{\sqrt{102.08n}} = 1.65$$

Solution is: $\{[n = 341.86]\}$, i.e., $n \approx 342$

`CDF[NormalDistribution[22.5 (342), Sqrt[102.08 (342)]], 8000]`

`0.948698`

39.

Let X_i = payoff to the ith gambler; $Y = X_1 + \ldots + X_{1000}$

$EX = \dfrac{18}{37}(10.) + \dfrac{19}{37}(-5.);$

$EX2 = \dfrac{18}{37}(10.^2) + \dfrac{19}{37}(-5.)^2;$ VarX = EX2 – EX^2; EY = 1000 EX; VarY = 1000 VarX;

SY = Sqrt[VarY];

PE = 1 – CDF[NormalDistribution[EY, SY], 2500.]

`0.196281`

41.

Key: to know the prob distribution of Y_n and then apply the CLM.

We see that $E(Y_n) = \frac{1}{n}(n\, E(X_1)) = p$ and $Var(Y_n) = \frac{1}{n^2}(n\, Var(X_1)) = \frac{pq}{n}$

The condition is $P(\,|\,Y_n - p| \geq 0.05) \leq 0.05$

Equivalently: $P(|\frac{Y_n - p}{\sqrt{\frac{pq}{n}}}| \geq \frac{0.05}{\sqrt{\frac{pq}{n}}}) \leq 0.05$ or $\Phi(\frac{0.05\sqrt{n}}{\sqrt{p(1-p)}}) = 0.975 = \Phi(1.96)$

This means that we require $\frac{0.05\sqrt{n}}{\sqrt{p(1-p)}} \geq 1.96$

or $n \geq \left(\frac{1.96\sqrt{p(1-p)}}{0.05}\right)^2$ Since $\frac{1}{4} \geq p(1-p)$

We require that $n \geq (\frac{1.96\,(1/2)}{0.05})\wedge 2$

$\left(\dfrac{1.96\,(0.5)}{0.05}\right)^2$

384.16

43. (a)

$$X_n = X_0(1 + f)^S (1 - f)^F$$

(b) From (a), we have

$$\frac{X_n}{X_0} = (1 + f)^S (1 - f)^F$$

or

$$\left(\frac{X_n}{X_0}\right)^{\frac{1}{n}} = \left((1 + f)^S (1 - f)^F\right)^{\frac{1}{n}}$$

Define

$$G_n(f) = \log\left(\frac{X_n}{X_0}\right)^{\frac{1}{n}} = \log\left(\left((1 + f)^S (1 - f)^F\right)^{\frac{1}{n}}\right)$$

$$= \frac{S}{n}\log(1 + f) + \frac{F}{n}\log(1 - f)$$

As $n \to \infty$, by SLLN, $\lim \frac{S}{N} \to p$. Thus

$$E[G_n(f)] \overset{a.s}{\to} p\log(1 + f) + q\log(1 + f) \equiv g(f)$$

(c) We see that when $p = 0.53$, $f^* = 0.06$. It is a global maximum

```
p = 0.53;
g[f_, p_] = p Log[1 + f] + (1 - p) Log[1 - f];
α* = Quiet[α /. Solve[D[g[α, p], {α, 1}] == 0, α]]

{0.06}

D[g[α, p], {α, 2}] /. α → α*

{-1.00361}
```

In other words, when $g'(f) = 0$, we have

$$f^* = p - q = 2p - 1 = (2)(0.53) - 1 = 0.06$$

Also,

$$g''(f) = -\frac{p}{(1+f)^2} - \frac{q}{(1-f)^2} < 0 \quad \Longrightarrow \quad \text{maximum}$$

(d) At $f^* = 0.06$, the expected growth rate is

$$g(f^*) = 0.53 \log(1 + f^*) + 0.47 \log(1 - f^*) = 0.001801$$

Thus, we have

$$\log\left(\frac{X_n}{X_0}\right) = 0.001801n$$

When $\frac{X_n}{X_0} = 2$, we solve $0.001801n = \log(2)$. This gives $n \approx 385$

45. Let $g : [0, \infty)$ be strictly increasing and non-negative function. Show that

$$P(|X| \geq a) \leq \frac{E[g(X)]}{g(a)}, \qquad a > 0$$

Proof. From the Markov inequality, we have

$$P(|X| \geq a) \leq \frac{E(|X|)}{a}$$

We see

$$\frac{|x|}{a} \leq \frac{g(|x|)}{g(a)} \quad \text{if } |x| \geq a \text{ as } g \uparrow$$

We take the expectation of the above equation and assert

$$\frac{E(|X|)}{a} \leq \frac{E(g(|X|))}{g(a)}$$

Hence

$$P(|X| \geq a) \leq \frac{E(g|X|)}{a} \qquad \qquad \Box$$

47. (a) Using the result obtain from Example 16 of the chapter, we have $a = -\frac{1}{2}$ and $b = \frac{1}{2}$. Hence the characteristic functions for X is

$$\varphi_X(u) = \frac{e^{\frac{iu}{2}} - e^{-\frac{iu}{2}}}{iu}$$

(b) Since Z is the convolution of X and Y, we find

$$\varphi_Z(u) = \left(\frac{e^{\frac{iu}{2}} - e^{-\frac{iu}{2}}}{iu} \right)^2 = \left(\frac{2\sin\left(\frac{u}{2}\right)}{u} \right)^2 = \frac{2(1 - \cos(u))}{u^2}$$

(c) We find

$$f_X(x) = \begin{cases} 1 - |x|, & -1 < x < 1 \\ 0 & \text{otherwise} \end{cases}$$

(d) See the following MATHEMATICA output:

```
         2 (1 - Cos[u])
cfn[u_] = ──────────────;
              u²

InverseFourierTransform[cfn[u], u, x, FourierParameters → {1, 1}]

    1               1                  1            1               1
2 (- - Sign[-1 + x] + - x Sign[-1 + x] - - x Sign[x] + - Sign[1 + x] + - x Sign[1 + x])
    4               4                  2            4               4

f[x_] = FullSimplify[%]

1
- (Sign[1 - x] + x Sign[-1 + x] - 2 x Sign[x] + (1 + x) Sign[1 + x])
2

Plot[f[x], {x, -1, 1}]
```

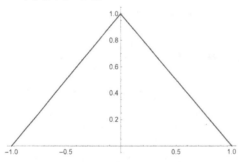

Index

Printed in the United States
by Baker & Taylor Publisher Services